Tropical pasture utilisation

Tropical pasture utilisation

L. R. HUMPHREYS

*Professor of Agriculture,
University of Queensland,
St Lucia, Australia*

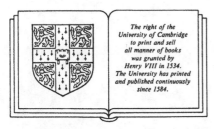

The right of the
University of Cambridge
to print and sell
all manner of books
was granted by
Henry VIII in 1534.
The University has printed
and published continuously
since 1584.

CAMBRIDGE UNIVERSITY PRESS

Cambridge
New York   Port Chester
Melbourne   Sydney

CAMBRIDGE UNIVERSITY PRESS
Cambridge, New York, Melbourne, Madrid, Cape Town, Singapore, São Paulo

Cambridge University Press
The Edinburgh Building, Cambridge CB2 2RU, UK

Published in the United States of America by Cambridge University Press, New York

www.cambridge.org
Information on this title: www.cambridge.org/9780521380300

First published 1991
This digitally printed first paperback version 2005

A catalogue record for this publication is available from the British Library

Library of Congress Cataloguing in Publication data

Humphreys, L. R.
Tropical pasture utilisation/L. R. Humphreys.
 p. cm.
Includes bibliographical references and index.
ISBN 0-521-38030-8
1. Pastures—Tropics. 2. Grazing—Tropics. 3. Forage plants—
Tropics. I. Title.
SB193.3.T76H86 1991
633.2′02—dc20   90-42849   CIP

ISBN-13 978-0-521-38030-0 hardback
ISBN-10 0-521-38030-8 hardback

ISBN-13 978-0-521-67341-9 paperback
ISBN-10 0-521-67341-0 paperback

To I.L.H.

Contents

Acknowledgements

I am indebted to Wolfson College, Oxford, Department of Plant Sciences, University of Oxford and the Institute for Grassland and Animal Production, Hurley, for support and to Ann Hansen for typing the manuscript.

I am grateful for advice or assistance from: D. E. Akin, L. C. Bell, D. A. Carrigan, R. J. Clements, T. M. Davison, D. G. Edwards, P. B. Escuro, H. Fujita, S. Fukai, D. L. Garden, C. J. Gardener, R. L. Hall, R. C. Gutteridge, J. W. Hales, R. L. Ison, A-M. N. Izac, O. R. Jewiss, C. Lascano, M. M. Ludlow, R. J. Jones, R. M. Jones, M. K. Murphy, B. W. Norton, F. Riveros, K. G. Rickert, C. W. Rose, R. Shawyer, H. M. Shelton, H. B. So, F. von Sury, J. K. Teitzel, M. Tjandraatmadja, J. M. Vieira, H. Wahab, F. R. Whatley, F. R. S., R. J. Wilkins, J. R. Wilson, L. Winks, and N. D. Young.

Abbreviations

ADF	Acid detergent fibre
AU	Animal unit
BF	Butter fat
CIAT	Centro International Agricultura Tropical
CSIRO	Commonwealth Scientific and Industrial Research Organization
DE	Digestible energy
DM	Dry matter
DMD	Dry matter digestibility
Et	Evapotranspiration
GDM	Green dry matter
hd	head
ILCA	International Livestock Centre for Africa
IVDMD	In vitro dry matter digestibility
L	Leaf area index, leaf area supported by unit ground area
LW	Liveweight
LWG	Liveweight gain
MAFF	Ministry of Agriculture, Forestry and Fisheries
ME	Metabolisable energy
NDF	Neutral detergent fibre
OMD	Organic matter digestibility
SNF	Solids-not-fat
SR	Stocking rate or animals per hectare
TNC	Total non-structural carbohydrate
WSC	Water-soluble carbohydrate

1

Introduction

1.1
The function of tropical pastures

Pastures feed grazing animals and contribute to the stability of landscapes and of some cropping systems. Farmers who improve forage supply in the tropics seek to convert effectively the added forage grown to animal products, and to devise production systems which sustain forage yield in a resilient ecosystem.

The savannas of the tropics and subtropics (Tothill & Mott, 1985) constitute an immense natural resource. Much of these lands is inviolate, in the sense that it is only suited for grazing by ruminants, which can convert the long chain structural carbohydrates in the vegetation to products that are useful to humans (R. J. Jones, 1988b). These lands are not inviolate from overgrazing, and this book is concerned with the principles of managing pastures which take account of the requirements of both plants and animals for growth and replacement.

Livestock in the tropics and subtropics function in agricultural systems which are mainly extensive, with a low level of inputs, but which include some intensive dairy operations. In many regions there is a close integration of livestock with cropping; crop residues provide the main sources of feed, and animal draught power and nutrient return are crucial to the success of cropping. Meat, milk, fibre, and perhaps skins and hides, are the conventional livestock outputs for western societies; power, fuel, manure and capital accumulation may be equally significant in the functioning of farm systems in the tropics.

The majority of the world's herbivores are in the developing countries: c. 66% of the cattle, c. 64% of the sheep and goats, c. 80% of the equines, and almost all the camels and the buffaloes (R. J. Jones, 1988b). However, production per head is much less than in developed countries. Only c. 30% of the world's beef, c. 18% of cow milk and c. 54% of sheep and goat meat are produced in developing countries (FAO, 1988).

Western scientists may view these figures as rep-

resenting inefficiency of animal production. The truth of the matter is that systems of animal production in Europe, North America and Japan are inherently inefficient, since they depend for their survival on (1) heavy subsidies from other sectors of the economy, and (2) heavy subsidies of support energy (Wilkins, 1982); they also generate persistent problems of environmental pollution for their communities. Students from tropical countries who seek training of relevance to their own farming systems in countries which have failed so dismally to produce self-sufficient and sustainable livestock production may have some disappointments. The tropical world has also had to bear the advice of western scientific tourists, accustomed to capital-intensive, subsidised agricultural systems, who have paternalistically advocated inappropriate technology and who have added to the burden of Third World debt.

There are many biological characteristics which distinguish tropical and temperate pasture. The C_4 photosynthetic pathway of the grasses provides efficient use of water and nutrients, but leads to a high content of structural elements which are poorly digestible. Sward density is lower than in temperate pastures, affecting the potential bite size for prehension. The higher growth potential of C_4 grasses relative to that of C_3 legumes leads to difficulties in the maintenance of a grass–legume balance. The use of shrubs and trees as feed sources requires technologies which have not developed in temperate regions. Mineral deficiencies, both for pasture and for animal growth, are more prevalent in the highly leached soils of the tropics, which are also more susceptible to erosion.

These distinctive pasture characteristics, together with features of the climatic environment, lead to options for management whose emphases are different from those encountered in the management of temperature pasture, and these are developed in this book.

1.2

The misleading legacies of past thought about pasture utilisation

Tropical pasture science is a new discipline, and the application of concepts developed in temperate or subtropical regions, where there is a longer tradition of scientific study, was inevitable in its emerging phases. Many of the ideas prevalent in the decade 1950–60 have since been evaluated in field practice. Some have been validated or have provided the basis for new developments of thought; others were erroneous, but are enumerated in this section since they persist in some segments of the scientific and agricultural advisory community.

(i)   Scientists, when faced with the degradation of the landscape caused by grazing, undervalued the role of stocking rate (SR, the number of animals grazing per unit area) and overemphasised the role of selective grazing. Continuous grazing was designated as sinful, and the intermittent nature of defoliation of pasture plants in paddocks where animals were always present did not receive sufficient weight. Thus Acocks (1953) referred to ' … the most fundamental principle of grazing management, viz. that grazing should be heavy for limited periods and must not be continuous'. Voisin (1959) believed (1) livestock should only occupy a paddock for a short period, since animals might otherwise defoliate a plant twice, and (2) there should be sufficient interval between grazings to replenish reserves of labile carbohydrate in the roots and crown and to enable the pasture to enter its 'blaze of growth' phase. There might also be an assumption that rotational grazing, in which animals are rotated about different paddocks, 'overcomes the disadvantages of under- and over-grazing' (McIlroy, 1964); the same author dismissed continuous grazing as a system where 'the stocking rate is relatively low'.

There were scientists sceptical of the claims for rotational grazing. Donald (1946), when reviewing experiments in the USA stated 'In no instance have consistent economically worthwhile differences been established in favour of rotational grazing'.

(ii)   Many scientists attached great importance to the effects of rest, or absence of defoliation, on (1) the accumulation of carbohydrate 'reserves', and (2) the development of leaf area to intercept radiation, in so far as these affected plant growth and persistence. It was not generally recognised that the variation in seasonal growth rate of continuously grazed pastures inevitably provided 'rest' when growth greatly exceeded consumption.

There were many studies in the USA and in southern Africa which dealt with the effects of season, defoliation, and nitrogen nutrition on the accumulation of non-structural carbohydrate in the roots and crown (Weinmann, 1961). There was strong teleological argument supporting the God-given role of carbohydrate reserves as purposefully providing for the persistence and regrowth of plants, if these plants were sufficiently rested to maintain adequate levels of reserve carbohydrate.

A certain minimum level of energy residual is necessary for plant persistence, but the influence of reserves on the rate of regrowth after severe defoliation appears to be limited to the first three days of regrowth. Plants accumulate surplus assimilate in the roots and crown when leaf growth is constrained by a shortage of nutrients or water; the excessive allocation of carbon to a root system inaccessible to livestock contributes to nutrient immobilisation through increased requirements for the eventual degradation of this material. The advocacy of root reserves as a management target was therefore a mixed blessing.

The concept of the leaf area index (L, or area of leaf supported by unit ground area) excited great interest in the decade under review. Brougham (1956) related the growth of pastures after defoliation to the development of L needed to intercept radiation. Davidson & Donald (1958) suggested for *Trifolium subterraneum* that sub-optimal L led to low growth rates because radiation reached the ground rather than the plant; supra-optimal L led to reduced growth because of inefficient sward structure. This emphasis on L has had continued utility in the management of humid zone or irrigated pastures, and in promoting plant cover as the defence against soil erosion.

The weaknesses of the concept are discussed later, but the main drawbacks which were not appreciated at the time were (1) manipulation of L had little effect on growth in circumstances where factors other than light, such as water and nutrients, limited growth (Humphreys, 1966a), and (2) maximisation of pasture utilisation and not maximisation of growth was the primary management objective – the significance of leaf senescence (Humphreys, 1966b; Wilson & Mannetje, 1978) was relatively unappreciated.

(iii)   The control of the botanical composition of natural pastures was understood in the secure terms of Clementsian succession, which worked well in the tall-grass prairies of the USA, where the theory evolved (Clements, 1920). This school believed that following a disturbance, such as overgrazing, drought or fire, the distance from a stable grazing climax is indicated within a linear species succession by the sort of species present. The stability of the ecosystem is the overriding objective, and 'the higher order, or climax, plants produce the most pounds of meat or milk per acre' (Archer & Bunch, 1953).

This view was challenged by scientists who recognised the greater productivity of the seral species intermediate in the succession in Africa (Davidson, 1962), which also provided the main cultivated tropical pasture grasses from the genera *Cynodon, Panicum, Paspalum* and *Setaria*. We also know that disturbance can lead to a new balance of botanical composition, in which the relative content of species or of growth forms or the appearance of new species may travel in many alternative directions according to the concatenation of abiotic and biotic circumstances (Humphreys, 1989). Reduction in grazing pressure may increase shrub invasion, or it may reduce it (Noy-Meir & Walker, 1986). Resilience, which is a measure of the capacity of the ecosystem to absorb changes and still persist (Holling, 1973), is a more realistic objective than stability, where we expect a return to an equilibrium state. Resilience requires a more open-ended approach to management. It is conceivable that different growth forms may coexist stably because particular weather sequences provide varying growth opportunities which favour individual species, according to their ontogenetic stage and their abundance at the time these occur (Westoby, 1980). Ecologists have come to recognise the overriding significance of pulses of plant recruitment (Noy-Meir, 1980), and grazing management needs an episodic character to manipulate the course of these events.

A new structure of knowledge to replace Clementsian succession is emerging from rigorous ecophysiological studies which have the necessary autecological bias; useful predictive frameworks may be developed from these to guide management decisions.

(iv) The concept of non-selective grazing (Acocks, 1966; Savory, 1983) has not weathered well. This has partly arisen from the failure of its application to control botanical change, but also because short-duration grazing systems reduce animal production (Denny, Barnes & Kennan, 1977). The attainment of satisfactory levels of individual animal performance is contingent upon the opportunity of the grazing animal to improve the quality of the average diet on offer by selectively consuming the plant organs present which have superior nutritive value.

Scientists were seized with enthusiasm in the first flush of success of many tropical plant improvement programs by the superior growth and persistence of cultivated grasses and the high level of nitrogen fixation of the tropical legumes. There was a slower awakening to the realisation of the intransigent character of the lower nutritive value of C_4 grasses, which limited long-term animal performance to *c.* 0.7 kg liveweight gain per head per day (LWG hd^{-1} d^{-1}) for beef cattle, or *c.* 12 kg milk hd^{-1} d^{-1} for dairy cows. These levels were only attainable where animals could select leaf rather than stem, and had access to young rather than aged pastures.

(v) Scientists accustomed to the housing of stock in European and North American winters and to the need to conserve forage for the winter assumed that the dry season shortfall in feed supply in tropical regions might be overcome by hay or silage conservation. This approach was in the main rejected by farmers, since (1) the feed shortage was less acute, (2) other approaches were available, (3) tropical forages were difficult to conserve and dependence upon the build-up and catabolism of animal body reserves was in some circumstances more efficient than fodder conservation, and (4) penalties to animal performance by withdrawing land for fodder conservation, substitutionary (rather than supplementary) feeding of hay for pasture, and compensatory weight gain after the period of shortfall in forage supply all reduced potential gains from fodder conservation.

(vi) The use of cut-and-remove systems of animal feeding is more prevalent than grazing in many mixed farming systems in the tropics, because of the integration of crops and pastures and the problems of controlling free-grazing animals in this situation. It was also considered that 'zero grazing or soiling would be an effective means of making the maximum use of intensively fertilised and managed grass' (Runcie, 1960). The heat stress suffered by European cattle in the tropics was thought to necessitate their housing during the day.

This system exacerbated problems of animal disease, caused immense movement of nutrients from pasture to the animal house and attendant pollution problems, reduced the opportunity for selective grazing and was expensive of labour. It has proved to be quite unnecessary for tropical dairying (Hongyantarachai *et al.*, 1989) except where it is dictated by farm convenience.

1.3
Current affirmations about pasture management
Tropical pasture science has moved slowly from the false assumptions enumerated above, which were not in any event accepted in all scientific communities. This book is not a manual and does not attempt an encyclopaedic coverage of the literature. It sets out a personal view of the central principles of the utilisation of tropical pastures,

illustrates these principles with concrete examples of research findings, and invites their validation in the diverse farming systems of the tropics. The basic affirmations are as follows:

(i)   The central objective of management is to effect a synchrony between the pasture available and the forage requirement of the animal. This is achieved primarily by adoption of a stocking rate (SR) which sustains animal production in the long term through (1) adequate forage allowance (the forage available per head), (2) plant growth and persistence, (3) maintenance of cover as a defence against erosion, and (4) desirable botanical composition.

(ii)   Continuity of forage allowance is desired to maintain animal production and to minimise animal stress. This is attained through (1) varying effective SR seasonally, through policies of purchase and sales and of mating, (2) providing a sequence of feeds of differing seasonal utility, (3) modifying the pasture environment through fertiliser or irrigation practice, and (4), as a last resort, conserving or purchasing feeds.

(iii)   The performance of individual animals attains satisfactory levels if there is maximum opportunity for selective grazing, little interference with the spatial or botanical preferences of the grazing animal, and young rather than aged pasture available.

(iv)   In humid and subhumid zones the legume plays a central role in the maintenance of nutritive value and in raising the N status of the ecosystem. Grazing management may be directed to the persistence and plant replacement of the legume component. Fertiliser inputs may be needed to ensure adequate levels of N fixation, and the mineral requirements of both plants and animals have to be addressed if the environmental resources for sustained production are to be well utilised.

(v)   There is a diversity of management options available to any farm system. The role of the scientist is to quantify the effects of grassland improvement and management practices, so that the pasture manager may choose options which are suitable for the goals, acceptance of risk, skills, intensity of production and diversity of output of the individual farm enterprise.

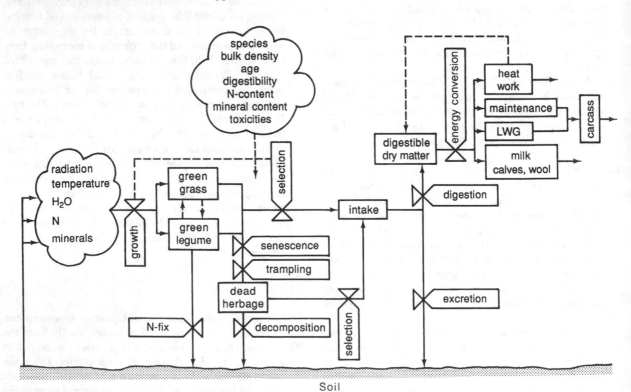

Fig. 1.1 A simplified model of the grazing system. (After Mannetje & Ebersohn, 1980.)

1.4
The content of the book

(i)   A brief overview of the livestock systems of the tropics is provided first (Chapter 2). These are categorised as those in which there is (1) close integration of animal production and the production of annual or perennial crops, (2) intensive ruminant production, (3) sedentary grazing based on planted or native pastures, and (4) nomadic pastoral activity.

(ii)   The effects of grazing animals on the pasture are then described. Their influences on the environment in which the pasture grows (Chapter 3) are described in terms of (1) edaphic changes (erosion, compaction, nutrient redistribution), (2) biotic effects (energy flow, pests and disease), and (3) modifications to the climatic micro-environment (temperature, water and light). Defoliation influences pasture growth (Chapter 4), which is considered in terms of (1) the balance with senescence and respiration and of the utilisation of light, (2) the restoration of the leaf surface after defoliation, (3) the distribution of assimilate to plant organs, and (4) its effects on nitrogen fixation in legumes. The modification of botanical composition by grazing (Chapter 5) occurs through (1) effects on plant replacement, and (2) plant interference, especially in the light relations of the mixed sward in humid zones, and through the effects of selective grazing.

(iii)   The latter half of the book is concerned with the response of the grazing animal to the pasture available, and the system with its interacting elements is schematically shown in Fig. 1.1 (Mannetje & Ebersohn, 1980). Pasture attributes of nutritive value and sward structure which influence animal performance (Chapter 6) are described, together with grazing behaviour and the effects of selectivity on the animal. The overriding significance of stocking rate (Chapter 7) on output per head and per unit area and its characterisation in terms of forage allowance are outlined; the interaction of SR with fertiliser input influences the success of intensive systems.

Continuity of forage supply is maintained through attention to seasonal variation in animals requirements and the integration of different pastures (Chapter 8), the use of special-purpose legume stands, fodder crops and crop residues or the modification of the environment through fertiliser application and irrigation. The alternatives (Chapter 9) are pasture conservation and the use of supplements.

Systems of rotational and deferred grazing, mixed grazing and the role of fire in pasture management are dealt with briefly (Chapter 10).

(iv)   Some remarks about the balance of management objectives and the priorities for research are made in Chapter 11.

2

Pastures in tropical farming systems

The nature of the feed resource determines the type of livestock system which evolves in particular regions, and this in turn reflects the species of grazing animals which are best adapted to the physical and biological environment, and whose use meets the social and economic objectives of the farmers in that society. The level and quality of the feed supply is determined by climatic and edaphic factors, as modified by the intervention of the farmer and the plant species available.

The diverse livestock systems of the tropics may be categorised as those in which there is (1) close integration of animal production and the production of annual or perennial crops, (2) intensive ruminant production, (3) sedentary grazing based on planted or native pastures, or (4) nomadic pastoral activity. Grassland plays little part in pig and poultry production, which are not treated in this book.

2.1
Pastures in annual cropping systems

2.1.1
Significance

Payne (1976) estimated that *c.* 290 million cattle and buffalo are carried on mixed farms raising annual crops in the tropics. These farms are mostly operated by small-holders; in south-east Asia *c.* 90% of the ruminant population is found on mixed farms (Zandstra, 1983) which are less than 5 ha. Subsistence farming predominates; farmers' attitudes to risk are more cautious than those of more entrepreneurial farmers and are directed to ensuring that the family is fed.

The emphasis on the outputs from the crop or from livestock shows different patterns, and the objectives of the livestock operation have a varying balance between draught power, the rearing of young animals for sale, and the production of meat, milk and dung. The processing of crop residues and crop by-products through the ruminant is an essential feature of the system, and adjacent grasslands, shrubs and trees contribute fodder in varying degree.

2.1.2
Types of enterprise

The predominant farming systems in South-east Asia (Zandstra, 1983) are:

(1) irrigated wetland farm types with intensive land use and high density of human population;
(2) rain-fed wetland farm types with less population density and a pronounced dry season;
(3) upland (dryland) farm types of varying moisture status and varying incorporation of trees;
(4) mixed land farm types, in which farmers grow wetland rice and a variety of upland crops, with or without rice.

Three descriptive studies are used to illustrate the main features encountered.

(i) A village in the Subang district on the north coast of West Java (lat. 6° S) is selected to represent low altitude (< 100 m) use of land for irrigated rice (which was 44% of the area), rain-fed ricefield (18%) and mixed gardens (36%, for coconut, bananas, pawpaw, jackfruit, cassava, maize and beans); the average farm size was 0.4 ha (Sumanto *et al.*, 1987). The main wet season occurs from October to April. Land preparation (ploughing and levelling) by ruminant rearers occupied *c.* 40 days ha^{-1}, but only 15–30 days ha^{-1} when ruminants were hired. The average herd size was 2.0 cattle or 1.7 buffalo; 70% of farmers who did not own ruminants hired cattle or buffalo for traction, and 80% hired non-family labour. Calving rates were 68% for cattle and 45% for buffalo.

In this district forage is not preserved. Rice straw is fed in the green state, and considerable quantities of rice straw are burnt or incorporated in the soil. Bean or groundnut hay, green maize leaves, whole maize stover and sweet potato leaves are the other crop residues fed. Roadsides and canal banks bearing *Polytrias amaura*, *Ischaemum timorensis*, *Brachiaria* spp., *Chloris acicularis*, *Cynodon dactylon*, *Imperata cylindrica* and *Calopogonium* spp. are grazed or cut. Forest verges of *I. cylindrica*, *Axonopus compressus*, *Paspalum conjugatum*, *P. amaura*, *Setaria* spp., *Panicum* spp. and *Centrosema pubescens* are cut and hand-fed, and weeds from gardens include *Amaranthus*, *Bidens*, *Brachiaria* and *Eleusine indica*. Hand feeding is practised predominantly in the wet season, and rearers may convey forage loads of over 50 kg for 5–15 km to supplement locally available forage. Labour supplied by cattle owners occupies 4.5 h d^{-1} for cutting and carrying feed, and 8.7 h d^{-1} for herding.

(ii)   North-east Thailand (lat. 14° to 19° N) is a slightly elevated plateau (100–300 m) which contains *c.* 2.7 million cattle and *c.* 2.4 million buffalo. It has a tropical savanna climate; at Khon Kaen 85% of the annual rainfall of 1260 mm falls from mid-April to mid-October. Three villages were surveyed (De Boer, 1973) in which *c.* 57% of the cropping area, both rain-fed wetland and upland, was planted to rice. Kenaf, corn with mung bean as a following relay crop, and sugar cane were the other main crops. The livestock enterprise in each village varied in its emphasis on breeding or draught. This was reflected in the age distribution of the herd: animals 0–2 years comprised 23% of the mature cattle biomass in a village directed to breeding and 6% in a village directed to draught. Similarly the male:female ratio (biomass basis) varied from 0.6:1 in the former to 5.8:1 in the latter, since castrated males were preferred for draught work. The annual birth rate (live births/cows) was 71% and 60% for cattle and buffalo respectively.

Animals were herded from May to December to avoid damage to crops, and used communal pasture, forested areas and fallow land which comprised 14% of the total area.

Short grasses, such as *Dactyloctenium aegyptium*, *Brachiaria miliiformis*, *Chrysopogon aciculatus*, *Digitaria*

Fig. 2.1. Rice straw saved for draught livestock in north-east Thailand.

adscendens and *Eragrostis viscosa*, were the principal grasses close to settlement, whilst taller grasses such as *Arundinaria ciliata* and *Arundinaria pusilla* occurred further from dwellings (Robertson & Humphreys, 1976). Hand-cut grass was fed during the early part of the cropping season, especially in July and August, when feed availability was restricted. Animals grazed the whole area of crop stubbles and weeds from December onwards, and were supplemented with rice straw in the latter part of the dry season (Fig. 2.1). The overall annual stocking rate (300 kg animal unit = AU) was 1.1 AU ha^{-1}. The village with the highest crop production had the highest stocking rate (1.3 AU ha^{-1}), indicating the complementary (rather than competitive) character of crop and livestock production.

(iii)   An African illustration is taken from the Ethiopian highlands at Debre Birhan, (lat. 10° N) where the annual rainfall is 980 mm and has a bimodal distribution (Gryseels & Asamenew, 1985). The average farm size was 3 ha, of which 0.7 ha was left fallow; communal grazing was additionally available. The mean livestock holding comprised 6.7 cattle, 10 sheep, 2 donkeys, 1.2 horses and 3.3 poultry. Farm cultivation depended heavily on animals, and the average annual input per farm was 440 ox-pair hours, 40 h by other cattle, and 80 h by donkeys and other equines. Donkeys also spent *c.* 300 h yr^{-1} walking to and from market.

The area cultivated was positively linked to the size of the livestock holding, as illustrated in Fig. 2.2. This is a further example of the complementarity of crop and livestock production; as more land is cropped so there are generated more crop residues which contribute to feeding the expanded livestock population. In this district cash income mainly arose from livestock: 31 % from livestock products and 52 % from trade in animals. Subsistence inputs were also dependent upon livestock, which contributed 53 % of gross farm margins (excluding draught but including home consumption).

2.1.3
Characteristics of the feed supply
The above specific illustrations are partly given because of the unfamiliarity of many western readers with this type of farming system. Farmers show considerable ingenuity in using such diverse feed sources to maintain continuity of forage supply to their animals. A concrete example (Moog, 1980) of seasonal change in feed source is given for backyard cattle-raising in a mixed farming rice–corn uplands area in the Batangas district of the Philippines (Fig. 2.3). The bulk of the feed constituted volunteer species taken from cultivated fields, roadsides and uncultivated wastelands. The predominant grass was

P. conjugatum, but volunteer weed grasses in rice and corn fields were preferred during the main growing season from July to November, as well as broadleaf weeds such as *Synedrilla nodiflora* and *Ipomea triloba*. Green corn was fed in July and August. Crop residues, especially corn stover and fresh corn stalks, increased in significance during the early season in December and January, and green leaves of trees and shrubs, including *Gliricidia sepium* and *Acacia*, played a significant role from March to June. Concentrate feeding was insignificant as a production source. Volunteer weeds in the maize crop were perceived as a positive feed resource, and crop husbandry directed to weed control may not attune with farmer objectives.

(i)   The intensity of cropping and the suitability of land will determine the extent of the companion uncropped areas which may be used for forage (Humphreys,

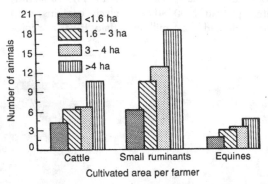

Fig. 2.2. Livestock holding and cultivated area per farmer in the Debre Birhan area of Ethiopia. (From Gryseels & Asamenew, 1985.)

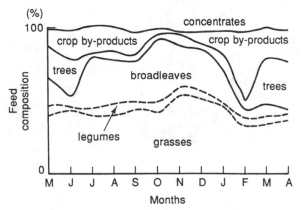

Fig. 2.3. Composition of feeds given to cattle in rice/corn cropping system in Batangas, Philippines. (From Moog, 1980.)

1986*a*). In wetland areas the paddy bund or mound demarcating the rice bay may be sown to crops such as cowpea, may be planted to a line of edible shrubs, may carry volunteer herbage, or may be oversown to a forage legume such as *Stylosanthes hamata* cv. Verano (Gutteridge, 1983). Grazing reserves adjacent to the cropping areas are heavily used during the main growing season, when crop land is unavailable for grazing, and poorly drained areas have a special role at the beginning of the dry season, since they retain green forage for a longer period. A narrow road 20 m wide provides 2 ha of forage per km; if the tarmac is paved the pasture area is reduced but the added run-off extends the grazing season on the run-on areas. Shrub legumes such as *Leucaena leucocephala* and *G. sepium* are planted on roadsides in many countries, and there is increasing interest in the establishment of areas of reserve fodder, especially adjacent to homesteads. In Bali a 'three-strata' system of feeding occurs in which cattle are fed successively on pasture, shrubs, and tree leaves as the season advances (Nitis, 1985). Fully improved pastures, in which the existing vegetation is replaced by planted

species and soil nutrient deficiencies are rectified, account for only a small segment of production, but are increasingly used in the house backyard situation. Conserved pasture hay or silage has found little place in the system.

(ii) The interplanting of upland crops with fodder legumes represents an attempt to maximise output from unit land resource, and recognition of the value of this practice in the maintenance of soil fertility may increase. The use of alley-cropping, in which food or cash crops are grown between contour-planted lines or shrub legumes, is being quite widely evaluated in South-east Asia and in Nigeria (Kang, Grimme & Lawson, 1985). Opportunist relay-cropping with forage legumes, such as *Crotolaria juncea* (Kessler & Shelton, 1980) is feasible, either by using the stored soil moisture available at the end of the main growing season, or by taking a short season catch crop during the uncertain opening monsoon rains. The rotation of short-term ley pastures with annual crops is rarely practised in the tropics.

Fig. 2.4. Drying dung cakes for fuel in Ethiopia.

(iii)   The considerable increases in the production of the tropical cereal grains rice, maize and sorghum have mainly occurred through the greater diversion of assimilate from leaf and stem to grain, and farmer resistance to the adoption of some new varieties has been associated with the poorer yield and nutritive value of the straws used in animal production (Reed, Capper & Neate, 1988). This whole question is discussed in Section 8.2.3. The increasing incorporation of pulses in farming systems, as intercrops, relay crops or sole crops in a rotation, contributes substantially to improving the nutritive value of crop residues.

In addition to these fibrous residues, farmers may use crop residues or crop by-products which are high in available carbohydrate, such as reject green bananas, banana stem, molasses, sugar cane, sweet potato, citrus pulp and pineapple pulp. Feeds with high N content include cassava leaf, copra meal, cottonseed meal, spent tea and meals made from peanut, rapeseed, sesame and soybean (Sastradipradja, 1981).

2.1.4
Crop–pasture–animal relations
Tropical livestock production from smallholdings is criticised by western observers for the low rates of gain, milk yield and reproductive performance; in fact these represent high rates of return from efficient, low input systems.

(i)   Crop residues and volunteer pastures on roadsides and residual uncropped areas constitute a biological resource which may be used or discarded. A maize crop might represent 127 KJ of metabolisable energy (ME) ha^{-1} and 620 kg protein ha^{-1} in the tops (McDowell, 1988). Use of the grain alone for human consumption delivers 39% of the energy and 20% of the protein; use of the bran and stover for animal feed raises these figures to 56% and 28% respectively. The efficiency of conversion depends upon the ruminant species used, as discussed in Section 8.2.3. Buffalo utilise rice straw more efficiently than cattle, due to their larger gut capacity

Fig. 2.5. Ploughing with oxen in Ethiopia.

Fig. 2.6. Coconut producing areas of the world. (From Reynolds, 1988.)

and greater extent of rumen fermentation. Cattle respond better to higher quality diets than buffalo, and their greater retention time of feed in the rumen relative to that for sheep and goats enables them to use high fibre diets better than the latter. The selective grazing of sheep and of goats, and the capacity of goats to browse, lead to other specific advantages.

(ii)   The nutrients returned to the cropping system as animal excreta are more readily available to plants than crop residues, whose breakdown otherwise immobilises a good deal of nitrogen, as discussed in Section 3.1.3. Spatial transfer of nutrients is also effected; animals which are housed at or near the dwelling concentrate nutrients for use on the adjacent high-value garden crops or backyard pastures. The return of nutrients to the cropping land may be integrated between separate crop and livestock producers; in West Africa sedentary crop farmers welcome the Fulani and their cattle onto their crop stubbles in the dry season.

Animal manure may have a significant cash value, either as a fertiliser for intensive crop production, or as fuel. In an eastern district of Ethiopia where fuel is scarce the output of manure as sun-dried fuel returned $c.$ US\$80 cow^{-1} yr^{-1} (1988 figure) (Fig. 2.4). Manure also contributes to family fuel supplies, with some loss to the maintenance of soil fertility.

(iii)   Cropping systems operating with animal traction (Fig. 2.5) depend for their success on the working condition of the cattle or buffalo. If this is high at the end of the dry season, the animals are ready for ploughing at the onset of the rains; the area prepared and the timeliness of preparation hinges upon an adequate feeding policy for the dry season. Farmers considering the adoption of new crops with a higher cash value may reject these if the feeding value and yield of stubble does not accord with their fodder needs. The introduction of small farm tractors in wetlands has altered the balance of animal feed requirements. Small tractors have been less successful in upland areas.

(iv)   Animals may constitute a family banking system, a means of capital accumulation, a hedge against inflation and an insurance against disaster. In many societies animal sales relate to the occurrence of specific family needs or social requirements.

Interventions designed to increase output from annual crop/livestock systems only succeed if the inter-relationships that exist among the farm system, and between the constituent elements and the farm environment are recognised. The complementarity of the component animal and crop factors may be used to promote sustained productivity of both types of output.

2.2
Pastures in plantation agriculture

2.2.1
Significance

Plantation agriculture in the humid tropics incorporates a significant ruminant population which is increasing as the thrust to diversification of enterprises and the benefits from this combination are recognised. There is a high estimate of $c.$ 15 million cattle and buffaloes in the perennial crop areas of the tropics (Payne, 1976); most of these animals are associated with coconut production. About 90% of the world's coconuts ($c.$ 9 million ha total) are grown in Asia and Oceania (Reynolds, 1988), but there are appreciable areas on the south-east and central western coasts of Africa, in the Caribbean, and on coastal Brazil (Fig. 2.6). Considerable scope exists for the expansion of livestock production into rubber and oil palm areas, where ruminants are found on a low proportion of holdings. Animals are used in orchards, such as the lum-yai in northern Thailand and the cashew in southern Thailand, in other plantations such as kapok and clove, and in forest plantings.

(i)   Cattle grazed in coconut plantations were traditionally seen as contributing to weed control, and as a means to the maintenance of short vegetation, which facilitates nut collection. Current thinking emphasises the plantation area as a developed resource which does not require further clearing for pasture establishment, and whose use may be intensified to increase and diversify farm income (Shelton, Humphreys & Batello, 1987). Plantation products are often export commodities, sold on widely fluctuating markets. The production of milk or meat helps to stabilise farm income, especially if animal products are sold to meet a rising domestic demand and facilitate import substitution. They provide a source of income from land being replanted to new crop varieties or to replace old trees. The possible secondary benefits to soil fertility, soil stabilisation and weed control are discussed subsequently.

(ii)   A primary difficulty is the added managerial and technical skills needed to run a dual enterprise. There is also an entrenched concern in estate management that livestock will damage rubber or oil palm plants, or reduce the growth of young coconuts; there may be an antipathy between the needs of the domestic cows of plantation workers and the need of management to protect the palms or trees from their perceived depredations. This problem does not arise in smallholder production systems. Heavy grazing also leads to trampling damage, and some pasture systems provide competition for the main crop.

In estate plantations the continuous replanting programmes ensure the availability of young plantation areas with potentially good carrying capacity. This situation is less common for smallholders.

2.2.2
Pasture performance in shaded conditions

(i)   Pastures and palms do not compete for light in the classical sense, since they occupy different strata; palms modify the light environment in which the pasture grows, and the level of shade is the most significant factor influencing pasture output. Coconuts are especially suited to understorey pastures, since there is high light transmission in both the very early and in the late phases of palm production. A study in the Solomon Islands (Litscher & Whiteman, 1982) of smallholder coconuts planted 9 to 19 years previously showed light transmission in the range 40 to 70%. Palm density (X_1, palms ha^{-1}) and palm height (X_2, m) negatively influenced percentage light transmission (Y) in a linear fashion:

$$Y = 102.6 - 0.195X_1 - 1.47X_2$$

Palm height and age was positively correlated; in later years a positive relationship between age and light transmission is expected (Nelliat, Bavappa & Nair, 1974). Mature oil palms provide less light (6–16% transmission; Chen & Othman, 1984) and opportunity for forage production in rubber plantations is also limited to the early years, since deep shade develops.

Net photosynthesis may be viewed simply as reflecting the size of the photosynthetic system and the efficiency of unit green surface. Shading reduces leaf area, which reduces light interception. The efficiency of the conversion of the light which is intercepted is increased under shade (Ludlow, 1978), despite the lesser photosynthetic capacity of shade leaves relative to leaves developed in full sunlight. There is less radiant energy to be fixed as carbohydrate. Plants exhibit various modifications in shade: the shoot to root ratio increases, the leaves become thinner, shoots are taller, and seed production of grasses is reduced (Oliveira & Humphreys, 1986).

The tropical grasses, with their C_4 photosynthetic pathway, show a greater decrease in relative growth rate in shade than the C_3 tropical legumes (Ludlow, 1978). This suggests that the competitive advantage of C_4 grasses over legumes is reduced by shading. In Section 5.4.2 an illustration is given of the development of legume dominance under shade; this may also be related to selective grazing of grasses in an environment with no pronounced dry season.

(ii)   Pasture species vary greatly in their shade tolerance (Table 2.1). Some species which are true shade species, such as *Axonopus compressus*, exhibit little reduction in growth or persistence in shaded plantation conditions, but are less productive than truly vigorous sun species, such as *Brachiaria decumbens* or *Calopogonium mucunoides*, except in deep shade (Wong, Rahim & Sharudin, 1985). Mixed pastures are established to cater for the range in shade conditions occurring spatially in the plantation, and the changed shade conditions as the trees age. In coconut plantations, plants of short stature are favoured over tall plants, since the latter cause difficulties for nut collection.

(iii)   The nitrogen nutrition of shaded pastures has unusual features which are not well understood. The nitrogen fixation of well-nodulated pasture legumes is positively associated with the rate of photosynthesis, in so far as this reflects the supply of carbohydrate to the nodule, and this is reduced by shading. However, the nitrogen fixation of shade-tolerant legumes such as *Desmodium intortum* and *Leucaena leucocephala* is well maintained in shade (Eriksen & Whitney, 1982).

Growth and the quantity of light energy are usually closely related, but there are now well-documented instances of pasture grasses, such as *Panicum maximum* var. *trichoglume*, exhibiting higher total biomass production under moderate levels of shade than under full sunlight (Wong & Wilson, 1980; Wilson, Catchpoole & Weier, 1986). This has occurred where N availability was limiting, and is associated with higher concentrations of tissue N and higher total N and K uptake.

2.2.3
Interactions between plantation crops and pastures

Yield of plantation crops may be positively or negatively affected by the pasture system, depending on the nature of the interference which develops and the net effect on the crop environment. Positive gains from the N fixation of leguminous cover crops are reflected in increased growth of rubber (Broughton, 1977) and oil palm (Agamuthu *et al.* 1981). These gains are also associated with the accumulation of a pool of nutrients in the pasture which are slowly released to the roots of the plantation crop. The content of legume in the pasture and its mineral nutrition condition the extent of the N increment.

Coconut yield is reported to increase at higher stocking rates (Santhirasegaram, 1967; Rika, Nitis & Humphreys, 1981). Yield of nuts (Table 2.2) was higher from coconuts grown with a pasture of *B. decumbens / Centrosema pubescens* than from an adjacent area of natural pasture in west Bali; it was also greater in paddocks continuously grazed at 4.8 and 6.3 yearlings ha^{-1} than at the lighter stocking rates of 2.7 and 3.6 yearlings ha^{-1}. This may be due to more efficient nutrient cycling, as discussed in Chapter 3. Fertility transfer under cut-and-remove systems increases fertiliser needs, but physical damage to young trees is avoided.

Monospecific grass swards may cause a reduction in coconut yield, especially under conditions where the fertiliser needs of the pasture are not met. These effects are associated with particular species; in Sri Lanka the yield of nuts with *Brachiaria brizantha* or *B. miliiformis* showed a positive trend after 10 years and was twice that of the yield with *P. maximum* (Ferdinandez, 1972). This has been attributed to the greater yield of *P. maximum*. However, interference is not simply related to grass yield: higher yielding grasses gave less reduction in the girth of rubber than lower yielding grasses in another Sri Lankan study (Waidyanatha, Wijesinghe & Stauss, 1984). Negative effects of pasture on crop yield are expected in areas where

Table 2.1. *Shade tolerance of some grass and legume species*

Shade tolerance	Grass species	Legume species
High	*Axonopus compressus*	*Calopogonium caeruleum*
	Brachiaria miliiformis	*Desmodium heterophyllum*
	Ischaemum aristatum	*Desmodium intortum*
	Ischaemum timorense	*Desmodium ovalifolium*
	Ottochloa nodosum	*Flemingia congesta*
	Paspalum conjugatum	*Mimosa pudica*
	Stenotaphrum secundatum	
Medium	*Brachiaria brizantha*	*Centrosema pubescens*
	Brachiaria decumbens	*Desmodium canum*
	Brachiaria humidicola	*Leucaena leucocephala*
	Imperata cylindrica	*Macroptilium axillare*
	Panicum maximum	*Neonotonia wightii*
		Pueraria phaseoloides
		Vigna luteola
Low	*Brachiaria mutica*	*Calopogonium mucunoides*
	Digitaria decumbens	*Macroptilium atropurpureum*
		Stylosanthes guianensis

Source: Shelton *et al.* (1987).

Table 2.2. *Effect of stocking rate on coconut yield*

Treatment	Nut number (No. ha^{-1} month^{-1})	Nut yield (kg ha^{-1} month^{-1})
Sown pastures		
Stocking rate (beasts ha^{-1})		
2.7	263a*	507a
3.6	287a	516a
4.8	439b	713b
6.3	454b	779b
Natural pastures	291	483

* *Values in the same column followed by different letters differ at $P < 0.05$.*
Source: Rika *et al.* (1981).

rainfall is marginal for crop adaptation (Plucknett, 1979), but a comparison with clean cultivation systems is unrealistic, since this is rarely practised. The favourable effects of a close sod (such as occurs with *Stenotaphrum secundatum*) on moisture infiltration and the control of soil erosion should be taken into account.

2.2.4
Animal production

Shade usually reduces the nutritive value of grass. This may be reflected in lowered animal intake, and decreased *in vitro* digestibility, especially of stem tissue. It is attributed to reduced digestibility of cell wall constituents, to higher concentrations of lignin and crude fibre, and to lower soluble carbohydrate (Wilson & Wong, 1982). The relative proportions of digestible mesophyll tissue and indigestible vascular tissue are involved in this effect.

This question has been raised relatively recently, and requires more research; the generalisations stated above may not apply in all circumstances. It may be that the nutritive value of shade-tolerant species, such as *S. secundatum*, is less adversely affected by shade than that of sun species, such as *Pennisetum clandestinum*, for which reduced animal intake of shaded grass has been reported (Samarakoon, 1987). Animal liveweight gains are lower in plantations than on open pastures in the same district. This is sometimes confounded with differences in forage availability, but in situations where intake is not limited by forage presentation yields, individual animal gains are reduced in shade, despite the cooler microclimate of the plantation. For example, lightly stocked pastures on the Guadacanal Plains of the Solomon Islands gave average cattle liveweight gain of 0.53 kg hd^{-1} d^{-1} (Smith & Whiteman 1985*b*), relative to 0.38 kg hd^{-1} d^{-1} on lightly stocked pastures under coconuts (Smith & Whiteman, 1985*a*).

Despite this stricture, levels of animal production are high from pastures containing legumes. For example, in Western Samoa annual liveweight gain of cattle was 350 kg ha^{-1} on planted grasses with *Centrosema pubescens* under well-spaced coconut palms, and 180 kg ha^{-1} on *Axonopus affinis* with *Mimosa pudica* (Reynolds, 1981). In the Solomon Islands production was 360 kg ha^{-1} on pastures containing *C. pubescens* and *M. pudica* and growing under coconuts with light transmission of 62% (Smith & Whiteman, 1985*a*), whilst in west Bali a high figure of 550 kg ha^{-1} obtained under old palms (Rika *et al.*, 1981). In Malaysia production was 210 kg ha^{-1} under 1–3 year old oil palm, and 114 kg ha^{-1} under 5–7 year old oil palm.

2.2.5
Management considerations

The primary management decision, as for all pasture systems, revolves about the synchrony between stocking rate and pasture availability, as discussed in Chapters 6 and 7. For plantation agriculture pasture availability is conditioned by age and density of trees, and stocking rate is adjusted accordingly. For example, recommended stocking rates in the Solomon Islands for differing levels of light transmission might be: 35%, 0.7 beasts ha^{-1}; 45%, 1.0 beasts ha^{-1}; 50%, 1.3 beasts ha^{-1}; 60%, 1.6 beasts ha^{-1}; and 80%, 2.5 beasts ha^{-1}.

Rotational grazing is not recommended in this book, except in special circumstances which are discussed in Chapters 5 and 10. A rotational grazing system is favoured for convenience by some managers of coconut plantations; this is arranged so that labourers collecting nuts and brushing weeds follow the cattle. In Malaysia many grass and broadleaf weeds of rubber plantations are accepted by sheep, and short-duration grazing systems are being adopted to facilitate inexpensive weed control and to give dominance of legume covers.

Considerable scope exists for pasture improvement and livestock intensification in plantations, to the mutual benefit of both sectors of production.

2.3
Intensive pasture systems

2.3.1
Significance

Intensive pasture systems in the tropics are characterised by the replacement of native pastures with planted species and the use of fertiliser to mitigate the constraints of mineral deficiencies on pasture growth. Animals with a high genetic potential to respond to good quality feeds are used. Special interventions are directed to maintaining continuity of forage supply; the farm system often incorporates some irrigation, the planting of short-term fodder crops, conservation of fodder, and purchase of concentrates and grains from off-farm sources.

These systems occur in isolated pockets of the tropics and subtropics. Intensive dairy pasture systems are more common than intensive beef-systems. Dairying is often located in the cooler highlands of equatorial regions, as in Malaysia, India, Sri Lanka, Tanzania, Kenya and Mexico, or in the subtropics of Brazil, Argentina, southern Japan and eastern Australia. Some intensive rearing of young beef animals occurs in Florida and southern Japan, and intensive systems of both beef and dairying are found in Cuba. Lot fattening of beef with grain is outside the scope of this book; monogastric conversion of grain is more efficient than conversion by ruminants, and the latter is not indicated in countries where grain is in marginal supply for human needs. The dairies close to cities which operate almost wholly on external feed sources are also not considered.

2.3.2
Characteristics of the feed supply

(i)   The development of improved pastures has followed the clearing of forest and the planting of exotic species capable of exploiting the environmental growth factors newly available to herbage. The agronomic characteristics and adaptation of these plants have been widely reviewed (Bogdan, 1977; Skerman, 1977; Whiteman, 1980; Crowder & Chheda, 1982; Humphreys, 1981*b*, 1987). The main tropical grasses used in intensive systems are: *B. decumbens, Chloris gayana, Cynodon dactylon, Cynodon niemfuensis, P. maximum, Paspalum dilatatum, P. clandestinum, Pennisetum purpureum,* and *Setaria sphacelata* var. *sericea.*

Tropical pasture legumes are successfully used in some extensive systems; with the exception of some shrub legumes mentioned in Section 2.1 they have rarely persisted in intensive farm practice. Twining legumes such as *Neonotonia wightii* are capable of giving good levels of milk production; unsupplemented Friesians milking on *N. wightii* with *P. maximum* var. *trichoglume* near Atherton, Queensland (lat. 17° S, 1300 mm rainfall, 700 m altitude) produced 4100 kg milk per lactation of 330 days, when grazed at 1.3 cows ha^{-1} (Cowan *et al.*, 1974). It is unusual to record levels

of milk production greater than *c.* 12–13 kg hd^{-1} d^{-1} (averaged over the whole lactation) from tropical legume-based pastures, unless energy supplementation is practised or temperate pasture species employed. However this level of production is only feasible at modest stocking rates where animals are able to graze selectively to maintain a satisfactory dietary intake. Thus in the Queensland dairy industry the twining tropical legumes *Macroptilium atropurpureum* cv. Siratro and *N. wightii*, and the scrambling legumes *Desmodium intortum* cv. Greenleaf and *D. uncinatum*, which were widely sown in the 1960s and 1970s, have not persisted; this is associated with their poor resistance to heavy grazing and to the ravages of disease. These legumes are well maintained on lightly stocked properties, or on sections of the holding moderately grazed by growing or dry stock; they have a further specialised use as constituents of the fodder banks established for dry season use, as in Cuba and other parts of the Caribbean (Paterson, 1987). The use of shrub legumes in fodder banks is further discussed in Section 8.2.2.

(ii)   The increasing use of fertilisers and of soil amendments to overcome constraints to pasture growth occasioned by mineral deficiencies is a significant feature of intensive pasture systems in the tropics. On many dairy farms nitrogen fertiliser is now applied routinely at levels of 200–400 kg N ha^{-1} yr^{-1}, and stocking rates are increased to 2–3 cows ha^{-1}. This practice does not increase milk yield per cow above that attainable from legume-based pastures, but enables more animals to be grazed on the farm (Lowe & Hamilton, 1986).

Concurrently soil pH levels decrease as N fertiliser is applied, and although most tropical grasses are tolerant of soil acidity, lime application may be desirable if the pH moves below 4.5. Lime requirement is greater for clover-based systems. Other mineral deficiencies, commonly S or P, less usually K, together with micronutrients such as Mo, Zn, Mn, Cu or B, are applied to provide a balanced nutrient programme; the need for these is accentuated as the major constraint of N is overcome, and as nutrient removal is accelerated through increased milk production and animal sales.

(iii)   Various interventions are used to maintain continuity of forage supply. The timing and level of applications of fertiliser N can be manipulated in order to ameliorate the seasonality of pasture growth, as mentioned in Section 8.2.4. In subtropical and in highland tropical locations temperate clovers and grasses are grown, commonly in conjunction with irrigation. These are designed both to exploit the capacity for growth in the cool season and to provide a higher quality diet than is available from tropical species.

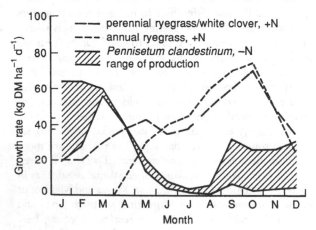

Fig. 2.7. Seasonal growth rates of perennial rye grass/white clover + N; annual rye grass + N; and the range of production (25–75 % of years), of *P. clandestinum* − N. (From Murtagh & Moore, 1987.)

The differences in seasonal growth of these components of the feed supply are illustrated (Fig. 2.7) for the Kyogle district (lat. 29° S, 1230 mm rainfall) of northern New South Wales (Murtagh & Moore, 1987). The growth of *P. clandestinum* pastures exceeds 60 kg dry matter (DM) ha^{-1} d^{-1} in January and February in 25% of years, but decreases to c. 3 kg DM ha^{-1} d^{-1} in the July winter. Irrigated pastures of *Lolium perenne/Trifolium repens* or *L. multiflorum* fertilised with N are capable of producing c. 40 kg ha^{-1} d^{-1} in July, and c. 75 kg ha^{-1} d^{-1} in October. Dairy farmers can budget their monthly feed supply according to the balance of irrigated *Lolium* and of rain-grown *P. clandestinum* on the property (Murtagh & Moore, 1987). The expectation is that a cow producing 15 kg milk d^{-1} from *Lolium* requires 73 MJ metabolisable energy (ME) d^{-1}, plus 54 MJ ME d^{-1} for maintenance energy, which will be met by 12.1 kg DM d^{-1} or 15 kg DM cow^{-1} d^{-1} at a level of 80% utilisation. Similarly, on *P. clandestinum* pastures with a lesser ME value the lower expectation of 11 kg milk d^{-1} may be met with a grazing requirement of 18 kg DM cow^{-1} d^{-1} at 67% utilisation.

A 50-ha farm carrying 125 milking cows with 30 ha of irrigated *Lolium* and 20 ha of *P. clandestinum* could meet most of the cow requirements if the frequency of irrigating *Lolium* and the fertilising of the *P. clandestinum* pastures with N were adjusted as required by the prevailing seasonal conditions; these management tools may be used to prevent surplus feed accumulation and feed deficiency. The combination of irrigated temperate pastures and fertilised tropical grasses has proved highly profitable to dairy farmers in subtropical Australia.

(iv) Some intensive dairy systems are heavily based upon annual forage crops. In southern Japan summer maize crops grown for silage or haylage and annual *Lolium* pastures grown in the cool season provide the main on-farm feed sources.

The rationale for using annual fodder crops rather than perennial pasture species revolves about several issues. (1) A high genetic potential for growth is found in annual summer-growing grasses with the C$_4$ photosynthetic pathway such as *Zea mays*, *Sorghum* spp. and *Pennisetum americanum*. These plants are larger seeded than equivalent perennial species and this is reflected in their superior seedling vigour; early growth also benefits from the mineralisation of soil N following cultivation. (2) Seasonal feed shortages may be overcome by the storage of soil moisture in fallowed land and the planting of cool season species such as *Avena sativa* and *A. strigosa*. (3) Erect forage crops have been selected for ease of harvesting in cut-and-carry feeding systems. (4) The integration of grain and forage production is facilitated by the flexibility attaching to a system incorporating annual fodder crops. (5) Crops such as *Zea mays* have superior characteristics for ensiling relative to perennial tropical grasses. This topic is expanded in Section 8.2.3.

(v) Despite the misplaced efforts of extension workers, pasture conservation has a minor role in tropical intensive systems. The problems of ensiling bulky perennial grasses of low sugar content and the climatic difficulties associated with producing hay from young pasture of reasonable nutritive value are described in Chapter 9. Forage conservation in the tropics is based on annual fodder crops and is mainly used to provide maintenance energy, especially for periods of seasonal feed scarcity. Conservation systems are rarely employed on farms with sizeable access to irrigation (Lowe & Hamilton, 1986) as the latter is a preferred option.

Significant advances in the technology of forage conservation in the tropics have engendered more confidence in the feasibility of conservation in capital-intensive enterprise. These features include: the erection of permanent storage and processing structures for haylage (the 'Harvestore'), the 'Silopress' system in which silage is compressed into plastic 'sausages' up to 40 m long, the production of round bales each containing up to 1000 kg and the improved mechanical handling of processed feeds.

(vi) Supplementation of grazing animals with high energy feeds is conventional practice for intensive dairy systems since the low ME intake from unsupplemented tropical pastures places a ceiling on milk production per cow which is unacceptably low for most farming economies; energy supplementation may also be used to overcome seasonal scarcity of forage. Grain feeding is additionally indicated where the milk from cows grazing pasture alone does not meet quality standards for concentration of Solids-Not-Fat (SNF).

In many dairy systems farmers feed perhaps one tonne of grain per lactation to each cow; grain is fed at a higher rate in the early phase of lactation (Davison, Jarrett & Martin, 1985) when higher response levels occur, and according to the genetic potential of the individual animal. Molasses is a common energy supplement on farms not too distant from sugarcane production mills; the response levels may be of the order of 1.25 kg molasses equivalent to 1 kg maize grain (Cowan & Davison, 1978).

Protein supplementation is not usually indicated on high-producing tropical pastures but mineral supplementation, as mentioned in Section 6.1.3, is often mandatory.

High energy supplementation of beef animals is practised in some specialised market situations (Tierney,

Evans & Taylor, 1983), and is appropriate where there is high internal price support, as in southern Japan.

The evolution of animal production systems in many tropical countries is towards intensification, since the growth in economic demand is outstripping the rate of increase in production. In Latin America economic demand for beef increased at an annual rate of 5.3% (1971–81) whilst beef production increased at 2.2% per year (CIAT 1985).

2.4
Sedentary ranch systems

2.4.1
Improved pastures

(i)   Most of the planted pastures in the tropics are on lands which were formerly forest or scrub, and occur in humid and subhumid zones. African grasses introduced to Latin America were planted or became naturalised in areas to which settlement expanded; *Panicum maximum* was present in Barbados in the seventeenth century and was spread to Brazil, and *Hyparrhenia rufa* was another early introduction which was widely used in Brazil, Venezuela, Colombia, Peru and the Caribbean region. *Melinis minutiflora* and *Brachiaria mutica* were used in Latin America in the eighteenth century, whilst *P. clandestinum* and *Digitaria decumbens* were more recent acquisitions from east and southern Africa (Parsons, 1972). These grasses were mainly planted on the more fertile forest soils, and their productivity was poorly sustained in the long term on the more infertile oxisols and ultisols. In the 1970s *Brachiaria decumbens* (and to a lesser extent *B. humidicola*) were widely planted on the acid soils of the cerrados and other regions on some 15 million ha; subsequently problems with spittlebug (*Zulia columbiana*) led to the use of other grasses, and *c.* 0.3 million ha of *Andropogon gayanus* were established by 1987 (CIAT, 1988). Concurrently the extension of cattle production in the Amazon basin following the destruction of rainforest was based initially upon *P. maximum*, sometimes planted with *Pueraria phaseoloides* and *C. pubescens*; these pastures have suffered considerable run-down in phosphorus (Serrao *et al.*, 1979).

In northern Australia coastal forest lands have been cleared and developed for dairying and beef production, whilst further inland predominantly *Acacia harpophylla* woodland has been replaced by *c.* 1.5 million ha of planted pastures of *Cenchrus ciliaris*, *P. maximum* var. *trichoglume* and *Chloris gayana* (Weston *et al.*, 1981), which are used for beef production. There are also substantial areas of introduced species which have become naturalised; *C. ciliaris* has colonised the Cloncurry river valley, and *Bothriochloa pertusa* predominates in the Bowen basin of Queensland.

The balance of improved and natural pastures in sedentary ranch systems hinges on the nature of the original vegetation; replacement of closed forests requires introduced herbage plants, whilst in open woodland situations the degree of reliance upon planted or native pastures is a question for managerial decision.

The degree of intensification on ranch properties reflects local social and economic conditions, as well as the availability of feasible technology. Many large enterprises based on planted pastures maintain their competitive efficiency through the use of low input systems with high product output per worker, whilst smallholder production uses high labour inputs. Three concrete illustrations of diverse enterprises are given.

(ii)   The first example is of dual cattle production (meat/milk) from 'family' farms in the Andean Piedmont of Caqueta (lat. 1° S, 3500 mm rainfall). Amazon forest was cleared and planted mainly to *H. rufa* and *Axonopus scoparius* (CIAT, 1988). On these low pH ultisols the pastures degraded to less productive species of *Paspalum* and *Axonopus*, locally called 'criaderos'. Farm size averaged 131 ha, of which 95 ha were pasture: 65% criaderos, 27% *B. decumbens*, 3% *H. rufa* and 5% other species. The average cropping area of 4 ha was mainly devoted to subsistence crops of rice, plaintains and cassava, and to minor cash cropping with maize. Primary forest had disappeared on 20% of farms and averaged 9 ha per farm.

The replacement of criaderos with *B. decumbens* proceeded slowly from 1969, when it was first planted in the district, to 1978 when it was grown on only 15% of farms; by 1986 it was grown on 97% of farms.

Average herd size was 121 head (49 cows) of which 72 were owned by the operator. Labour units per farm averaged 3.8, of which half were family. The balance of milk/fattening was higher on the smaller farms. Mean calving rate was 61%, milk output averaged 580 kg per lactation, and mean cattle growth was 83 kg h^{-1} yr^{-1}. In this high rainfall area relatively high stocking rates of 1.3–1.5 beasts ha^{-1} apply in the farm situation but the low output levels per head must reflect the constraints to animal nutrition which apply.

(iii)   The cerrados of central Brazil (lat. 10° S to 24° S) occupy *c.* 150 million ha and provide the major resource for cattle production in that country. The vegetation is categorised as campo limpo or grassland, campo sujo or grassland with occasional shrubs, campo cerrado or open savanna, cerrado or closed savanna with low, open tree forms, and cerrado or closed savanna with higher, more closed tree forms. Oxisols, which are deep, well-weathered soils of good physical structure but with high aluminium saturation and P fixation, predominate together with entisols.

Planted grasses are gradually replacing the native vegetation, often following a pioneering cropping phase with rice. A simulation study (Barcellos *et al.*, 1979) contrasts the situation of a ranch following the traditional native pasture system (total area 6590 ha, 300 ha *H. rufa* pasture, 4970 ha rangeland, and 1320 ha reserve areas) and a ranch of the same characteristics in which 1040 ha (19% of total area) were cultivated to pasture, using P inputs. The ranch is used for breeding operation; the traditional system provides a maximum of 1731 animal units (AU, where breeding cows and three-year old heifers rate as 1 AU), whilst the improved system gives 2469 AU. Stocking rate was estimated as 0.3 AU ha^{-1} for rangeland, 0.8 AU ha^{-1} for *H. rufa*, and 1.3 AU for *B. decumbens*.

Production parameters after 10 years for the traditional and the improved systems respectively are: mortality rate for calves 8% and 5%, for other stock 5–6% and 2–4%, calving rate 45% and 70%, bull:cow ratio 1:15 and 1:25, liveweight production 17 and 38 kg ha^{-1} overall. Beef production in the improved system showed little increase until the seventh year, since young females were retained to build up the breeding herd; thereafter excess heifers contributed substantially to sales. The economic parameters used predicted an internal rate of return of 15% on the investment, which is considered satisfactory; other economic studies in the cerrados have been less sanguine (Monteiro, Gardner & Chudleigh, 1981).

(iv) This example is of a large-scale breeding and fattening cattle enterprise in which native pasture is the predominant feed source and in which the development of planted pastures on a small area of the property has been associated with a substantial increase in output of cattle.

'Wrotham Park' is a leasehold property in north Queensland (lat. 16° N, 960 mm rainfall, wet/dry climate) which operates on an aggregation of 1.0 million ha (Edye & Gillard, 1985). The vegetation is open *Eucalyptus* or *Melaleuca* savanna with *Shizachyrium fragile*, *Sorghum plumosum*, and *Chrysopogon fallax* as the predominant understorey grasses; the soil is a P-deficient alfisol. Over the period 1971–82 an average of 45 000 cattle (*c*. 26 000 breeders) were carried at 0.045 beast ha^{-1} on native pasture.

Pasture development was carried out on 16 000 ha (1.6% of the property) from 1967–83 at a total cost of A\$0.7 million. The annual legume *Stylosanthes humilis* was initially planted, but was superseded in 1974 by *S. hamata* cv. Verano because the disease *Colletotrichum*

gloeosporioides attacked *S. humilis*. Timber was cleared and burnt and the area roughly ploughed for the first phase of development on 8000 ha; subsequently sowings were made on uncleared areas. Single superphosphate was applied at 220 kg ha^{-1}, and paddocks of *c*. 2400 ha were fenced and supplied with stock water. Pastures were grazed at 0.5 beast ha^{-1} immediately after sowing, and thereafter at *c*. 0.4 beast ha^{-1} these high stocking rates reduced the grazing pressure on the unimproved pastures. *S. hamata* has maintained dominance, and the perennial grasses have been replaced by the annuals *Digitaria ciliaris* and *B. miliiformis*.

Animal output from the property has fluctuated with seasonal conditions and cattle prices. In the period 1964–69 sales averaged 5280 head yr^{-1}, and these increased to 7730 head yr^{-1} in 1978–83. Slaughter weights also increased, giving a net increase in total output of 70%; cattle were sold at 3–4 years of age relative to 5–6 years in 1964–69. The pasture development and improved management have led to a substantial increase in output per unit of labour. This is a striking example of how pasture legumes planted on a small fraction of a property and directed at a key constraint in the production process (the rate of fattening) lead to a significant increase in both output and efficiency.

2.4.2
Natural pastures

(i) A description of the natural pastures in different tropical regions of the world and their ecological relationships is outside the scope of this book, but many reviews are available (Whyte, 1968; Eiten, 1972; Coupland, 1979; UNESCO/UNEP/FAO, 1979; Huntley & Walker, 1982; Boulière, 1983; Sarmiento, 1984; Tothill & Mott, 1985). Most of the world's tropical livestock production comes from natural rather than planted pastures; for example, sedentary ranching in Queensland derives an estimated 11.7 million beef equivalents or 89% of the total carrying capacity from natural pastures, whilst sown pastures provide only 1.4 million beef equivalents (Weston *et al.*, 1981). Whilst gradual further transition to planted pastures and crop sources of feed may be expected in humid and subhumid zones, the semi-arid and arid regions will be retained as natural grazing in varying degrees of disturbance.

It is convenient to categorise natural grazing lands as follows (Humphreys, 1981*a*). (1) Swamp grasslands, as for example in the Bajo Llano of Venezuela and Colombia, are a valuable resource, since they extend the season of green forage availability; parasite problems may restrict their utilisation in the wet season. They form an edaphic grassland climax in which the soil conditions of salinity, impeded drainage or flooding are more suited to pastures than to trees or to arable systems. (2) Watershed grasslands occur as a grassland disclimax in the humid and subhumid tropics on lands which were formerly forest. They are often the end point following slash and burn cropping systems, in which increasing frequency of fires and cultivation together with declining soil fertility have led to the invasion of tall grasses, such as *Imperata cylindrica* in south-east Asia. (3) Montane grasslands, as in east Africa and in the Andes, occupy relatively small areas of the tropics but usually receive good rainfall and have a high potential for dairy development. (4) Savanna is found in every continent, and is a category of tropical and subtropical vegetation which grades from woodlands with a sparse understorey of herbage, to grassy open forests, shrublands, and to 'true' grasslands, in which shrubs and trees are virtually absent (Tothill & Mott, 1985); all forms have a continuous graminoid stratum.

(ii)   The ruminants and equines used in sedentary ranching on natural pastures reflect varying adaptation to climate and to feed characteristics; their distribution is modified by social factors (Table 2.3).

The water buffalo (*Bubalus bubalis*) is not found in the arid and semi-arid zone (Devendra, 1987); it is a larger sized (400–600 kg) stocky ruminant with low hair density, limited sweat glands and a grey skin which reduces heat load and which is indented. Cattle (*Bos indicus*) in the dry zones are taller, larger, have a loose skin and appendages, and are able to walk long distances, whilst cattle in humid areas tend to be more compact, with thicker skin, higher hair density and more sweat glands than buffaloes. Goats (*Capra hircus*) are also larger, taller and have longer legs and ears in the dry zone than in the humid regions; their fat reserves are less but their capacity to withstand dehydration and to survive in dry areas is superior to that of sheep (*Ovis aries*). Sheep tend to be larger and to have longer legs in the dry than in the wet zones. Camels *Camelus*

dromedarius) have long legs and neck, a single hump, and special mechanisms for resisting dehydration; they are mainly husbanded in the arid and semi-arid regions (Table 2.3). Equines (*Equus asinus*) are of small stature in the dry zone and have a relatively small stomach. Camels are used for milk, meat, hair, draught and dung; the other species provide these in varying degree. Herds often comprise a mixture of animal species. Sedentary ranching of game animals has its advocates but is only practised in isolated instances, although game animals generally make a significant contribution to food, hides and recreation. The following points are concerned with management interventions used on ranches.

(iii)   The number of animals grazed on unit area, or the stocking rate, is the most significant management decision which determines the profitability and sustainability of the ranch enterprise. Effects of stocking rate on botanical composition and on animal production are discussed in Chapters 5 and 7 respectively.

The attempt to synchronise animal appetite with forage supply is difficult in pastoral situations where seasonality of pasture growth is marked, and pastures are usually underutilised, since the stocking rate is set by the rate at which animals will survive the dry season. Grazing pressure (animals per unit of feed availability) fluctuates greatly, since animal density usually changes less rapidly than pasture availability. In extensive production systems the nutritional stresses occasioned by dry seasons regulate animal numbers through their effects on animal mortality and reproductive performance independently of managerial decision. It is claimed that the lag in the build-up of animal density following the breaking of a drought provides a greater opportunity for recovery of the pastures than may occur under more controlled management. Certainly abiotic factors predominate in determining botanical change in semi-arid and arid zone pastures, as described in Section 5.2.8.

A herd structure based on breeding and young stock is more vulnerable to seasonal nutritional stress than is

Table 2.3. *Distribution of ruminant animals in the arid and humid tropics*

Climatic zone	Water buffaloes	Cattle	Goats	Sheep	Camels
Distribution (%)					
Arid/Semi-arid	0	13	28	34	89
Subhumid	85	77	62	63	11
Humid	15	10	10	3	0
Total population ($\times 10^6$)	79	543	271	213	14
% of world population	63	47	59	19	79

Source: Devendra (1987).

a mixed herd, in which growing animals may be disposed of at varying ages, according to feed availability and market opportunity. The policies concerning purchase and sale of animals, culling of breeding females, and time of mating are used by managers to minimise discontinuity in forage supply.

(iv) The effective occupation of the property requires a spatial distribution of water points that allows animals to graze all the pastures for most of the year; this also minimises the destruction of vegetation which occurs on stream frontage country and about isolated stock watering points (the 'piosphere', Lange, 1969). On large holdings a developed infrastructure of water points enables animals to graze on the areas of the property which may have received unusually favourable rains, or where the moisture retention capacity of the soil sustains a longer period of green feed availability. The installation of water points usually has priority for capital investment, as this increases stock carrying capacity and minimises animal stress; fencing to provide stock control has an important but secondary priority.

The distance animals will walk to and graze out from water points varies with animal species, breed, stock condition, and with the location of preferred feeds and their level of availability, since animals tend to optimise the energy used involved in grazing (Squires, 1982). In western Queensland the upper limits for distance to water are regarded as *c*. 16 km for cattle and 5 km for sheep; on properties where the stocking rate is 0.8 sheep equivalents per ha graziers might attempt to develop stock watering so that sheep walked no more than 1.5 km to water, or cattle no more than 5 km.

(v) The reduction of tree and shrub density, and the control of tree and shrub encroachment form a central preoccupation for many graziers. In Africa this may arise from the need to reduce the occurrence of trypanosomes, associated with the forest habitat of the tsetse fly; the use of trypanotolerant breeds, the eradication of the fly or the introduction of vaccines (Murray *et al.*, 1979) would modify this requirement. More commonly trees are seen as competitors with herbage.

The objectives of woodland management (Burrows, 1985) are: (1) to maximise pasture production by reducing tree density. The negative relationship between tree density and herbage availability is a powerful one, as illustrated for two sites in central Queensland (Fig. 2.8); at site 2, where the productive potential is low, a density of only 150 trees ha^{-1} reduces the herbage to 17% of the yield attained in the absence of trees. The negative effect of trees on herbage is more severe in the subtropics, where competition for water

during the growing season is more acute ·than in monsoon tallgrass savannas (Mott *et al.*, 1985). Loss of herbage production is of less moment where browse is edible and provides a longer season of green feed availability. (2) Tree regrowth may be minimised by adoption of suitable arboricides, choice of season for clearing, pattern of clearing (strips or clumps retained), and by appropriate grazing and firing management (Section 10.3). (3) Tree removal is avoided on steep slopes greater than *c*. 20% and on land susceptible to salting. Reserves of timber are retained on the property as required for fire breaks, fuel, fencing and construction of shade and shelter for animals. (4) Trees (preferably edible) are re-established to maintain a varying age structure in the tree population, or if overclearing has occurred.

(vi) Fire is a simple management tool which is widely used on native pastures in the humid and subhumid zones; it is less prevalent in drier areas where the fuel continuum is less and where surplus feed is viewed more positively by pastoralists. Fires are lit (1) to increase the accessibility of green pasture; (2) to control weeds which are fire susceptible; (3) to provide more equitable competitive relations between preferred and non-preferred species; (4) (hopefully) to stimulate out-of-season growth; (5) to assist ease of mustering; (6) to moderate future risk of fire; and (7) to minimise pests and disease. The use of fire has decreased in pastoral communities which recognise the value of low quality roughage when fed with supplements such as urea/molasses, or who have become conscious of the environmental damage associated with some burning systems. The topic is treated more fully in Section 10.3.

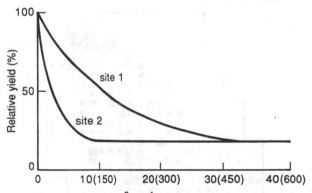

Fig. 2.8. Effect of tree basal area (m² ha^{-1}) and tree density (number ha^{-1}) on relative pasture yield (site 1, yield in open 3500 kg ha^{-1}; site 2 yield in open 1000 kg ha^{-1}). (From Burrows, 1985.)

(vii) Other aspects of ranch management include the provision of mineral supplements (Section 6.1.3), energy and protein supplementation (Section 9.3.1), measures to preserve animal health, and action to control predators and theft.

2.5
Nomadic pastoral systems

The term 'nomad' derives from the Greek *nomos* = pasture, so that the seasonal use and deferment associated with nomadic grazing are central to the origin of our conception of pasture. The biological boundary between sedentary and nomadic systems is gradual, since on large

sedentary holdings animals will move within that holding according to feed quality and availability (Ridpath, Taylor & Tulloch, 1985) or there may be considerable animal traffic within an aggregation of pastoral leases which are spatially separated.

Pastoral systems may be totally nomadic, in which case the animals owners have no permanent place or residence and cultivate no crops; the families move with the herd (Ruthenberg, 1980). Transhumance, or semi-nomadism, implies that stock owners have permanent or semi-permanent places of residence, usually near to land on which the family may cultivate crops, but travel with the herd for long periods away from their settlement area. In many countries government policies are directed to fostering more sedentary systems where defined grazing

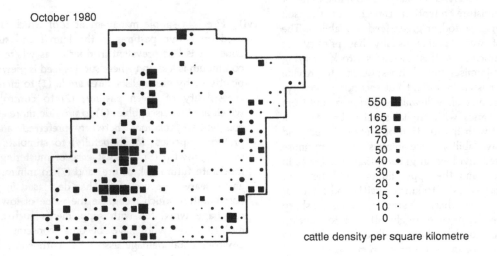

October 1980

550 ■	
165 ■	
125 ■	
70 ■	
50 ▪	
40 •	
30 •	
20 ·	
15 ·	
10 ·	
0	

cattle density per square kilometre

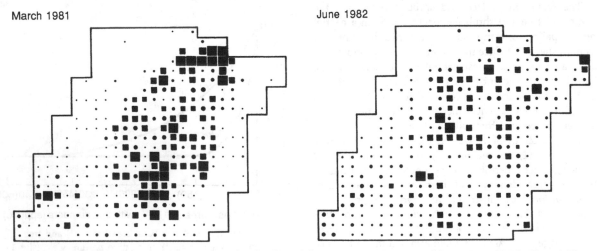

March 1981

June 1982

Fig. 2.9. Seasonal distribution of cattle density (no. km^{-2}) at different times in the Niger delta region of Mali. (From Milligan & de Leeuw, 1983.)

areas are adjacent to the settled areas. Pastoral systems may be defined as those where households are more than 50% dependent upon livestock products, whilst in agropastoral systems families are 10–50% livestock dependent.

The first objectives of the pastoralist are (1) 'survival in a risky environment, and (2) maximising his output' (Cossins & Upton, 1988). The main characteristics of nomadic systems are as follows:

(i) The movement of herds is dictated by the spatial variation in feed availability and by the seasonal accessibility of stock water. Operations are conducted on land which is communally used. This does not imply unrestricted access for all comers; there are defined rights of use or a hierarchy of rights of access which are periodically adjusted between particular communities. Animals in milk may be retained closer to settlement whilst dry stock are herded on more remote pastures. Market considerations may also modify stock movement. In the Fezzan and southern Algeria camel herders may trek 500 km yr^{-1}, or 700–800 km if forage is scarce. Camels may graze 30–80 km from water points, where 4–8 km might be more normal behaviour for cattle. In many arid and semi-arid regions the occurrence of rain may be locally restricted, and mobility of herding contributes to efficient pasture use.

In regions with a unimodal rainfall distribution the dry season migration may be to lands where cropping residues are available, as in the case of the Fulani in West Africa, or to perennial floodplain grasslands with an extended season of green feed availability, as on the floodplains of Benin or Nigeria, or the shores of Lake Tchad. Cattle in the Niger delta region of Mali amount to *c.* 1.2 million in the period of greatest use, and *c.* 0.5 million sheep and goats are also grazed (Milligan & de Leeuw, 1983). During October (Fig. 2.9) high cattle densities occur in the transition zone between the delta and the higher country, and few cattle are in the central delta where the flooding is deep, or in areas where water is absent. In March *c.* 70% of the cattle have moved to the floodplain, in two distinct north and south aggregations, but by June many cattle have migrated elsewhere.

(ii) Herds of mixed species composition are the norm in East Africa (Cossins, 1983). More uniform utilisation of mixed vegetation occurs if animal species using browse or herbage in varying degree are combined. Camels and goats are less susceptible to drought than sheep or cattle, and changing seasonal patterns will alter the balance of these animals. In some environments small ruminants are regarded as being more susceptible to disease and to predators than are large ruminants.

The production system of the transhumant Borana of southern Ethiopia illustrates the flexibility of herd composition and its adjustment according to seasonal conditions and to herd size (Cossins & Upton, 1987). An average household (Fig 2.10) comprises a man, his wife and two children or three active adult male equivalents, whose total energy requirements approximate 11 600 MJ yr^{-1}. The basic food is milk from cattle, which have *c.* 920 kg total production per lactation; the offtake is *c.* 312 kg per lactation, which represents *c.* 60% of household energy needs. As in most pastoral systems, the needs of the family and of the calf for milk are in competition. The reduced milk supply to the calf increases calf mortality and delays maturity; first calving of *c.* 70% heifers occurs at more than four years of age. This penalty is reduced in larger herds where owners usually milk 35–45% of cows, relative to 65–70% of cows in smaller herds. 'Fallen meat' or cull cows and animals which die contribute *c.* 5% of household energy requirements, and surplus cattle are sold. The ratio of females:males varies with the degree of dependence on

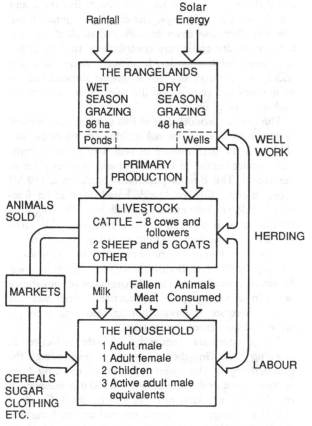

Fig. 2.10. The Borana pastoral system. (From Cossins & Upton, 1987.)

milk, and decreases as trading requirements are perceived as more important. The herd (Fig. 2.10) also includes goats and sheep, and perhaps camels for milk and carriage, and equines. The small stock is raised for sale, and this provides grain and other food, which supplies the residual 35% of family energy needs, together with cash income for the other household requirements, averaging perhaps US$100–150. At the time of this study 1 kg animal liveweight purchased 2.5 kg grain, which is a favourable energetic transfer for the pastoralist. Grain is cultivated in some favoured situations, and householder production of grain could be used to substitute partially for milk to the advantage of calf production; alternatively the growing of supplementary fodder crops for young stock might benefit herd performance (Cossins & Upton, 1987).

In this region primary plant production varies from 1.5 t ha^{-1} yr^{-1} in the western zone to 2.7 t ha^{-1} yr^{-1} in the northern zone. Families each have access to c. 86 ha of pasture during the wet season, when surface ponds provide stock water. During the dry season the grazing area is restricted to c. 48 ha, and considerable family labour is required to water stock from deep wells: cattle every three days, sheep and goats every five days, and camels every 8 to 14 days. The dry herd is grazed more distantly than the cows in milk and small stock, and blood from dry cattle may contribute up to 25% of the food needs of the herder. In drought years there is a wider migration of dry stock, greater dependence on small stock and camels, and the sale of many animals to exchange for grain.

The overall productivity of this pastoral transhumant system based on communal grazing compares favourably with the performance of sedentary ranch systems with similar rainfall in East Africa and in other parts of the world. The Borana pastoralist produces c. 119 MJ gross energy (GE) ha^{-1} yr^{-1}, based on 6.25 kg liveweight (LW) offtake and 21 kg milk ha^{-1}; this might be contrasted with c. 96 MJ GE ha^{-1} from 19 kg LW from the Laikipia ranches in Kenya, and with still lower figures from ranches in northern Australia (Cossins & Upton, 1987). The Borana level of production does, however, provide only a small margin above subsistence needs in an average year. A reduction in stock numbers would therefore aggravate human need, unless people are withdrawn from the system.

Range scientists frequently advocate reduction in stocking rate in the interests of improving the productivity of the vegetation and of reducing bush encroachment and soil erosion; the social consequences of adjusting stocking rate need recognition. Sandford (1983) also suggests there is no evidence that varying stocking method ('rest and rotation' systems) in an attempt to improve the productivity of vegetation has been successful in dry tropical Africa. On the other hand, the development of more and better watering facilities extends the access of stock to forage; again, uncontrolled access extends range degradation.

Many governments seek to impose sedentary systems on nomadic or transhumant populations. This may improve the quality of drinking water for the household, and provide better opportunities for health care, other forms of welfare, and education. It also favours taxation and military conscription; the goals and preferred lifestyle of the pastoralist may be at variance with government policies. It has also to be recognised that to this point the development of sedentary systems involving a restriction of grazing rights has usually led to more rapid degradation of the range resource and a decrease in animal output from the range.

In contrast to the situation for pastures in humid and subhumid regions, there are few technical inputs generated by pasture research which can be applied in semi-arid and arid regions. The problems of people in these regions are acute, and the sustainability of some of the pastoral systems are endangered. The solutions to these problems are essentially social, cultural, economic and organisational; there is no intention to minimise their seriousness by regarding them as lying outside the scope of this book.

In this chapter pastures in tropical farming systems have been broadly grouped to 'reflect fundamental structural differences' (Fresco & Westphal, 1988) and according to major variations in the technological problems encountered. A system is 'an arrangement of components or parts that interact according to some process and transform inputs into outputs'. There are many ways of viewing systems concepts applicable to human activity systems (Pearson & Ison, 1987). The segments of this chapter have provided descriptive material to illustrate the main features of different farming enterprises, and to introduce the central but different technological problems which bear on the efficiency of each type of enterprise.

The work of grassland scientists in seeking to improve the operation of farming systems and the welfare of rural communities functions at three interacting levels (Humphreys, 1986b): (1) understanding the farm situation; (2) describing and developing the ecosystem; and (3) removing constraints to plant performance and utilisation. Efforts to improve the situation of farmers require true assessments of the goals, skills and resources of farmers as well as comprehension of the physical and biological constraints to farm production and of changing market incentives. The grassland scientist seeks to quantify options for pasture improvement and management. These are designed for farmers who have varying skills, are engaged in production with varying intensity of input such as stock, capital and fertiliser, who have different production outputs, and who differ in their acceptance of risk.

The succeeding chapters fall into two groups: those concerned with the pasture response to utilisation, and those which focus on the animal response to the feed supply. The component processes discussed need to be viewed in the holistic context of their interaction with other processes in the farming system. The treatment is intended to be sufficiently general to provide principles which may be tested by the reader in diverse but particular tropical pasture ecosystems.

3

Grazing and the environment for pasture growth

Tropical and subtropical environments have distinctive features which control the growth and persistence of pastures. Many tropical soils are acid, of low base exchange capacity, highly weathered and leached, and subject to greater rates of breakdown of organic matter and erosion than occur in temperate regions (Sanchez & Tergas, 1979; Whiteman, 1980).

The warmer climatic conditions lead to the predominance of grasses with the C_4 dicarboxylic acid photosynthetic pathway, and these have a greater photosynthetic output than the C_3 grasses, especially in the high radiation conditions of the subtropics (Whiteman, 1980). The amplitude of variation in daylength between seasons is positively related to latitude, but many tropical pasture plants are sensitive in their flowering responses to small changes in photoperiod (Humphreys & Riveros, 1986). The rainfall regime varies from the truly arid desert areas of north Africa and north-west Australia, through the Köppen Aw wet and dry climates which encompass much of the savanna grasslands of Latin America and central Africa, to the truly humid equatorial regions of south-east Asia and the Amazon. The higher insolation of the tropics causes greater evapotranspiration, and acute moisture deficits can occur in soils which receive a high annual rainfall of intermittent or seasonal incidence. The combination of edaphic and climatic factors leads to a biotic environment of tropical organisms which exhibit special features with respect to the cycling of nutrients and the flow of energy between plants, animals and soil. This chapter deals with the ways in which grazing or cutting management modifies the environment in which the pasture grows.

The grazier perceives the primary activity of the ruminant as eating, and the effects of defoliation on pasture growth are discussed in Chapter 4. We should not neglect the lesser understood action of animal hooves on pasture growth and on soil characteristics. Stocking rate and the method of grazing modify plant cover, which influences the infiltration of water into the soil and the degree of erosion. Animal excretion, described as 'the shower of fertility', effects a spatial redistribution of mineral nutrients which is influenced by grazing management and which creates problems of environmental pollution if animals are confined or housed in cut-and-remove feeding systems.

Stocking rate influences the rate of nutrient cycling; it also determines the degree of litter accumulation and the partitioning of assimilate to subsurface plant organs. These factors influence the decomposer industry and the amount of energy which is captured in animal products, and change the microenvironment which influences the development of insects, the larger soil-inhabiting organisms, fungi and bacteria. The microclimate with respect to light interception and light quality, temperature of the soil and at the base of the sward, and moisture status are also modified by defoliation practice.

The ways in which management interventions affect the (1) edaphic, (2) biotic and (3) climatic environment are considered.

3.1
The edaphic environment

3.1.1
Soil compaction and animal treading

(i) Animal hooves have direct effects on pasture growth, and indirect effects on soil structure which influence subsequent pasture performance. These effects are modified by animal species and behaviour.

The sharp-edged hooves of most grazing animals are deleterious to the pasture and to the soil. Grazing animals influence the pasture environment by their activities in walking and resting, and also cause disturbance by pawing and digging the ground or, in the case of buffaloes, by wallowing.

The static load on the pasture depends on the ratio of animal biomass to total hoof area; common values are 0.7–0.9 kg cm^{-2} for sheep and 1.3–2.8 kg cm^{-2} for cattle (Spedding, 1971). The effects on the pasture are greater when animals are moving, and the dynamic pressure may be twice the static load. Pressure is greatest when animals are accelerating; in addition to vertical compression there is horizontal rotary force when the hoof leaves the ground, and there are shear and kick components. It follows that management directed to moving animals slowly and peacefully about the pasture is less detrimental.

Analogies may be drawn between the animal treading and the influence of wheel traffic associated with activities of crop protection, fertiliser application, conservation and general farm movement.

(ii)   The direct effects of treading on the pasture are evident in the bruising, crushing and tearing of leaves and stems; in extreme cases plant crowns and roots are damaged. Tall pasture is knocked down and trampled, so that plant material may be buried, especially in wet conditions.

The negative effects of treading on plant growth are illustrated for a susceptible legume, *Lotononis bainesii* (Pott, Humphreys & Hales, 1983). It is difficult to quantify these effects in farm practice, since treading is usually confounded with defoliation and with differences in plant cover. In this study a mixed *Lotononis bainesii/D. decumbens* pasture was mown every 3 months to 3 cm height and the cut material removed. A flock of sheep of 1000 kg biomass was then walked up and down the pasture a varying number of times, to simulate different stocking rate (SR) equivalents, in which 3 months' normal walking was applied within a single day. This produced very similar changes in soil bulk density (0–7 cm) to those which occurred in a contiguous grazing experiment in which sheep were grazed throughout the year at the same stocking rates (Pott & Humphreys, 1983).

Shoots of *D. decumbens* were badly damaged, but growth recovered rapidly from underground buds on rhizomes and from persisting stolons; regrowth was independent of treading intensity. By contrast the cumulative regrowth of *L. bainesii* (Fig. 3.1) was reduced from 178 g m^{-2} in the zero treading treatment to 45 g m^{-2} at 28 sheep equivalents ha^{-1}; on some occasions the proportionate effect was more severe, especially in the first 30 days of regrowth. Young seedlings died by the rupture of the exposed hypocotyl and cotyledons or by plant uprooting and root exposure; more advanced seedlings with stolons produced regrowth from coronal axillary buds on the buried crown and from stolons. Treading of *P. dilatatum*

(Hunt, 1979) caused death of tillers, but this was followed by increased tillering. The relative yield of 68% at 60 sheep equivalents ha^{-1} was due to reduced growth per tiller and not reduced tiller density.

(iii)   Effects of treading on the population dynamics of swards are more problematical than the negative effects observed on pasture growth.

Treading is expected to increase plant mortality and to reduce the life-span of seedlings, although this does not occur at every treading occasion. Negative effects on plant growth may decrease seed yield and *L. bainesii* seed reserves (0–4 cm) decreased from 21 000 seeds m^{-2} at zero treading to 13 500 seeds m^{-2} at 28 sheep ha^{-1} (Pott *et al.*, 1983).

On the other hand, soil disturbance by treading may favour seedling regeneration. Hoof cultivation is used to create a 'gap' by the destruction of competing vegetation; it may be used deliberately to produce a seedbed for the establishment of pastures (Mohamed-Saleem, Suleiman & Otsyina, 1986). Improved soil/seed contact, which improves the moisture supply to the

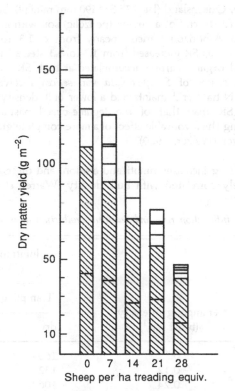

Fig. 3.1. Summation of *Lotononis bainesii* yield of 90-day regrowth periods following 6 successive treading occasions (plotted from the base up) of varying intensity. (From Pott *et al.* 1983).

seed, and seed burial to favourable depths have positive effects on seedling regeneration, provided a compacted topsoil is not inimical to coleoptile emergence or the penetration of the soil by the emergent radicle. Following one treading occasion there was a positive association of *L. bainesii* seedling density with treading intensity which varied from 32 to 58 seedlings m^{-2} respectively over the range 0–28 sheep ha^{-1} (Pott *et al.*, 1983).

(iv)   Treading which causes a packing of the soil particles and a loss of the larger pores in the soil mass results in an increase in bulk density. The damage resulting from the mechanical disturbance of a wet soil, known as 'poaching', 'pugging' or 'puddling', arises when the bearing strength of the soil and vegetative cover is exceeded. These changes in total porosity, in pore size distribution and in aggregate stability reduce aeration, moisture infiltration, moisture retention and drainage. Resistance to root penetration is increased and the biological and chemical activity of the soil is decreased.

Soil bulk density increases with SR (Fig. 3.2). After 4 years' grazing of a mixed pasture of *S. sphacelata* var. *sericea/M. atropurpureum/S. guianensis* at Tedlands near Mackay, Queensland (lat. 22° S, 1490 mm rainfall), bulk density (0–10 cm) of a duplex (podzolic) soil with light textured A-horizon varied linearly from *c.* 1.3 to *c.* 1.5 g cm^{-3} as SR increased from 1.2 to 3.3 steers ha^{-1}. Less soil organic carbon accumulated at high SR. Pure grass pastures of *S. sphacelata* var. *sericea* receiving 300 kg N ha^{-1} yr^{-1} maintained a lower bulk density at higher SR than that of the legume-based pastures, indicating the favourable effect of a more complete grass sod cover (Walker, 1980).

(v)   Treading increases runoff and erosion, and these are positively correlated with bulk density (Warren *et al.*,

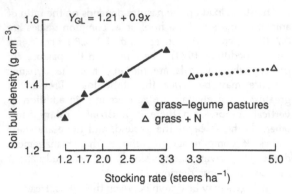

Fig. 3.2. Effect of SR on soil bulk density near Mackay, Queensland. (From Walker, 1980).

1986). Proponents of short-duration grazing systems (Savory, 1983) may claim that hoof action will enhance infiltration and reduce erosion, but there is no good evidence to support this.

The confounding effects of plant growth and cover on soil responses were removed in an experiment at Sonora, Texas (Warren *et al.*, 1986), where the vegetation was killed with herbicide and variation in cattle trampling of a silty clay soil studied during the dry summer period. In a series of trampling events plots were left in a dry condition or irrigated with 10 mm water. Simulated rainfall was then applied at 203 mm h^{-1} for 45 minutes. Infiltration rate after trampling (Table 3.1) was negatively related to stocking equivalents whilst the amount of soil erosion (sediment production) increased sharply with SR. Some recovery of soil characteristics occurred during the rest periods between trampling but effects on soil characters persisted. Trampling when the soil is moist reduced aggregate stability, and this has the effect of increasing runoff and erosion.

Table 3.1. *Infiltration rate and sediment production after simulated rainfall in relation to trampling intensity*

Trampling intensity (beast ha^{-1})	Infiltration rate (mm h^{-1})		Sediment production (kg ha^{-1})	
	Trampling condition		Trampling condition	
	dry	moist	dry	moist
0	166	160	980	2010
0.12	132	130	3820	2750
0.24	106	83	4610	5050
0.37	101	82	7080	7470

Source: Warren *et al.* (1986).

(vi) Treading damage may be contained and minimised by management. Heavy textured soils are more vulnerable than light soils, and may receive special treatment with respect to pasture species composition and grazing management.

Sod-forming grasses, as illustrated with *P. dilatatum* by Hunt (1979) and *D. decumbens* by Pott *et al.* (1983), evade treading damage. The botanical composition of the highveld at Frankenwald, South Africa, changes to dominance by stoloniferous and seral grasses such as *C. dactylon* and *Digitaria tricholaenoides* at the expense of erect, tussock grasses in response to increased simulated treading (garden roller; Gillard, 1969). The regrowth of *P. americanum* is less sensitive to wheel traffic than that of *Sorghum bicolor* cv. Sudax (Fribourg, Overton & Mullins, 1975).

Animals may be moved away from treading susceptible soils when these are wet to hard-standing or to areas of well-drained light soils. The last resort is to

Fig. 3.3. Soil erosion resulting from overgrazing and upland cropping.

concentrate animals on 'sacrifice' areas which will be subsequently cultivated and sown to green manure crops or to pasture for environmental repair.

3.1.2
Soil erosion

(i) In the tropics soil erosion (Fig. 3.3) and its attendant accelerated runoff, which cause serious environmental degradation and loss of agricultural production, are of little consequence in well-managed grasslands which provide adequate ground cover.

Wind erosion has great significance in arid and semi-arid regions, but the focus of this book is on the subhumid and humid regions where water erosion predominates. Tropical soils are often exposed to high intensity rainfall, and soil loss may amount to 115 t ha^{-1} yr^{-1} from unprotected ploughed land in western Nigeria (Lal *et al.*, 1975). This has an immediate effect on farm yields and farm income. Organic matter and nutrients are concentrated in the top soil, depending upon soil type; in western Nigeria crop yield may decrease *c.* 40–50% following the loss of 2.5 cm soil. The hydrological characteristics of farm watersheds are changed. Reduced rates of infiltration lead to greater surface run-off and a drier farm environment; ground-water is depleted and stream flow becomes more intermittent, with a greater incidence of flash-floods. These lead to damage to public utilities and human communities. The increased sedimentation rates in streams shorten the life of water reservoirs. Near Dodoma, Tanzania (lat. 6° S, 570 mm rainfall) the rates of soil accumulation in four reservoirs were 195, 405, 601 or 729 m^3 km^{-2} catchment yr^{-1}, representing an annual catchment area soil loss of 0.2–0.7 mm yr^{-1}; these figures do not include soil lost and deposited upstream from the reservoirs (Rapp *et al.*, 1972).

Soil erosion arises firstly from the kinetic energy of falling raindrops, which slakes and disperses the surface soil, splashes soil particles downhill, and may cause 'surface seal' as the smaller dispersed particles block the soil pore space and reduce the rate of infiltration. These actions have most effect if rain drops strike unprotected bare soil. The destructive energy of falling raindrops is graphically illustrated by a study at Mazoe, Zimbabwe (lat. 17° S, 910 mm rainfall, 1200 m altitude). Soil loss from bare soil (fersiallitic sandy clay loam, 4.5% slope) covered by two layers of wire gauze suspended *c.* 15 cm above the soil surface averaged 1.02 t ha^{-1} yr^{-1} relative to 119 t ha^{-1} yr^{-1} from unprotected bare soil (Hudson, 1957). Runoff from adjacent grass plots growing *Cynodon swazilandensis* was similar to but slightly higher than from the protected bare soil. In this situation the transporting and eroding power of runoff was a minor factor relative to the effects of direct raindrop on bare soil, which was prevented by the wire gauze screens.

The movement of runoff over the soil surface both transports the soil displaced by raindrop action and scours additional soil particles from the soil surface, according to the velocity of flow, which is influenced by the slope and length of the land surface, the depth of flow, and the characteristics of the surface which impede flow, and according to the entrainment of sediment and its balance with sediment deposition. The efficiency of entrainment of sediment (denoted η, where $0 < \eta < 1$) expresses the proportion of stream power which is used in entraining and transporting sediment. This is discussed subsequently in relation to cover in Fig. 3.5 (Rose, 1988); it is highest for vertisols (Pellusterts and Chromusterts), silt loams (mesic Fragiudalfs) and loess (Rose, 1988). Sod-forming grasses such as *P. clandestinum*, which provide a high level of anchored cover, even when well grazed, provide good soil resistance to entrainment.

(ii) Runoff, expressed as a fraction of the rainfall received (the runoff coefficient), has a negative curvilinear relationship with percentage soil vegetative cover.

The effect of vegetative covers on runoff at Sungei Tekam in peninsular Malaysia from a Munchung series soil of 10% slope is illustrated in Table 3.2 (Hong, 1978). Runoff coefficient averaged *c.* 0.21–0.23 on bare soil, whilst in the first year of establishment for the twining legumes *Pueraria phaseoloides* and *Centrosema pubescens* it decreased from 0.17 to 0.06–0.08 as the canopy developed. The established natural covers of the sod forming *Paspalum conjugatum*, *Mikania cordata* and *Passiflora foetida* gave high infiltration throughout and negligible soil loss, despite the occurrence of rainfall events with an intensity of up to 68 mm h^{-1}.

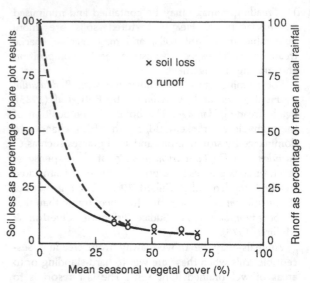

Fig. 3.4. Effect of cover on annual soil loss in Zimbabwe. (From Elwell & Stocking, 1976).

Cover is defined variously; the projected foliage cover clearly influences the proportion of rain striking the soil surface directly, but the rate of infiltration and the damage occasioned by water flow depends upon the vegetal cover in contact with the soil, which in turn is composed both of litter and of plant material anchored in the soil. Differences in definition (foliage cover plus litter or basal cover plus litter) may in part account for the wide discrepancy in the suggested values for cover which are critical for runoff and soil loss; these vary from 75% (Lang, 1979) to 30% (Elwell & Stocking, 1976). The latter study, which is illustrated in Fig. 3.4, was conducted at Matopos, Zimbabwe (lat. 20° S, 610 mm rainfall) on a medium-grained fersiallitic sand on 4% slopes.

Table 3.2. *Effect of covers on runoff coefficient and soil loss (t ha^{-1}) in Malaysia*

	Bare soil	Legumes	Natural cover
Nov–Dec			
Runoff	0.21	0.17	0.01
Soil loss	13.5	9.0	.01
Mar–May			
Runoff	0.23	0.06	0.01
Soil loss	30.2	1.8	0.01
June–Aug			
Runoff	0.22	0.08	0.01
Soil loss	11.2	0.01	0.01

Source: Hong (1978).

Vegetal cover increases infiltration by maintaining aggregate stability, and thus preventing the formation of 'surface seal', which is a function of the capacity of the soil to withstand the disruptive slaking action by raindrop impact and entrapped air; this occurs when soils are wetted quickly (Cernuda, Smith & Vicente-Chandler, 1954). Mulch prevents the maximum slaking action by reducing dryness (which may also have negative effects on infiltration) and the rapid wetting of the soil surface. Grassland also increases infiltration through its effects on organic matter content, and on soil aggregate size and stability, as influenced by root activity, decomposition and exudation.

Dead cover prevents soil loss but may be associated with surprisingly high runoff, as reported in Natal by Scott (1951) from ungrazed, unburnt plots; he described this as a 'thatch effect', but it is also linked with reduced evapotranspiration from predominantly dead vegetation and high levels of antecedent moisture restricting infiltration of the main summer rains.

I have been unable to locate critical tropical studies of the effects of stocking rate on cover, runoff and soil loss. A grazing study at Kluang, Malaysia, which is discussed in Section 5.4.3 (Eng, Kerridge & Mannetje, 1978), found that bare ground after three years' grazing averaged 5, 10 and 30% respectively at SR 2, 4, and 6 beasts ha^{-1}. The rate of infiltration at Mazoe, Zimbabwe (Rodel & Boultwood, 1981b) on planted pastures was independent of summer SR over the range 9.9 to 17.3 beasts ha^{-1}; however, the mean rate of infiltration on these pastures was only 43% of that on adjacent lightly grazed *Hyparrhenia/Hyperthelia* veld. Soil bulk density at the two sites averaged 1.40 and 1.29 g cm^{-3} respectively.

(iii) Reduced cover associated with heavy grazing leads to unacceptable levels of soil loss.

Early conceptual models emphasised the impact of raindrops on bare soil and the transport of the displaced soil by overland flow; more recent work adds to these models the scouring action of overland flow in augmenting sediment level. Vegetal cover in contact with the soil effectively prevents the latter process, even at low levels of cover (Rose, 1988). Fig. 3.5 shows the relationship between the efficiency of entrainment of soil particles of two vertisols on the Darling Downs, Queensland, and the fraction of soil area which is bare; the point of inflexion below which efficiency of entrainment increases rapidly is only *c.* 10% cover.

Relationships between cover, soil loss and the amount of erosive rain at Matopos, Zimbabwe, were developed from the same study from which Fig. 3.4 was developed. Various models were tested (Elwell & Stocking, 1974) which assessed the energy impact of rainfall and of varying categories of rainfall amount and intensity; Fig. 3.6 is based on a definition of erosive rain where daily fall exceeded 25 mm and at least 20 mm fell at a minimum intensity of 20 mm h^{-1}. It will be noted that soil losses in excess of 4 t ha^{-1} yr^{-1} only occur at less than 33% mean vegetal cover. The cover values represent the mean of 10 years' estimates for particular pasture defoliation treatments; the decreased soil loss occurring above 150 mm erosive rain reflects a flaw in the model, since increased cover (higher than the long-term mean value) occurred in years of high rainfall incidence.

The primary question to be determined as a basis for management is the level of soil loss which is tolerable. This may be related to the rate of soil formation, which is determined by the nature of the parent material and the climate of the site, the depth of soil available, and its fertility down the profile. Estimates of tolerable soil loss vary from 1–5 t ha^{-1} yr^{-1} in the subhumid subtropics to as much as 13 t ha^{-1} yr^{-1} in humid tropical agricultural areas. Land-use planning for vulnerable, sloping land may prescribe a crop–grassland rotation in which the low erosion losses in the years under pasture are balanced by the higher erosion losses in the years under crop to give an acceptable mean figure.

In the Nogoa River catchment area of central Queensland the soil loss (Y, t ha^{-1} yr^{-1}) at two sites (lat. 24° S, 690 and 730 mm rainfall) was related to % vegetative cover (X) by the equation $L_n Y = 1.32 - 0.053 X$ (Ciesiolka, 1987). The recommendation for catchment management is that grazing

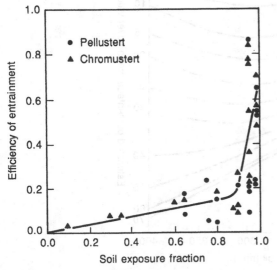

Fig. 3.5. Effect of soil exposure on efficiency of entrainment of soil particles. (From Rose, 1988).

might be controlled so that mean vegetative cover is maintained at minimum value of 30%; at this level soil loss would approximate 1 t ha^{-1} yr^{-1}.

There is considerable controversy concerning the effects of tree clearing on environmental quality (World Commission on Environment and Development, 1987). Much depends upon whether cleared land continues to be cultivated, or whether pastoral activities ensue which maintain grassland cover. The impact of grazing management on soil erosion from gently sloping land may be greater than the removal of tree and shrub cover (Reece & Campbell, 1986).

3.1.3
Nutrient cycling and the spatial redistribution of nutrients under grazing

(i) The pasture ecosystem may be managed so that nutrients are cycled between soil, plant and animal pools with minimum loss. The objectives are to sustain productivity at chosen levels by balancing inputs and outputs and to avoid both the drainage of excess nutrients which would cause environmental pollution and the depletion of the soil resource.

The pasture ecosystem is potentially conservative. The amount of nutrients retained as animal product

which is sold from the farm is small relative to the size of the nutrient pools, and pasture cover minimises soil loss and runoff. This situation is in stark contrast to that of many annual cropping systems, especially on sloping land.

A simplified flow diagram (Fig. 3.7, Wilkinson & Lowrey, 1973) shows the three main pools: plant, animal and soil, and the main sources of inputs and losses. Ingestion of plant material by grazing animals may be augmented by supplements from off-farm sources. Animal products are progressively sold from the farm; in some subsistence economies human consumption of these products leads to human wastes which may be retained in the farm system. Excretory products are added to the soil pool; urine enters the available nutrient pool or is lost to the atmosphere whilst dung is more slowly incorporated into forms available to the plant. Nutrients taken up by plant roots and channelled to the shoots are either consumed by animals or returned to the soil as litter. Plant litter enters the soil pool through the activities of micro-organisms and soil-inhabiting animals; their biomass gradually contributes residues which are incorporated in the soil humus complex. It is convenient to think of the soil in three interacting phases: the immobile and unavailable pool of minerals, the organic matter complex and the nutrients available to plants in the soil solution.

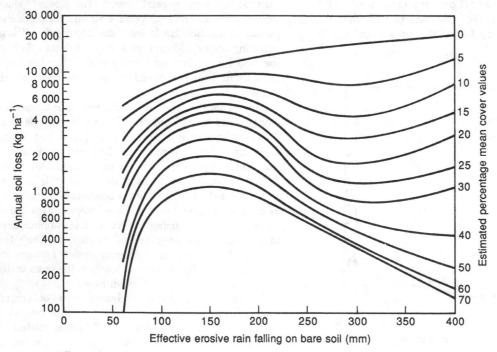

Fig. 3.6. Effects of amount of erosive rain on annual soil loss, as modified by long-term % mean cover. (From Elwell & Stocking, 1976).

Nutrients are also added from atmospheric sources and from fertiliser, and are lost by leaching, erosion, volatilisation and immobilisation.

The size of these nutrient pools fluctuates during the year according to the balance of processes operating, even in a system which is in long-term equilibrium. A large pool size may cushion the impact of unfavourable environmental circumstances, but productivity is determined not by pool size but by the rate of nutrient and energy flow between the pools.

(ii)   Nutrient cycling is illustrated in more detail in Fig. 3.8 (Robbins, 1984) for nitrogen, the nutrient of major interest. Pastures of *P. maximum* var. *trichoglume* at Gayndah (lat. 26° S, 730 mm rainfall) grown on a shallow basaltic soil were reserved from summer grazing

to provide stand-over winter forage and spring pasture; they were grazed each year from May/June until the end of November at 2.4 weaner steers ha^{-1}. Urea was applied in October at 58 kg N ha^{-1} yr^{-1}. In Fig. 3.8 the year is considered in three periods for a first year pasture. Period 1 is the phase from the onset of winter until the commencement of the growing season in September/October; little pasture growth occurs and litter accumulation is significant, representing an increase from 8 to 22 kg N ha^{-1}, whilst the shoot N pool is depleted. Warmer, moist conditions from October onwards (period 2) lead to a net loss of litter and a gain in shoot N. Dietary N concentration was *c.* 1.0%, and intake was 16 and 12 kg N ha^{-1} respectively in periods 1 and 2. Retention of N in animal tissues (removal) was estimated as 17% of ingested N. Excreted N was added to the soil pool, but it was assumed that 50% urinary N and 20% faecal N were lost to the atmosphere. During the summer–autumn of period 3 when the pastures were deferred from grazing, the shoot and root N pools increase whilst a net loss of N occurs in the litter pool, but little of this is recycled.

In this study the plant and animal N pools are very small in relation to the total N in the soil pool, which is estimated as 6000 kg N ha^{-1} in the 0–50 cm layer; only a tiny fraction of the N in the large soil organic matter pool is available to plants. In these circumstances small changes in litter accumulation and the immobilisation of N which this represents, together with N accretion in the root system, limit the availability of N to plant shoots. Reduced rates of animal liveweight gain occur as the pastures age after planting and these appear to reflect a reduction in %N of green leaf and hence dietary %N (Robbins, Bushell & Butler, 1987); this indicates a need to increase available N input if productivity is to be maintained.

Individual components of the nutrient cycle are considered next.

(iii)   The nutrient pool represented by domestic animals is most significant in dairy systems, since nutrient removal in milk is greater than in meat production, whilst nutrient removal in wool is insignificant. A high producing, intensively managed dairy pasture selling off 8000 litres ha^{-1} yr^{-1} loses *c.* 42, 8, and 11 kg ha^{-1} respectively of N, P, and K. By contrast 500 kg cattle LWG ha^{-1} represents 12, 4, and 1 kg ha^{-1} respectively of N, P, and K.

The biomass of feral animals may need to be taken into account in some pastoral situations. It is expected (from studies of temperate pastures, King & Hutchinson, 1980) that the biomass of invertebrate animals inhabiting the soil and the litter surface is negatively related to the SR of domestic animals. High SR which reduces litter

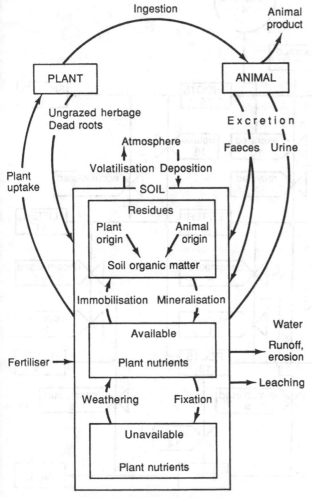

Fig. 3.7. Simplified nutrient cycle for pasture ecosystems. (From Wilkinson & Lowrey, 1973).

accumulation decreases the availability of food for microarthropods such as *Collembola* and *Acarina*. In subhumid lightly grazed native pastures near Charters Towers, Queensland, the biomass of termites such as *Amitermes vitiosus* is estimated to be between 40 and 120 kg ha⁻¹, which exceeds the cattle biomass present (Holt & Easey, 1984).

(iv) Excretion of N varies typically from 90–96% of ingested N for beef cattle, 87–95% for wethers and 72–87% for dairy cattle (Henzell & Ross, 1973;

Sugimoto, Hirata & Ueno, 1987a). Excretion of most of the 'ash' constituents is 90–96% of the ingested amounts.

The division of nutrients between faeces and urine influences their spatial distribution and losses (Watkin & Clements, 1978). Most of the excreted N (50–80%), K (70–90%), S (60–90%) and B are in the urine, and the other minerals predominate in the faeces. Faecal N approximates 0.8 g per 100 g of forage eaten, so a higher proportion of N appears in the urine from high N feeds (Barrow, 1987); similarly faecal S averages c.

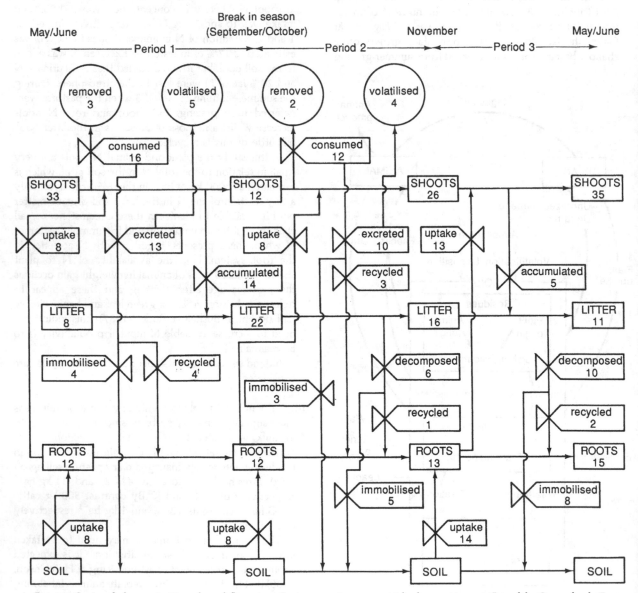

Fig. 3.8. Seasonal change in N pools and flows in a *Panicum maximum* var. *trichoglume* pasture at Gayndah, Queensland. (From Robbins, 1984).

0.11 g per 100 g of forage eaten. The N:S ratio is *c.* 20:1 in urine and *c.* 7:1 in faeces.

The organic P content of faeces averages *c.* 0.06 g forage eaten; for high P feeds a higher proportion of inorganic P (mainly as dicalcium phosphate) is excreted (Barrow, 1987). Ca, Mg, and the micronutrients Fe, Mn, Zn, Cu and Co are also excreted in the faeces. The dung pat, which typically covers 0.06–0.15 m² for cattle, causes a considerable concentration of nutrients; sheep pellets are dispersed more widely. Soil changes adjacent to the dung pat on an ultisol at Quilichao, Colombia (Fig. 3.9, CIAT 1982) show substantial increases in exchangeable K and Ca and available P, followed by a decrease with time, associated with leaching and plant uptake. The increase in inorganic N was followed by a steep decrease, and volatilisation of ammonia is a common source of loss, depending upon the activity of coprophagous beetles (Gillard, 1967).

Urine patches vary from *c.* 0.25–0.5 m² in area for cows, depending upon the wind and upon cow movement, whilst a single urination of 150 ml from a wether may occupy 0.03 m². Since wether urine commonly contains 0.9% N (Doak, 1952) the delivery of N may be as high as 45 g m⁻²; cow urine is more dilute and N concentration varies negatively with volume from *c.* 0.25 to 0.83 % N. Soil inorganic N, S and K are shown to increase adjacent to a urine patch at Quilichao (Table 3.3, CIAT, 1982), but great nitrogen loss subsequently occurs. About 75% of the N in urine occurs as urea, and this is readily volatilised and lost as ammonia, especially if deposited on dry soil in hot conditions. At Katherine, Australia (lat. 14° S, 950 mm rainfall), ¹⁵N-labelled urine was applied during the dry season. Nitrogen was lost rapidly in the first three days after application, and was then lost at a slower rate; the losses accumulated to 46% of the urea N applied (Vallis

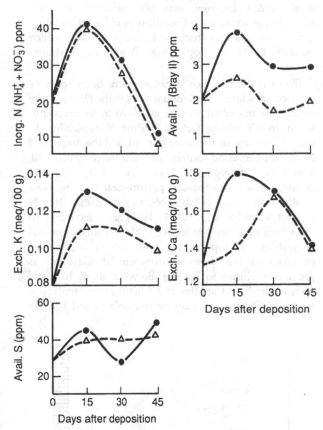

Fig. 3.9. Time changes in soil nutrients according to distance from dung pat (closed symbols, 20 cm; open symbols; 100 cm) at Quilichao, Colombia. (From CIAT 1982).

Table 3.3. *Effect of urine deposition on* Brachiaria decumbens *pasture on soil mineral content at Quilichao, Colombia*

Time after urine deposition (days)	Distance from urine deposition (cm)	Inorganic N (NH₄⁺ + NO₃⁻/ppm)	Available S (ppm)	Exch. K (meq./100 g)
0	20	20	25	1.20
	100	21	26	1.24
15	20	65	36	1.39
	100	35	33	1.17
30	20	28	37	1.61
	100	27	38	1.61
45	20	13	42	1.59
	100	9	40	1.56

Source: CIAT (1982).

et al., 1985). Leaching was not involved, and volatilisation as ammonia is the pathway of loss. This figure of 46% was greater than the losses of 16–32% which occurred in the dry tropics near Townsville, Australia.

(v)   Shoot material which is not eaten or burnt senesces and becomes litter. The amount of N in the litter and the amount of microbial protein involved in its decomposition are significant factors limiting N availability to plants. The size of the litter N pool, and the amount of N in aboveground organs are illustrated (Fig. 3.10, Mears & Humphreys, 1974*a*) for a May (autumn) occasion in a *P. clandestinum* pasture near Lismore, New South Wales (lat. 29° S, 1660 mm rainfall). Litter N present varied from *c.* 15–75 kg N ha^{-1}, according to SR, which negatively influenced litter accumulation, and to fertiliser N input. This pool size is larger than occurred in the drier environment at Gayndah as previously shown in Fig. 3.8. Between about 10% and 33% of the N in the litter of a *Chloris gayana* pasture in southern Queensland may be mineralised and become

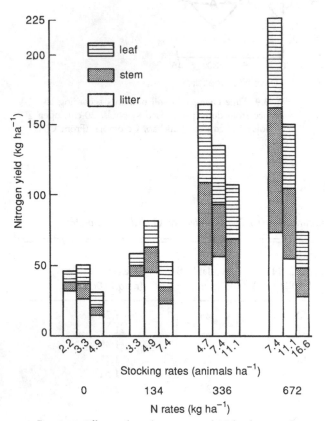

Fig. 3.10. Effects of stocking rate and N fertiliser rate on plant N pool for *Pennisetum clandestinum* pasture. (From Mears & Humphreys, 1974*a*).

available in the first year after deposition, but thereafter less than 5% of the residual N in litter is recycled each year (Vallis, 1983).

(vi)   The primary focus of interest in the soil nutrient pool is the rate at which available nutrients are replenished to meet the nutrient losses. Augmentation of N comes from legume N fixation, fertiliser N, inorganic nutrients in excreta and litter, and mineralisation from organic residues (Till, 1981; Floate, 1987; Steele & Vallis, 1988); the contributions from atmospheric sources and from free-living N-fixing organisms are more problematical. The losses from the available segment of the soil system arise from volatilisation of ammonia, as discussed above, denitrification, which is active in warm, wet conditions, erosion, chemical and biological fixation and leaching; N loss due to the last factor is *c.* 40 kg N ha^{-1} yr^{-1} under legume crops in Malaysia, or 100 kg ha^{-1} yr^{-1} from bare soil (Agamuthu & Broughton, 1985). Plant nutrient uptake, as modified by defoliation, is discussed in Section 4.5.3. Some examples of the levels of nutrient uptake in tropical pasture systems may be found in Billore & Mall (1976), Husz (1977), Christie (1979), Pemadasa (1981), Medina (1982) and Lamotte & Bourlière (1983).

(vii)   The rates of flow between these nutrient pools are accelerated as SR increases. A higher level of forage utilisation increases the total nutrients of the animal biomass and the rate of product output, except at excessively high SR (Chapter 7). The greater eventual partition of plant nutrients to animal excreta rather than to the more slowly available litter pathway (as suggested by Fig. 3.11) increases mineral availability. On the other hand increasing SR increases spatial transfer of nutrients and losses from volatilisation and leaching. Nutrient availability is increased at high SR in circumstances where nutrient inputs meet the additional losses. The reduced size of the root system at high SR (Section 4.5.1) decreases the level of nutrients immobilised in roots.

The adverse effects of high SR on nutrient losses have been emphasised by some reviewers (Steele & Vallis, 1988) but circumstances occur where nutrients are efficiently cycled at high SR (Tierney & Goward, 1983*b*). Good N retention was evident in Aracatuba, Sao Paulo, Brazil, where *P. maximum* pastures were grazed at 3.8 and 1.5 steers ha^{-1} in summer and winter respectively. When annual dressings of 200 kg N ha^{-1} were discontinued, residual effects gave a loss of only 50% yield in the next two years (Mott, Quinn & Bisschoff, 1970). At Overton, Texas (lat. 32° N), Rouquette, Matocha & Duble (1973) unexpectedly claimed an actual increase in total soil Kjeldahl N at 4.7 beasts ha^{-1} relative to the estimate under 2.7 beasts ha^{-1} (Fig. 3.11). The amount of recycled nutrients per beast from grazed *C. dactylon* pastures was similar at both SR, but more forage per ha was consumed at high SR and less material entered the litter fraction. Clearly this question requires more study.

(viii) The spatial transfer of nutrients is considerable, even under normal grazing practice. This was shown in Fig. 3.9 and Table 3.3 for dung and urine deposition. It is exacerbated by animal movement to shade and watering points. This is illustrated (Fig. 3.12, Sugimoto *et al.*, 1987b) for Holstein heifers grazing a *Paspalum notatum* pasture at Miyazaki, Japan (lat. 32° N); animals could move freely between a resting place with shade trees and open pasture. On warm days in July and September when the temperature exceeded 27 °C, 44–53 % of urinations and 26–29 % of defecations occurred in the shaded area; in October when maximum temperature did not exceed 23.5 °C only 11 % of urinations and defecations occurred in the shaded area, so that less nutrient transfer and concentration took place. Over the whole grazing season 22 % of the urine N and 14 % of the dung N were deposited on the resting area where little pasture grows (Sugimoto *et al.*, 1987c).

In dairy systems it is common practice to graze animals overnight in paddocks near to the milking shed, and transfer of nutrients to 'night paddocks' occurs if these paddocks are small. Horses may also be kept close to the dwelling for ease of use. A further concentration occurs if animals are corraled or housed overnight for protection against theft or predators. Some workers have associated the transfer of nutrients to night paddocks with increased excretion at night (Gillingham, 1987). However, Harker (1960) found that there was little difference throughout the day and night in the frequency of defecation of Zebu cattle in Uganda, whilst Fig. 3.12 shows a pronounced day-time pattern of excretion, with 80 % of urinations and 73 % of defecations occurring in the sunrise–sunset period.

Purchase from off-farm feed sources should be taken into account when assessing nutrient transfer. Feeding grain to animals at pasture increases soil fertility in the paddock, and Benacchio, Baumgardner & Mott (1970) were able to relate increased P and K status of the soil to previous levels of grain feeding. Husz (1977) details a nutrient budget for mixed farming in the Brazilian cerrado which incorporates nutrient transfers from cropping areas to grazing land upon which crop products are fed.

Cut-and-remove systems represent a potent source of nutrient transfer; increased fertiliser requirement is a concomitant consideration. A 400 kg cow consuming 2.5 % of body weight each day, and eating pasture containing 2 % N, 0.2 % P and 1 % K, accounts for 73 kg N, 7 kg P and 36 kg K in a year. A hay crop of the same nutrient composition yielding 6 t ha^{-1} removes 120 kg N, 12 kg P and 60 kg K from the field. Efforts to collect and return the excreta from housed animals to the fields are rarely successful in avoiding nutrient loss. The practice of cut-and-remove utilisation is entrenched for valid reasons in some mixed farming agricultural systems (Chapter 2) but is based on quite false premises for many

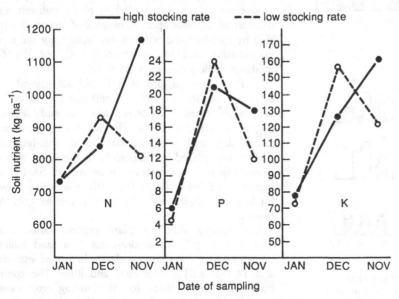

Fig. 3.11. Time changes in soil N, P and K as influenced by stocking rate. (From Rouquette, Matocha & Duble, 1973).

situations. The lowered opportunity for selectivity of feed relative to that presented by grazing may decrease individual animal performance, housed animals are more vulnerable to disease, and cross-bred animals are sufficiently heat tolerant to perform well while grazing in the tropics. At Muaklek, Thailand (lat. 15° N, 1090 mm rainfall, 220 m altitude), milk production from 5/8 or 3/4 Holstein/Friesian or Red Dane cows grazing *P. maximum/S. hamata* cv. Verano pastures during the day did not differ from that of housed cows fed on a cut-and-remove basis, as discussed in Section 6.3. The saving in labour costs in moving to a grazing system is considerable, and the undesirable concentration of nutrients at the barn is minimised.

Excess nutrients from the area where animals are concentrated or housed causes stream pollution. The leaching of nutrients may produce levels of nitrate in drinking water which are unacceptable, especially for infant consumption. Nitrate together with increased phosphate stimulates the growth of filamentous algae. As these die and accumulate at the bottom of the stream or pond, the bacteria involved in the decomposition of this organic matter consume the available oxygen and cause the death of fish and other marine life. This phenomenon of eutrophication is a common accompaniment to this type of intensification of animal production. Pasture systems need to be managed to avoid excess nutrient loss, both in order to sustain pasture productivity and to avoid environmental damage.

3.2
The biotic environment

3.2.1
Energy flow and the decomposer industry

The management of energy flow in grasslands is not directed to maximising the capture of solar energy in net primary plant production as discussed in Chapter 4. It is concerned rather with maximising a sustained energy flow to domestic grazing animals. This objective has to be consistent with a requirement to maintain a balance between protecting the soil environment through adequate accretions of litter and minimising the nutrient immobilisation associated with the decomposition of litter. Tropical pastures differ from temperate pastures in exhibiting higher levels of net photosynthesis but lower levels of litter accumulation, due to the faster rates of decomposition under warm, moist conditions (Lamotte & Bourlière, 1983).

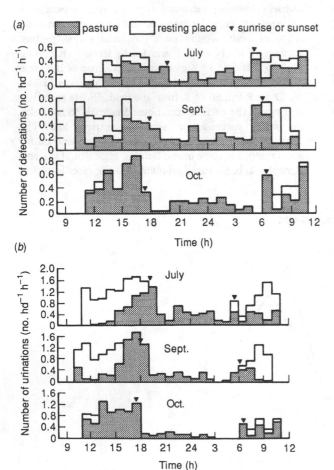

Fig. 3.12. Seasonal effects on diurnal changes in (*a*) defecations and (*b*) urinations by cattle on pasture or on shaded resting place. (From Sugimoto *et al.*, 1987*b*).

(i) The efficiency of energy capture as net primary plant production (which is regarded as photosynthesis less respiration) depends upon a complex of factors dominated by moisture and nutrient availability, by the level of radiation and by whether the C_4 or the C_3 photosynthetic pathway is involved.

A *P. notatum* pasture at a humid subtropical site (Miyazaki, Japan, lat. 32° N, 2680 mm rainfall in the year of observation) is selected as a case study (Hirata, Sugimoto & Ueno, 1986). The winters are cool and rather dry, and the long-established well-fertilised pasture was grazed intermittently in the warm season from May to October to give an overall SR in this period of 5.8 heifers ha^{-1}. The pasture was also mown for hay towards the end of the main summer growing period.

The energy value of plant organs varied from 18.8–19.9 KJ g^{-1} for inflorescence, leaf and stolon, 16.6–17.8 KJ g^{-1} for stem and attached dead material, and 14.4–16.4 KJ g^{-1} for root and litter. The contribution of each of these to the standing crop varied considerably during the year; the standing crop (including stolons and roots to 30 cm depth) increased to 33.0 MJ m^{-2} in summer and decreased to 22.2 MJ m^{-2} in winter. The addition of surface litter to these figures gave values for total organic matter of the sward of 42.0 and 27.5 MJ m^{-2} respectively. The relative energy distribution of the sward components (Fig. 3.13) shows the predominance of stolons and roots as the main energy pools in this sod-forming, grazing-resistant grass; these represented 24–40% respectively of the standing crop.

The figures for net primary production (Table 3.4) reveal a completely different pattern which emphasises the capture of solar energy for leaf and stem production and little net increase in stolon and roots; the latter organs contribute greatly to sward respiration and this continues during the winter when growth is slow. Total net primary production for the year was 29.0 MJ m^{-2}, which represented 0.57% of total short-wave radiation. However, the seasonal values for efficiency of energy capture varied considerably from 1.33% in July–August to −0.38% in December–February.

The flow of energy fixed as net primary production to the principal energy pools (Table 3.4) shows a high value of 41% consumed by the grazing animal, and 24% additionally conserved as hay. This illustrates an intensively managed, high producing and seasonally grazed system. Consumption by domestic animals in other tropical pasture ecosystems reflects much lower efficiencies of energy transfer and is often below 10% of net primary production in monsoonal rainfall climates. The energy fixed in animal product relative to energy ingested then becomes the key question; in this respect tropical grasses are at a disadvantage compared with temperate species (Okubo et al., 1985).

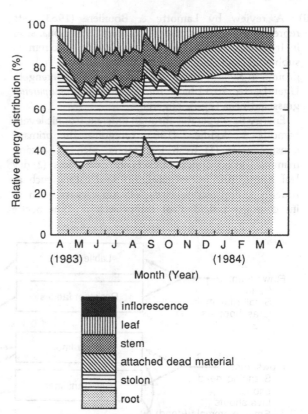

Fig. 3.13. Seasonal changes in relative energy distribution of organs in grazed *Paspalum notatum* pasture. (From Hirata *et al.*, 1986).

Table 3.4. *Energy transfer and efficiency in a grazed* Paspalum notatum *sward*

Component	Energy (MJ m^{-2} yr^{-1})	Efficiency Percentage solar radiation	Efficiency Percentage net primary production
Solar radiation (total short wave)	5090		
Net primary production	29.0	0.57	
Grazed herbage	12.0	0.24	41
Mown herbage	6.9	0.14	24
Litter production	8.9	0.17	31
Storage in plant (stolons and roots)	1.2	0.02	4

Source: Hirata *et al.* (1986).

(ii) A review by Lamotte & Bourlière (1983) lists representative values of peak biomass for regional sites in tropical savannas; these cover a wide spectrum in subhumid and humid zones from 0.5 t ha^{-1} above ground in *Kylinga–Sporobolus* communities in Serengeti, Uganda, to 20.8 t ha^{-1} for a *Pennisetum purpureum* grassland near Lagos, Nigeria.

Energy flow operates through different trophic levels. Lamotte & Bourlière (1983) categorise the primary consumers as (1) fresh leaf eaters, such as grazing mammals, grasshoppers, crickets and caterpillars; (2) dry leaf eaters, such as termites, (3) seed eaters, such as rodents and passerine birds; (4) sap suckers, such as insect bugs; and (5) root eaters such as some beetle

larvae and nematodes. Dead material is then consumed by detrivores or saprophages which also feed on living micro-organisms associated with decomposition; these include cockroaches, some beetle larvae and termites. Further decomposition is accomplished by humus and soil consumers, such as earthworms, termites, fungi and bacteria.

The decomposition system may be compartmentalised according to the source of substrate, and one such classification (developed as a subsection of the total-system ELM grassland model) is illustrated (Fig. 3.14, Hunt, 1977). The substrates are described as faeces, dead plant and animal tissues (subdivided into labile and resistant fractions) and humic material. The decomposers

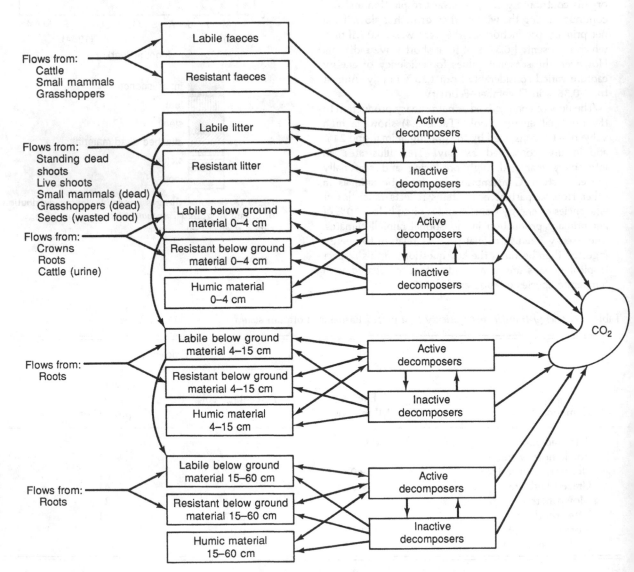

Fig. 3.14. Components of decomposition model and direction of flows. (From Hunt, 1977).

are either active or dead; in the latter case they form part of the organic substrate.

(iii) The significance of litter for soil conservation, infiltration and nutrient cycling has been discussed earlier in the chapter, and its effects on the microclimate and on seedling establishment are mentioned later. The rate of litter accumulation varies seasonally, and litter commonly builds up over the winter to reach peak values of litter present in early spring, and then decreases in late spring and summer to low values in early autumn.

At Coolum, Queensland (lat. 27° S, 1500 mm rainfall) the extreme seasonal values were 902 and 284 g m^{-2} for lightly grazed pastures (Bruce & Ebersohn, 1982). The average value of 558 g m^{-2} was similar to the level (600 g m^{-2}) present under a tropical legume cover crop in Malaysia (Broughton, 1977). Peak litter accumulation was 232 g m^{-2} in ungrazed *Bothriochloa/Imperata* grassland at Imphal, India (lat. 25° N, 1180 mm rainfall, Yadava & Kakati, 1985), whilst in the more arid cleared mulga shrublands of south-west Queensland litter present varied from 7 to 26 g m^{-2} (Christie, 1978). It is convenient to express the rate of litter accumulation in relation to the amount of standing crop present, since this provides the material available for senescence and detachment. At Coolum this varied from 3 to 6 mg g^{-1} d^{-1} in absolute terms.

Litter accumulation obviously depends upon how favourable is the climate for growth, and a higher potential rate of accumulation occurs in well-watered grasslands (Bruce & Ebersohn, 1982). Litter accumulation is greater on fertile soils. Litter yield in south-west Queensland decreased in the rank order: *Cenchrus ciliaris* on brown cracking clay, *Astrebla* grassland on grey cracking clay, and *Thyridolepis mitchelliana* grassland on sandy red earth; this reflects relative net primary production and soil fertility (Christie, 1979). SR exerts a major influence on absolute litter accumulation, as illustrated in Fig. 3.10; in this study of continuously grazed *P. clandestinum* pastures litter present on a year-round basis averaged 427 g m^{-2} in the lowest SR treatments and 230 g m^{-2} in the highest SR treatments (Mears & Humphreys, 1974a). Litter yield in semi-arid *Astrebla* grassland at Augathella, south-west Queensland was 68, 66 and 45 g m^{-2} respectively at SR of 1.3, 2.5, and 5 sheep ha^{-1}; however, the relative rate of litter accumulation was *c.* 1.5 mg g d^{-1} and independent of SR.

(iv) Litter disappearance is positively related to temperature and to available moisture, since these determine the activity and density of the decomposer organisms. In the Miyazaki (lat. 32° N) study with *P. notatum*, the rate of litter disappearance was only 0.7 mg g^{-1} d^{-1} in winter, but increased to 6.3 mg g^{-1} d^{-1} in autumn (Hirata *et al.*, 1986). Annual rates of litter loss were found to be similar at Gayndah, Queensland (lat. 26° S, 730 mm rainfall, 4.9 mg g^{-1} d^{-1}, Robbins, 1984), at a *Sehima* grassland in India (875 mm rainfall, 4.7 mg g^{-1} d^{-1}, Mall & Billore, 1974) and at an *Andropogon–Paspalum* pasture in Texas, USA (750 mm, 4.5 mg g d^{-1}, Britton, Dodd & Wiechert, 1978). In the moist Coolum environment litter disappearance was more rapid, and averaged 12 mg g^{-1} d^{-1} (Bruce & Ebersohn, 1982).

The rate of litter disappearance depends upon the quality of the substrate; soluble carbohydrates will be consumed more readily than cellulose, which in turn is less resistant than lignin. The presence of tannin-like polyphenols in residues may be responsible for the slower decomposition of litter of *Desmodium intortum* relative to that of *Macroptilium atropurpureum* cv. Siratro (Vallis & Jones, 1973). It is expected that litter with a low C:N ratio (< *c.* 20–30) will decompose more quickly; Robbins (1984) suggested that for residues of *P. maximum* var. *trichoglume* at a subtropical site the critical C:N ratio may be as high as 70. Dry matter is lost more rapidly than N; the N concentration of litter commonly increases with age and is greater than that of the standing dead material from which it derives.

The loss of root material has rarely been monitored. In the humid forest zone at Abidjan, Ivory Coast (5° N) *P. maximum* swards were cut every six weeks. Picard (1979) estimated that individual root turnover was rapid and varied from 2.6 to 4.8 months.

The decomposer industry in the tropics and subtropics encompasses a large range of species, which are outside the main focus of this book, but which are reviewed by Lamotte (1982), Morris, Bezuidenhout & Furniss (1982) and Lamotte & Bourlière (1983). A single example of earthworm activity is given to illustrate the significance of these invertebrates in regulating decomposition by micro-organisms. At Lamto, Ivory Coast (1250 mm rainfall) the geophagous species *Millsonia anomala* feeds on the organic debris found in the soil it ingests (Lamotte, 1982). This and other earthworms may consume 20 to 309 times their own weight of soil each day, if the soil is sufficiently moist, which may amount to a turnover of 1000 t ha^{-1} yr^{-1}. This assimilation and comminution of all soil fractions, together with the microbial activity in the gut of the earthworm and the mucus deposited by it lead to substantial mineralisation of organic matter, amounting to *c.* 800 kg ha^{-1} yr^{-1}. This also has beneficial effects on soil aggregate stability and size, and upon infiltration.

3.2.2
Pests and disease

The incidence of the pests and disease which reduce pasture productivity is sometimes modified by grazing management through its effects on (1) the pasture microclimate, (2) the removal of infective organisms or inoculum, and (3) the susceptibility of the host plants.

Tropical pasture plants provide the substrate for a wide spectrum of viruses, bacteria, pathogenic fungi, nematodes and insects. These reduce plant productivity and persistence, and may also decrease nutritive value; the digestibility of organic matter and nitrogen of *D. decumbens* leaves is reduced by infestation with the rust *Puccinia oahuensis* (Davis & Norton, 1978).

The major thrust of control is directed to identifying or breeding pasture plants which are resistant to attack or which recover resiliently after attack, and there are other approaches to control through the management of the balance of the whole biological population and the introduction of new organisms antipathetic to the organism which is reducing pasture productivity. Chemical control measures have restricted adoption on tropical pastures used for grazing, partly because of the high cost of such inputs in relation to the price of animal products, but also because of the persistence of harmful pesticide residues in meat and milk.

(i) Instances where pasture management alters the incidence of pests and disease may be related to a variation in pasture microclimate according to the amount of pasture material present. Twining legumes such as *M. atropurpureum* cv. Siratro develop a closed canopy and a humid microenvironment under lenient management which favours the fungal disease *Rhizoctonia solani*; this causes severe leaf and stem damage. Grazed Siratro may be grown in regions too humid for seed crops, which reach a high leaf area index L and succumb to *Rhizoctonia*. Walker (1980) found that high density Siratro swards grazed at low SR (1.2 beasts ha^{-1}) developed more *Rhizoctonia* than low density Siratro swards grazed at higher SR, and subsequently a moderate SR resulted in a higher yield of Siratro, due to its favourable effect on the disease factor.

The root-feeding weevil *Amnemus quadrituberculatus* is damaging to many pasture legumes in coastal southern Queensland and northern New South Wales. Attempts have been made to control this insect by heavy grazing in the late summer and early autumn to reduce the incidence of oviposition sites (Braithwaite, Jane & Swain, 1958).

Organisms in these two disparate examples flourish under lenient SR. On the other hand dense plant cover is inimical to the leaf hopper *Orosius argentatus*, which transmits the 'little leaf' mycoplasma to which many tropical pasture legumes are susceptible (Hutton, 1970). Mound building by the funnel ant *Aphaenogaster pythia* buries pasture swards on the Atherton Tableland, Queensland. Saunders (1968) found that the percentage of ground area covered by soil mounds was negatively related to an index of sward 'density' derived as the product of pasture ground cover and pasture height. The latter might be varied by SR and by the introduction of tall bunch grasses such as *P. maximum*. Similarly the incidence of the spittle bug *Deois incompleta* is reduced in *B. humidicola* swards in the Amazon region if pasture height is maintained no lower than 20 cm during population peaks of *D. incompleta* (Silva & Serrao, 1985).

(ii) The infection rate of a fungal organism is a function of spores initiating lesions, the latent period, the abundance of spore production and the rate of removals (Van der Plank, 1968). It is modified by the extent of the leaf canopy which provides substrate for the organism; heavy grazing also removes inoculum.

Disinfestation by fire is a further option. The fungal disease anthracnose (*Colletotrichum gloeosporoides*), which is capable of the rapid evolution of new races, has become a major limitation to the use of *Stylosanthes* spp. in tropical pasture development; considerable resources are being directed to plant breeding programmes designed to provide resistance. Lenné (1981) showed that fire reduced pathogenic inoculum, and had persistent effects; the burnt plots had 60% less anthracnose than unburnt plots one year after burning *Stylosanthes capitata* pastures. *S. capitata* regenerates well after burning, and this led to higher pasture yields. Biennial burning is suggested.

(iii) The development stage of the plant, which is modified by grazing, influences susceptibility to attack. Increased susceptibility with age is a common pattern of infectibility. Miles & Lenné (1987) monitored the abundance of anthracnose on 40 accessions of *S. guianensis* which were cut at 4-, 8- or 12-week intervals. Anthracnose severity was greater with less frequent than with more frequent defoliation, although no simple relationship with age of material or flowering abundance was observed.

The spectrum of disease and pest attack is usually local in character and of sporadic incidence. Field solutions involving grazing manipulations sometimes emerge from collaborative studies between pathologists, entomologists and pasture agronomists.

3.3
The climatic environment

The climatic environment in which the pasture grows is determined primarily by the weather, but grazing management which changes the level of pasture and litter availability may modify the sward microclimate sufficiently to influence the extent of frost damage, moisture loss or seed germination.

3.3.1
Temperature

(i) Pasture growth in the subtropics and highland tropics is limited by low soil temperatures during the autumn, winter and spring (Schroder, 1970). The removal of litter or plant material by severe defoliation or by burning causes the soil to warm up more quickly during the day.

This is illustrated for a site in south-east Queensland (Fig. 3.15; Tothill, 1969); the maximum soil temperature at 13 mm depth following burning or the removal of cover averaged *c.* 7 °C higher than under plots bearing unburnt *Heteropogon contortus*. Minimum temperature was similar in both treatments. Fig. 3.15 shows considerable daily variation in the difference between treatments. Under cloudy conditions at 1300 h Rickert (1970) found that the soil temperature at 12 mm depth was identical for bare soil and for soil bearing 5000 kg ha^{-1} mulch; on clear days the high surface albedo and low thermal conductivity of mulch caused

lower soil temperatures during the day and less amplitude of diurnal variation in soil temperature. Similarly daily maximum soil temperature at 25 mm depth under a *Paspalum plicatulum* sward was *c.* 4 °C higher for the month after burning than after cutting at 10 cm; however, minimum temperature in the latter treatment averaged *c.* 1 °C higher (Stür & Humphreys, 1987). The presence of litter which lowers soil temperatures reduces the rate of breakdown of legume hardseededness (Section 5.2.4).

(ii) The susceptibility of plants to radiation frosts and the severity of frosting are positively associated with the amount of litter or cover present. This finding arises because more heat is lost by nocturnal radiation from a bare soil surface than from a soil with a cover of litter. For example, on one occasion minimum air temperature 10 mm above the ground surface was −0.8, −1.4 and −2.4 °C respectively for bare soil, and for mulch treatments of 2500 and 5000 kg ha^{-1} (Rickert, 1974). The loss of heat to the leaves above the ground surface reduced the amount of frost damage. Fig. 3.16 shows a linear relationship between a score for percentage leaf kill of *M. atropurpureum* by frosting and a score for amount of soil cover provided by litter (zero to 100% cover, maximum litter 1400 kg ha^{-1}) on both burnt and unburnt areas. This also indicated that the effect of fire in reducing subsequent frosting was mediated through its effects on litter cover (Ludlow & Fisher, 1976).

The temperature profile against depth in the sward for radiation frosts significantly determines the extent of

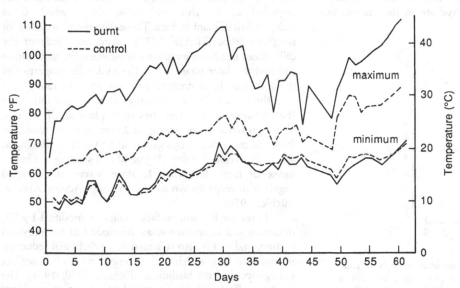

Fig. 3.15. Daily maximum and minimum temperature at 1.3 cm soil depth below burnt and unburnt *Heteropogon contortus* pasture. (From Tothill, 1969.)

damage. Temperatures are warmer close to the soil surface and decrease sharply in the zone 0–3 cm above the soil surface; low temperatures are also maintained above this zone. At Narayen, south-east Queensland, a temperature gradient of 6 °C between the soil surface and a height of 25 cm has been noted (Cameron & Ludlow, 1977). The height of the lowest bud is negatively associated with plant survival in *Stylosanthes guianensis* and in *Centrosema virginianum* (Clements & Ludlow, 1977); selection for frost avoidance is therefore feasible. This also suggests that grazing or cutting management which results in elevated apices or buds would increase the vulnerability of swards to frost damage. High above this zone the air is warmer, and shrubs which are 'managed tall' escape frost.

Unusually heavy frosts in July and August caused severe damage one year to swards of *Setaria sphacelata* var. *sericea* cv. Nandi and *Chloris gayana* cv. Samford at Samford (lat. 27° S) in south-east Queensland (R. J. Jones, 1988*a*). These pastures were well fertilised with nitrogen (336 kg N ha^{-1} yr^{-1}). The amount of pasture present in the autumn varied according to SR, and according to whether the pastures had been conserved for hay in March or not. Regrowth of pastures in September was greater in swards which had been previously cut for hay or stocked heavily. By early October yield of green pasture was 1360 and 610 kg ha^{-1} in the previously conserved and non-conserved treatments respectively, and averaged 880 and 1090 kg ha^{-1} respectively at 3.8 and 5.0 beasts ha^{-1}; this led to steers gaining weight more rapidly at the high than at the low SR.

It would be expected that lenient autumn grazing would increase labile carbohydrate in the crown, but

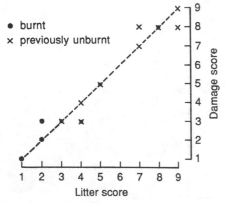

Fig. 3.16. Relationship between litter score (9 = 100% cover) and frost damage score (9 = 100% leaf kill) of *Macroptilium atropurpureum* cv. Siratro. (From Ludlow & Fisher, 1976.)

attempts to relate frost tolerance in tropical grasses to level of labile carbohydrate have not been successful (Hacker, Forde & Gow, 1974). Heavily utilised swards had superior frost recovery in the example quoted above, and this may well be related to reduction in litter or to the location of buds close to ground level.

3.3.2
Water

Solar factors predominate in the control of water loss from pasture, but the extent of the leaf canopy and the amount of litter modify the rate of evapotranspiration where foliage cover is incomplete and *L* (leaf area index, m^2 m^{-2}) is in the range zero to *c.* 3.

It is difficult to locate critical studies of the effect of SR on water balance in tropical pastures to illustrate this question. *Panicum maximum* var. *trichoglume* swards defoliated to *L* 0.1–0.6 in the autumn when growth was slow showed subsequent evapotranspiration (Et) of 33–78 mm respectively (Humphreys, 1966*a*). However, defoliation had no consistent effect on Et in the succeeding summer months when growth was faster, *L* values higher, and Et more rapid.

Similarly, evapotranspiration of summer crops increases as the leaf canopy grows, and Lascano *et al.* (1987) developed a model which predicts soil evaporation and transpiration from a cotton crop using soil and plant variables and daily weather data; rate of Et increased substantially after the crop was *c.* 55 days old. Evaporation from the soil surface accounted for *c.* 30% loss over the whole period to harvest. Evaporation from soil influenced Et of sorghum and cotton crops in Texas when *L* was below 2.7 (Ritchie & Burnett, 1971), and this value of *L* was regarded as the threshold value above which Et is independent of plant factors. These data were modelled for sorghum crops (Ritchie, 1972), taking into account the differences in rate of evaporation between the wet surface soil and the later phase when the hydraulic properties of the soil limit the movement of water to evaporating sites near the surface. The architecture of the canopy influences the partitioning of water loss from plant and from soil. Crops grown at a row spacing of 25 cm reduce evaporation from soil more than crops grown at 100-cm row spacing at the same overall density; degree of soil shading is a better index of this effect than *L*, since leaves are clumped together in crops grown at wide spacing (Adams, Arkin & Ritchie, 1976).

Litter on the soil surface, which is modified by SR, decreases soil temperature, as mentioned in the previous section, and this is also reflected in a substantial reduction in moisture loss, which may promote successful seedling emergence and establishment (Rickert, 1970, 1974). The positive effects of litter and plant cover on moisture infiltration were described in Section 3.1.2.

3.3.3
Light

The degree of canopy cover alters the spectral quality of the light reaching the soil surface. This affects tillering (Section 4.4.2) and the germination of light-sensitive seeds (Section 5.2.5).

The material in this chapter indicates that grazing management affects both the sustainability of agriculture and the types of environmental pollution which bedevil society. High SR, especially if applied to moist, heavy soils, alters soil structure and causes what may become permanent reductions in pasture output. The loss of topsoil and the increased runoff associated with reduced cover also decrease pasture output, and the higher incidence of flooding and dust storms, the damage to public utilities such as roads and bridges, and the decreased life of storage reservoirs have a social cost. The spatial transfer of nutrients associated with livestock concentration in green or feed lot systems represents a net debit to agricultural production, but also causes stream pollution and the nuisance to nearby residential communities of unpleasant odours and the increased fly population. Increasing population pressure on the lands of the tropics often forces forward strategies which provide short-term increases in rural productivity; these need to be evaluated in terms of their cost to a sustainable agriculture and to a benign environment which communities might otherwise enjoy.

4

Effects of defoliation on the growth of tropical pastures

4.1
Objectives of management

The primary objective of management is to maximise the intake of nutrients by the animals from the pasture on a sustained basis. This is achieved by optimising the balance between pasture shoot growth, losses due to tissue senescence, and the yield of pasture removed by grazing or cutting (Parsons, Johnson & Harvey, 1988).

Crop agronomists seek to maximise growth, and the yield harvested at the end of the crop growing season reflects previous photosynthesis. Maximum growth is a mistaken objective for scientists dealing with grass pastures since maximum rates of pasture growth only occur under conditions of suboptimal utilisation which are followed by high rates of tissue senescence. Plant material not eaten eventually senesces. The constant problem before the pasture manager is to reconcile the competing requirement of the animals for daily forage and the need to maintain a minimum sward leaf surface which will intercept and utilise current radiation.

The complexity of decision-making also arises from the nature of the feed requirements of the animal. Continuity of forage supply is sought. The seasonality of this is modified by defoliation practice; for example, close clipping of *Paspalum notatum* in Georgia, USA, reduces seasonal differences in feed availability relative to those which occur under more lenient defoliation (Beaty *et al.*, 1980). Alternatively forage consumption may be rationed in periods of abundance to provide for a seasonal shortfall. Systems involving a conservation cut differ in their effects from continuous grazing systems, as discussed later. The manager is also concerned with the nutritive value of the forage harvested, which decreases with long durations of regrowth and with advanced ontogeny, as treated in Chapter 6.

The literature of pasture science is scattered thickly with studies of the effects of cutting on pasture yield. These studies are useful in their local application to cut-and-remove pasture systems but have little relevance to grazing. Usually they do not distinguish between pasture growth and utilised yield, and lack the ancillary descriptive measurements which would assist the interpretation of the data through the light which might be shed on the processes involved. Our understanding of these is partial, despite many decades of research on pasture defoliation. This chapter will set out principles which should be tested in varying management situations in the tropics and subtropics.

4.2
Defoliation

Defoliation is used in the general sense of shoot removal, rather than in its more specific sense of lamina removal. It may be characterised in terms of its frequency, intensity, timing, selectivity and spatial distribution.

(i) The concept of frequency, or how often material is removed, is simple with respect to cut-and-remove systems. The time series of removal by grazing is more difficult to describe, since the frequency with which individual tillers are grazed varies widely, even at a constant stocking rate. The percentage of *M. atropurpureum* cv. Siratro runners (Fig. 4.1) which were grazed in any three-week period varied from 60% to 85% (Fig. 4.2; Clements, 1985) when a pasture at Samford, Queensland, was grazed continuously at 1.3 beasts ha^{-1}, a moderate stocking rate. The term 'continuous' grazing sometimes appears to be a misnomer, since an individual shoot has a substantial period of rest between grazing. A rotational grazing system of 4 days grazing, 17 days rest (Fig. 4.2) resulted in a similar frequency of runners being grazed, except for a period from March to May, when forage availability was less than earlier in the season.

Hyperthelia, *Themeda* and *Sporobolus* veld types

continuously grazed by steers at 0.27 beasts ha⁻¹ at Matopos, Zimbabwe (lat. 20 °S, 610 mm rainfall), exhibited a lesser frequency of defoliation than the previous example (Gammon & Roberts, 1978). Over the four months of the main growing season the cumulative percentage of tillers defoliated in the most heavily grazed veld type (*Sporobolus*) was: zero defoliation 26%, once 35%, twice 27%, thrice 8%, and four defoliations or more 3%. The interval between defoliations varied from 14 days to over 200 days, and *c.* 50% intervals between grazing exceeded 63 days. During the growing season there was a trend for previously defoliated tillers to be selectively eaten, whilst in the dry season previously undefoliated tillers tended to be eaten, perhaps since previously grazed tillers were short and closely appressed to the ground. An interesting feature of this study was that pastures rotationally grazed (mainly 12 days grazing, 58–9 days rest) exhibited

similar frequencies and intensity of defoliation to those under continuous grazing.

(ii) The intensity of defoliation, or how much pasture is removed, is described in terms of the proportion of material harvested, of the characteristics of the feed maintained on offer, or of the residual sward material after defoliation.

Some grazing systems are directed to adjusting animal numbers to the feed supply so that a relatively fixed percentage of yield is consumed. The amount of pasture present at the end of the growing season may be used to decide the future animal density in environments with a marked seasonality of production. For example, at Augathella, Queensland, 30% of the yield of *Astrebla* grassland in April may be the figure used to decide the recommended stocking rate (Orr, Bawly & Evenson, 1986), since the basal area of perennial grasses appears to be sustained at this level, and 30% use provides good wool production from individual animals, whilst 50 or 80% utilisation has adverse effects on both pasture and animal performance.

The forage allowance, or the amount of pasture maintained on offer per animal unit, has gained acceptance in American work, as illustrated by the study of Santillan (1983) in Section 5.4.4. It is convenient to adjust this figure for animal size, and to bear in mind the generalisation that maximum intake may approximate at least 2.5% body weight per day. Thus a range in forage allowance from 2.5 to 10 kg DM per 100 kg LW would suggest a potential range of 100 to *c.* 25% utilisation, as

Fig. 4.1. The vine-type legume *Macroptilium atropurpureum* cv. Siratro with exposed growing points.

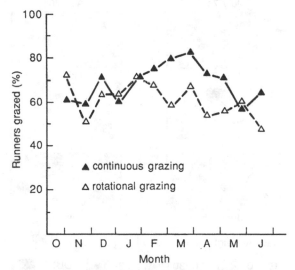

Fig. 4.2. Percentage of runners of *Macroptilium atropurpureum* cv. Siratro grazed in each 3-week period for continuously and rotationally grazed pastures at Samford, Queensland (From Clements, 1985.)

modified by the rate of pasture growth. Grazing pressure is a term which is best applied to suggest the relationship between animal intake or animal body weight per unit of pasture available, but is misleadingly used as a synonym for forage allowance; this usage is to be avoided.

The intensity of defoliation is sometimes described in rotational grazing systems by the residual forage present when the animals are removed from the field. This may serve as a guide to subsequent pasture growth, but poorly predicts animal performance, since this is better related to the rate of leaf removal, which is strongly influenced by the level of pasture presented at the beginning of grazing (Murtagh, Kaiser & Huett, 1980) and by the rate of leaf growth during grazing (Section 7.1.2). The utility of this index is also impaired by

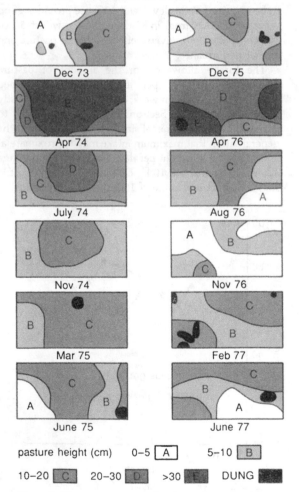

Fig. 4.3. Changes in pasture height within a fixed quadrat as related to deposition of dung pats at Beerwah, Queensland. (From Jones & Ratcliff, 1983.)

variation in degree of soiling by dung, which influences pasture acceptability.

(iii)   Selectivity of removal is minimised by mechanical cutting systems, although the height of cut will modify the proportions of leaf, stem and inflorescence material removed. Animals graze in a vertical plane downwards, but also graze clumped vegetation sideways in a horizontal plane. Selective grazing between plant species modifies competitive relations (Section 5.3.2); selectivity is less influenced by differences in plant structure as stocking rate increases (Stuth *et al.*, 1985). Grazing of particular plant species first removes the most detachable components, such as green lamina and flowers. Young shoots are removed in preference to old shoots, which reduces the potential photosynthetic output of the sward, since the latter have a lesser photosynthetic capacity.

The selectivity and intensity of defoliation varies with animal species. Material close to the soil surface is more accessible to the mouthparts of sheep and goats than of cattle and buffalo. Cattle use their tongues to roll leaf and other material into their mouths before it is bitten, and this tearing action results in residual plant organs having more irregular surfaces than plants grazed by sheep or goats. The direct cutting action of the teeth against the upper dental pad by sheep results in a cleaner cut surface; sheep will also prehend small twigs and stems and break them off relatively cleanly by head-jerking movements (Arnold, 1981). The wide muzzle of the buffalo gives less opportunity for selectivity than for cattle, and the latter make manipulative jaw movements to increase the legume content of their diet when, for example, grazing dense stands of *Aeschynomene americana/Hemarthria altissima* pastures (Moore *et al.*, 1985). The smaller mouths of sheep and goats enable them to remove material more selectively, and the split upper lip of the goat facilitates the selective prehension of small plant organs such as flowers. The split upper lip of the camel also facilitates selective browsing.

(iv)   The differences between cutting and grazing systems are most evident in the spatial heterogeneity of grazing. Animals vary their intensity of grazing according to the topography of the area, and seasonal or diurnal use of ridges and valley bottoms reflects differences in microclimate as well as in forage availability or quality. Past experience conditions animal behaviour, and previous pasture use which promotes accessible leafy material contributes to the persistence of patch grazing.

Fouling of pasture by excreta delays its utilisation. The spatial variation in height of a mixed pasture of sod-forming grasses in south-east Queensland which was continuously grazed at 2.5 beasts ha^{-1} was clearly related to the incidence of dung pats and their persistent effects on pasture acceptability and growth (Fig. 4.3; Jones & Ratcliff, 1983). Recurrent deposition of dung occurred within the fixed quadrats over the four years in which these were studied; each deposition influenced pasture height for c. 16 months within 20 cm of the deposition patch. Fig. 4.3 indicates how the location of patch avoidance changes with time, so that there is a continual alteration in the intensity of defoliation on a particular pasture microlocation.

The persistence of heavily grazed patches is reduced by burning, as mentioned in Section 10.3.

4.3
Growth of pastures

Growth of pasture is recorded as the net balance between increase in shoot tissue and the senescence and detachment of shoot tissue. Roots are not usually measured

and do not form part of the food supply of the grazing animal.

4.3.1
Growth and senescence

Senescence is the hidden factor in the equation for pasture growth and utilisation. Pasture managers are aware of senescence when a tropical pasture is frosted, or desiccated by unusual heat and drought. They are less aware of the gross magnitude of senescence which occurs continuously whilst pastures are growing rapidly, and which may exceed the biomass consumed by grazing animals, as mentioned in Section 3.2.1. Plants in dense swards of the annual S. humilis plants may be dying and disappearing unobserved at the rate of c. 80–90 plants $m^{-2} d^{-1}$ (Rickert & Humphreys, 1970)

The magnitude of senescence and the modifying effect of defoliation on the growth of the leaf surface is

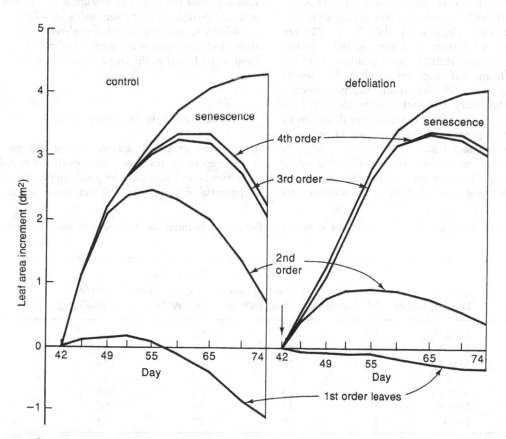

Fig. 4.4. Leaf area increment and leaf senescence for main shoot of undefoliated control plants and for last remaining leaf and its axillary growth of defoliated *Panicum maximum* var. *trichoglume*. (From Humphreys, 1966 b.)

illustrated (Fig. 4.4; Humphreys, 1966b) for seedlings of *P. maximum* var. *trichoglume* grown under good conditions at a low density of 1.6 plants m^{-2}. Floral initiation occurred 24 days after seedling emergence, and 42 days after emergence 60% of shoot material was removed, whilst control plants grew uninterruptedly. The defoliated plants were decapitated between the fourth and fifth last leaf on the main apex (1st order leaves), and Fig. 4.4 shows the subsequent leaf area increment for the fifth last leaf and its three axillary leaves on the defoliated plants, and for the undefoliated main shoot. Initially leaf expansion from the decapitated shoot was less than in the control, but increased tillering which produced more 3rd order leaves led to the production of a similar net leaf area in both treatments. Senescence of leaves 74 days after seedling emergence accounted for 48% of the accumulated true leaf expansion in the control shoots, whilst only 23% had senesced in the defoliated plants. This example shows the considerable extent of leaf senescence in young, well-spaced plants, and the decrease in potential utilisation if swards are undefoliated.

The seasonal trends in the rates of leaf development and senescence are shown from combined data of *P. maximum* var. *trichoglume* and *C. ciliaris* grazed swards at Narayen, south-east Queensland (lat. 26 °S, 710 mm rainfall) in Table 4.1 (Wilson & Mannetje, 1978). In this predominantly summer rainfall environment the rate of leaf development from leaf appearance until full lamina expansion was faster in the warm summer months. Senescence in this study is characterised by the time from leaf tagging to commencement of senescence (indicated by 95% of leaf being green) and the linear rate of decline in percentage area of green leaf.

The onset of senescence was usually delayed in autumn and spring. However, once senescence commenced, the rate of decline was generally faster in the main summer growing season, except when frosting occurred, as for the 23 May age class. An analysis of the effects of environmental variables showed that although high water stress following wet periods accelerated senescence, sustained water deficits delayed senescence. Similarly, leaf senescence of *Urochloa mosambicensis* and *U. oligotricha* was fastest under favourable moisture conditions in north Queensland (McIvor, 1984), whilst the rate of senescence of *P. clandestinum* lamina increased with temperature over the range 15–30 °C (Murtagh, 1987).

There is an effect of leaf position. Wilson (1976a) suggests that leaves of high insertion level (as occur in autumn) have intrinsically slower rates of senescence than leaves produced on new tillers appearing in the spring. It may be expected also that changing proportions of leaf and stem will alter the rate of shoot senescence, since lamina has a faster rate of senescence than stem. More work is needed to assess the effects of chemical composition on the degradability of tropical pastures; we have barely begun to monitor the influence of variations in grazing and cutting management on the senescence of tropical pastures. For *Lolium perenne* the height of pasture is well related to leaf area, and provides a convenient index for adjusting pasture availability (Grant *et al.*, 1983); the inferences we may draw from studies of temperate pastures (Parsons *et al.* 1988) need to be tested in the tropics and subtropics.

4.3.2
Growth and light interception

(i) Pasture growth is limited by the amount of light intercepted by the green surfaces of the sward, and if nutrients and water are in good supply the maximum potential growth rate will not occur unless light is

Table 4.1. *Seasonal trends in leaf development and senescence in* Panicum maximum *var.* trichoglume *and* Cenchrus ciliaris *swards*

Date of tagging leaves	Days to full expansion	Days to onset of senescence	Rate of decline in green tissue (% day^{-1})	Days to complete mortality
Jan 1	11	13	1.2	90
Mar 1	10	15	2.0	63
Mar 27	20	20	1.0	119
May 23	16	–	2.6	44
Sept 11	–	29	1.1	118
Nov 9	12	27	2.9	60
Jan 4	10	9	2.1	54

Source: Wilson & Mannetje (1978).

prevented from reaching ground level. Many scientists regard 95% interception (or a light value of 0.05 full sunlight) as a critical value. The angle of the sun to the horizon affects light interception, and in mid-summer perhaps twice the amount of leaf is needed to provide the critical value as would be required in mid-winter.

Following sward defoliation pasture growth usually exhibits a sigmoid function with time, and the instantaneous rate of growth (dW/dt where W is the weight of pasture present and t is time) increases as the size of the photosynthetic system, (which is described simplistically as the leaf area index, L, or the area of leaf surface supported by unit ground surface) increases to make more effective use of the photosynthetically active radiation (PAR, 400–700 nm wavelength) in fixing CO_2. Classical growth analysis regards the crop growth rate as the product of L and of a measure of the efficiency of L, the net assimilation rate, which is expressed as net growth rate per unit L.

Regrowth of a sward of *Paspalum notatum* at Fukuoka, Japan (lat. 33 °N) shows a sigmoidal increase in L with time, to a maximum value of 8.5 some 48 days after defoliation (Fig. 4.5; Agata, 1985a). However, net assimilation rate, which showed high values when all young leaves were fully illuminated, decreased as mutual shading increased, and as biomass developed to increase respiration. The net result was that absolute crop growth rate increased to a maximum value of 30 g DW m^{-1} d^{-1} when L was 5.5, but decreased subsequently as the rate of further increase in L did not match the decrease in net assimilation rate. The efficiency of conversion of radiant energy to dry matter, expressed as net photosynthesis, increased from 0.2% at the commencement of regrowth to 2.6% when crop growth rate was maximal (Agata, 1985b).

A further illustration is taken from a study by Ludlow & Charles-Edwards (1980) of a *Setaria sphacelata* var.

sericea/*Desmodium intortum* sward at Redland Bay, Queensland (lat. 28 °S), which had been cut for five years at intervals of three or five weeks at 7.5 or 15 cm height. Both shoot dry weight and L (Fig. 4.6a and b) showed a sigmoidal increase with time after cutting and a pronounced early lag phase. Light interception (Fig. 4.6c) was incomplete after three weeks, but reached the critical value after five weeks, when L was c. 5. After defoliation, interception immediately increased in the 3-week cutting treatments, whilst a short-term decrease occurred in the 5-week series. More light was intercepted by the 15 cm-series than the 7.5 cm-series, but gradually these treatment differences diminished as sward growth proceeded.

(ii)   Light is attenuated as it passes to the lower layers of the sward. The geometrical arrangement of the canopy influences the light environment and the photosynthesis of the green surfaces below the upper leaves and stems. The inclination of the leaf affects the amount of light passing down, as does the density of leaves. A clumped dispersion of leaves is less efficient than an even dispersion. The interception of light in representative profiles of these same *S. sphacelata* var. *sericea/D. intortum* swards is shown subsequently in Figs. 5.11 and 5.12, whilst the photosynthetic output from different layers of the sward is illustrated in Fig. 5.12.

There is robust relationship for particular swards between the PAR incident at the top of the canopy (I_o) and the PAR available in any particular horizon (I). Saeki (1960) showed that $I = I_o k e^{-KL}$, where L is the leaf area index above the particular horizon and k is the canopy extinction coefficient; low values of k indicate that more light penetrates to the lower layers of the sward than if high values occur. The extinction coefficient in the Ludlow & Charles-Edwards (1980) study was largely independent of treatment, and averaged 0.67; the *Setaria* leaves in the upper layers of the sward tend to fall into a horizontal position. By contrast, the *P. notatum* leaves in the Agata (1985b) study were relatively erect and the extinction coefficient was only 0.36.

(iii)   Leaves in the lower layers of the sward will be better illuminated as radiation increases; more of the leaves in a deep sward will be above the compensation point where photosynthesis exceeds respiration. The *P. notatum* study was conducted in a high radiation environment which averaged 17.9 MJ m^{-2} d^{-1} during the course of regrowth, but mean solar radiation decreased 28% from the beginning of regrowth to the final measurement period (calculated from Agata, 1985b). This may have influenced the shape of the curve for crop growth rate, which decreased as high L occurred (Fig. 4.5).

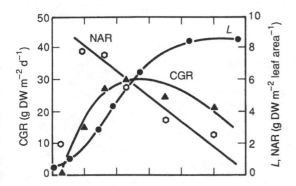

Fig. 4.5. Crop growth rate (CGR), leaf area index (L), and net assimilation rate (NAR) of *Paspalum notatum* after cutting at 10 cm height. (From Agata, 1985a.)

Ludlow & Charles-Edwards recalculated their data to show the relationship of L to net photosynthetic output (which is similar to growth rate) under constant but high radiation conditions (Fig. 4.7). Net photosynthetic rate (P_N) increases sharply over the range L 0–2, but then becomes asymptotic (in contrast to Fig. 4.5); increasing L produces the same net photosynthetic output.

(iv) The rate of net photosynthesis, which is the difference between gross photosynthesis and respiration, may increase, decrease, or remain stable at high levels of L. We need to understand what factors are involved in determining this variable response if predictive judgements are to be made.

The early work of Davidson & Donald (1958) with subterranean clover showed an optimal L of c. 4–5, above which growth rate decreased, as in Fig. 4.5. On the other hand Brougham (1956) under conditions of increasing radiation recorded a relatively constant growth rate over the range of L 5–9 (cf. Fig. 4.7), so this continuing dichotomy of responses has been evident since the topic was first studied. Fortunately the issue is not of great moment in the field, since management directed to efficient pasture utilisation avoids the accumulation of a large biomass in the field, especially as long periods of regrowth are associated with low nutritive value, at least for grasses.

The C_4 tropical grasses, which have the dicarboxylic acid photosynthetic pathway, show increasing photosynthetic output as radiation increases to the highest levels occurring in the tropics (Ludlow, 1978), since they have lower intracellular and stomatal resistances than C_3 legumes and the temperate pasture grasses; these have the phospho-glyceric acid photosynthetic pathway and show light saturation at 30–50% full sunlight illuminances of c. 2400 $\mu E\ m^{-2}\ s^{-1}$ quantum flux. The

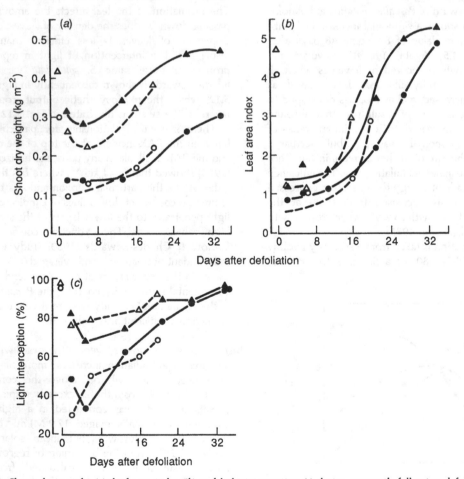

Fig. 4.6. Shoot dry weight (*a*), leaf area index (*b*) and light interception (*c*) during regrowth following defoliation of *Setaria sphacelata* var. *sericea*/*Desmodium intortum* swards cut at 7.5 cm (circles) or 15 cm (triangles) every 3 weeks (open symbols) or 5 weeks (closed symbols). (From Ludlow & Charles-Edwards, 1980.)

higher photosynthetic efficiency of the C_4 grasses may lead to a more sustained growth performance at high levels of L.

(v)   The rate of respiration has a component of 'maintenance' respiration, which is concerned with tissue repair and maintenance and is related to the biomass accumulated in the sward, and a component of 'growth' respiration, which is directly related to the rate of photosynthesis. The sum of these processes may or may not contribute to a reduced rate of net photosynthesis at high levels of L.

In Fig. 4.7 respiration is shown to increase as L increases to c. 2, and to remain constant at higher levels of L; it was thus proportional to the rate of photosynthesis and did not induce the optimum L type of response of Fig. 4.5. In assessing the total impact of respiration on growth, it is necessary to adjust the values in Fig. 4.7 for night respiration. At high L total respiration was c. 27–30 % of gross photosynthetic rate. In the study with *P. notatum* swards (Agata, 1985b) respiration was greater, perhaps associated with the massive subsurface crown and root development of 1.2 kg DM m^{-2}. Respiration increased to a maximum of 24.3 g m^{-2} d^{-1} when growth rate was maximal, and accounted for 46 % of gross photosynthetic rate. It decreased to 17.7 g m^{-2} d^{-1} in the final measurement period, preserving the same ratio to gross photosynthetic rate; decreased growth at high L was therefore not due to increasing respiration but was associated

with some other factor, such as reduced radiation during this period.

(vi)   There are many studies in which utilised yield increases with the duration of the period between cuts. This is interpreted as reflecting the positive effects on light interception of higher mean leaf area duration (Ludlow & Charles-Edwards, 1980). The effect is usually greater in plants with an erect habit. At Samford, Queensland, the annual yield increase for each increase of one week in the cutting interval over the range 6–12 weeks was 1310 and 2110 kg ha^{-1} respectively for the erect *Paspalum commersonii* and *P. plicatulum* cv. Rodds Bay, but only 110 and 280 kg ha^{-1} respectively for the more decumbent *P. dilatatum* and *P. notatum* cv. Pensacola. In all studies with grasses the extension of the interval between cuts results in decreased nutritive value; this consideration is less cogent for legumes in which the effect of age on nutritive value is less.

4.4
The restoration of the leaf surface after defoliation

4.4.1
Residual leaf area

The rate of restoration of the leaf surface after defoliation hinges upon the extent of the residual leaf area and its photosynthetic capacity, the density of leaves which are still capable of expansion, the density of buds present and their further expansion and differentiation.

The relative growth rate (RGR) of *Crotalaria juncea* in the immediate post-defoliation period was shown to be positively and linearly related to the residual leaf area by the equation RGR (g g^{-1} d^{-1}) = −0.022 + 0.044 L (Kessler & Shelton, 1980); RGR was negative if residual L was less than 0.5. Higher residual L also indicated more buds.

This topic is complex, and compensatory effects may occur which diminish the significance for growth of differences in residual L. For example, swards of *M. atropurpureum* cv. Siratro at Samford, Queensland with residual L varying from zero to 0.7 grew at the same average rate during the succeeding 50 days (Jones, 1974b). Similarly, in Fig. 4.7 there is a great deal of noise in the relationship between P_N and the values for L in the range 0–2; at L 1.3 P_N varies from zero to 1.1 mg CO_2 m^{-2} s^{-1}s.

(i)   The photosynthetic capacity of the residual leaves reflects the light environment in which they have developed. This is illustrated from a shade study in which *P. maximum* var. *trichoglume* and *M. atropurpureum* were grown at 100, 33 and 11 % full sunlight, which

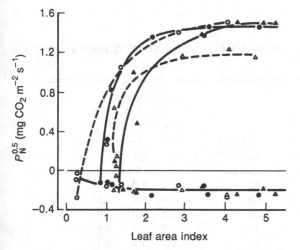

Fig. 4.7. Leaf area index and canopy net photosynthetic rate ($P_N^{0.5}$; measured at 0.5 mW m^{-2} photosynthetically active radiation) and dark respiration rate of swards described in Fig. 4.6. (From Ludlow & Charles-Edwards, 1980.)

averaged 15.2 MJ m^{-2} d^{-1} (Ludlow & Wilson, 1971a). Leaves formed in the lower levels of the sward have a higher leaf area/leaf weight ratio (or specific leaf area) as suggested in Table 4.2, but this advantage is offset by a lowered photosynthetic capacity. The C$_4$ grass exhibited a higher P_N than that of the C$_3$ legume, but shading greatly reduced photosynthetic capacity, which was associated with a decreased chlorophyll content. The decline in P_N was associated with increased stomatal (r_s), mesophyll (r_m) and, to a lesser extent, boundary layer (r_a) resistances in *P. maximum* var. *trichoglume*, and mainly with increased mesophyll resistance in Siratro. Shading changed leaf anatomy; shaded leaves had fewer, smaller and less densely packed cells, and were thinner.

Sudden changes in sward structure, as occurs in short duration grazing systems and with cutting, places residual leaves in an environment to which they may not be fully adapted. However, photoinhibition (Powles, 1984) was a minor factor only in the growth response of previously shaded leaves of *P. maximum* var. *trichoglume* exposed to full light (Ludlow, Samarakoon & Wilson, 1988).

(ii)   Grazing preferentially removes young shoots, and the basal leaves remaining after cutting often have an older age distribution than leaves higher in the sward. The significance of senescence in determining net growth was discussed previously; it should be recognised that photosynthetic capacity decreases before the onset of senescence is visually evident.

Net photosynthetic rate of leaves at various ages (Fig. 4.8; Ludlow & Wilson, 1971b) increased with illuminance to the highest value tested for *P. purpureum*, and

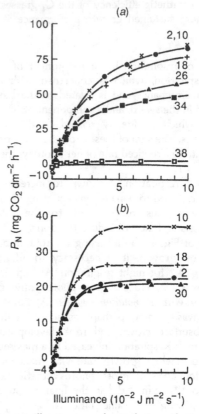

Fig. 4.8. Illuminance and net photosynthetic rate of leaves of (a) *Sorghum almum* and (b) *Calopogonium mucunoides* at various times (days) after unfolding. (From Ludlow & Wilson, 1971b.)

Table 4.2. *Specific leaf area (SLA), leaf net photosynthetic rate (P_N) and carbon dioxide transfer resistances of plants grown at three illuminances*

Species	Illuminance[a] (dm^{-2} g^{-1})	SLA (dm^{-2} g^{-1})	P_N (mg CO$_2$ dm^{-2} h^{-1})	r_a	r_s	r_m
Panicum maximum var.	100	3.3	72	0.8	1.0	0.8
trichoglume	33	5.0	40	0.9	1.9	2.3
	11	8.3	22	1.6	4.0	4.9
Macroptilium atropurpureum	100	2.8	36	0.7	1.1	2.9
cv. Siratro	33	7.1	20	0.7	1.5	6.3
	11	10.0	13	1.1	2.2	10.6

(Header note: CO$_2$ transfer resistances (sec cm^{-1}) spans the r_a, r_s, r_m columns.)

Source: Ludlow & Wilson (1971a). [a]100 = full sunlight.

r_a, boundary layer resistance; r_s, stomatal resistance; r_m, mesophyll resistance.

was greatest for young leaves. The maximum value of 80 mg CO_2 dm^{-2} h^{-1} decreased to 50 mg some 34 days after unfolding. By contrast, *Calopogonium mucunoides* showed light saturation at relatively low illuminance, and young leaves which were still unfolding had lower P_N than leaves 10 days of age. The unrolled lamina of grasses is fully expanded whilst the intercalary meristem is still promoting leaf extension; in legumes further expansion occurs after unfolding and maximum P_N is less than in C_4 grasses. P_N then decreased following complete leaf expansion to about half the maximum value at 30 days of age (Fig. 4.8*b*). As in the previous example concerning the effects of shade, the decrease in P_N with age is associated with increased stomatal and

mesophyll resistances and with reduced chlorophyll concentration. In *P. notatum* P_N is quite well maintained until leaves are 45 days old, and decreases sharply until senescence occurs at *c.* 60 days (Sampaio, Beaty & Ashley, 1976).

(iii) The removal of young grass laminae which are still extending is probably less detrimental to regrowth than the removal of young legume leaves, whose replacement is wholly contingent upon the differentiation of new leaves from axillary buds. Bud density affects regrowth in the absence of residual leaf area in legumes, as has been shown for *Aeschynomene americana* (Albrecht & Boote, 1985).

4.4.2
Apical dominance and tillering

The rate at which the leaf surface is restored after defoliation depends not only upon the residual leaf area present, but upon the rate at which new shoots are differentiated to replace the shoots removed.

(i) The dormancy and expansion of lateral buds are controlled by the balance and supply of specific plant hormones. It is generally considered that auxin produced in young leaves inhibits the expansion of buds (Phillips, 1975), and the application of indole-3-acetic acid (IAA) to cut tissues simulates this effect (Clifford, 1977), whilst abscisic acid (ABA) is also involved in bud suppression (Knox & Wareing, 1984). The presence of inhibitory levels of auxin in elongating stem internodes nearby may also be involved, and Jewiss (1972) considers that stem extension may be more important than flowering effects which occur at the apex in restricting lateral bud expansion, although the plant apex has long been considered the main site of dominance, especially in plants which are flowering (Pedreira, 1975).

Other growth regulators, and especially cytokinins, are involved in and necessary to the activation of lateral buds. For *Sorghum bicolor* there is an optimum concentration of cytokinin which activates the expansion of lateral buds, although low concentrations of cytokinin are also promotive (Nojima, Oizumi, & Takasaki, 1985). The subsequent growth of the bud after initiation of growth has occurred is further promoted by sucrose and gibberellin (GA_3) (Oizumi *et al.*, 1985).

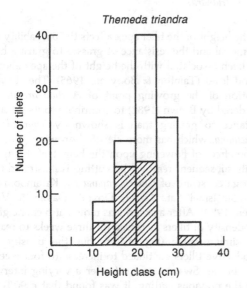

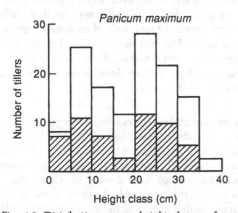

Fig. 4.9. Distribution among height classes of new shoots (shaded from the interior and unshaded from the exterior of the crown) during regrowth of *Themeda triandra* and *Panicum maximum*. (From Richards *et al.*, 1986.)

(ii) We do not always understand the precise mechanisms of hormonal control which are modified by defoliation management, although the above brief background gives a general conceptual framework against which the ontogenetic effects may be viewed.

The presence of young vegetative shoots inhibits further tillering in *P. maximum* var. *trichoglume*, and removal of these enhances the rate of appearance of new shoots (Humphreys & Robinson, 1966). The presence of young flowering shoots is also inhibitory and the continual removal of young inflorescences of *P. maximum* var. *trichoglume* as they appear promotes tillering and eventually increases inflorescence density once the process of removal is terminated. Tropical grasses differ from many bred temperate species, since shoot differentiation continues strongly during flowering; this process of continued tillering is more pronounced in *Cenchrus ciliaris* than in *P. dilatatum* (Masuda, 1976). Nevertheless, shoot removal during flowering increases tiller density. The strength of apical dominance diminishes as advanced fruiting stages are reached, and in *P. maximum* var. *trichoglume* the removal of inflorescences at the post-anthesis stage does not influence the rate of tillering (Humphreys, 1966*b*).

(iii)   Synchrony of tiller emergence increases the vulnerability of the plant to defoliation. The restoration of the leaf surface is likely to be slower if a high proportion of the tillers have elevated apices which are removed by defoliation, so that regrowth depends upon the activation of new basal tiller buds and not upon the continued growth of established tillers. *Themeda triandra* is regarded as less resistant to grazing than *P. maximum* var. *trichoglume*, and the latter has more short tillers, and shows a greater diversity of tiller height (Fig. 4.9; Richards, Mott, & Ludlow, 1986) and age of tiller than does *T. triandra*.

(iv)   The height of the tiller apex affects the probability of its removal and the resistance of grasses to grazing has long been associated with the height of the apex above ground level (Tainton & Booysen, 1965). The slower elevation of the growing point of *Aristida armata* is considered by Brown (1982) to contribute to its greater resistance to grazing than is shown by *Thyridolepis mitchelliana*, which has more elevated growing points.

The effect of flowering upon the height of the apex and its subsequent removal by cutting is illustrated for an irrigated sward of *Chloris gayana* cv. Katambora at Rehovot, Israel (lat. 32 °N; Dovrat, Dayan & Van Keulen, 1980). After a sward was mown at 6 cm height, total density of tillers increased for three weeks to reach a maximum value (Fig. 4.10*a*), but the density of reproductive tillers continued to increase to four weeks after cutting. Swards were cut after a varying interval from the previous cutting. It was found that *c.* 90% of tillers were capable of regrowth when 1-week regrowths were cut again (Fig. 4.10*b*) whereas less than 20% of tillers were capable of regrowth when 4-week regrowths were cut again. This was directly related to the density of tillers above 60 mm height which had reached a reproductive stage with an elevated apex at the time of the second cut (Fig. 4.10*c*).

(v)   The system of defoliation which is employed changes the long-term structure of the sward. The types of tiller produced are modified by defoliation. *Pennisetum purpureum* has both an erect habit and large rhizomes. Belyuchenko (1980) has shown in Cuba that the proportion of rhizomes and of crown basal shoots is negatively related to cutting height, whilst the incidence of apical buds below cutting height and of axillary buds developed on vertical shoots increases as cutting height increases.

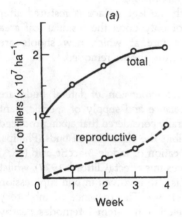

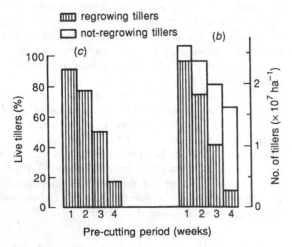

Fig. 4.10. (*a*) Density of total and reproductive tillers during regrowth of *Chloris gayana* cv. *Katambora*; (*b*) density of tillers regrowing and not regrowing; and (*c*) % live tillers after different pre-cutting periods. (From Dovrat, Dayan & van Keulen, 1980.)

Swards cut frequently at the same height develop a layer of green leaves immediately below cutting height; the average height of the junction between the blade and sheath of the lowest green leaf is located further above the soil surface in swards cut high than in swards repeatedly cut close to ground level.

The effect of a wide range of stocking rate (SR) on tiller density and tiller weight is shown (Fig. 4.11; Mears & Humphreys 1974a) for *P. clandestinum* pastures at Lismore, New South Wales (lat. 29 °S, 1660 mm rainfall). In this study an incomplete factorial design was employed in which increasing SR was allied with increasing N fertiliser inputs. Shoot density on these continuously stocked pastures varied from 1250 to 8500 m^{-2}. Increasing SR above the lowest SR employed at any level of N input greatly increased shoot density. There was a compensatory and negative relationship between tiller density and tiller weight. Increasing N had varying effects on tiller density which were probably associated with the level of shade involved, as related to pasture availability.

There are many studies which support the general proposition that defoliation promotes tillering or branching, since apical dominance is removed. The immediate effect of a single defoliation may be quite transitory since new patterns of dominance are soon established, as shown for *S. hamata* cv. Verano (Wilaipon, Gigir & Humphreys, 1979) but repeated cutting or grazing will modify the hormonal balance and result in the continual activation of the buds present. Early grazing of mixed *S. hamata* swards in Thailand at four weeks promoted a better residual *L* and a higher shoot density than grazing at eight weeks (Tudsri *et al.*, 1989); similarly early initial grazing of *Aeschynomene* spp. in Florida promoted better branching (Gildersleeve *et al.*, 1987). Tiller density of *P. notatum* at Nagoya, Japan, showed wide seasonal variation but increased after each grazing (Hirakawa, Okubo, & Kayama, 1985). Frequent cutting (4 or 8 week interval) of *B. decumbens* in north Queensland (McIvor, 1978) led to higher residual leaf density and *L*, faster regrowth and higher utilised yield than infrequent cutting (12 or 16 week interval).

(vi)   Height of cutting or grazing has diverse effects. It appears that a shortage of buds below cutting height restricts regrowth in some legumes, as mentioned previously. A similar effect occurs in erect grasses, and *P. purpureum* in Florida develops fewer axillary buds per tiller under high levels of use (Rodrigues *et al.*, 1986).

On the other hand close cutting enhances utilised yields in grasses capable of strong basal tillering. *Cynodon dactylon* cv. Coastal at Clayton, North Carolina, develops a more prostrate habit under close cutting, and growth arises from both lateral and terminal buds (Clapp, Chamblee & Gross, 1965). The most severe defoliation schedule tested (growth to 5 cm height reduced to 1.9 cm), which maintained *L* usually in the range of only 1.5–2.2, gave higher utilised yield than more lenient systems (for example, growth to 15 cm height reduced to 7.5 cm). Young stems of *C. gayana* show reduced tillering potential as the cut on the stem is raised, and Kipnis, Dovrat & Lavee, (1977) suggest that the stimulus of defoliation is negatively related to the distance between the site of action (the tiller base) and the site at which it operates (the height of cut). The bunch grass *P. maximum* develops a higher tiller density and gives greater regrowth when cut at 15 cm than at 30 cm in Jaboticabal, Brazil (Mecellis & Favoretto, 1981).

(vii)   Plant genetic constitution modifies the structural response to defoliation. Tussock-forming grasses with apogeotropic (erect) shoots having long internodes which arise from basal shoots with short internodes, such as *Andropogon gayanus* (Fig. 4.12) may be contrasted in habit with sod-forming grasses with predominantly diageotropic (horizontal) stems, such as *P. clandestinum* or *P. dilatatum* (Fig. 4.13); there are many intermediate forms. Belyuchenko (1976) divides the renewal buds of tropical grasses into three groups: (i) those on the surface of the soil or immediately below, such as *C. gayana* and *Eragrostis curvula*; (2) those not deep

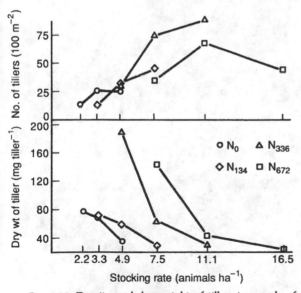

Fig. 4.11. Density and dry weight of tillers in swards of *Pennisetum clandestinum* continuously grazed at different stocking rates receiving different N inputs. (From Mears & Humphreys, 1974a.)

seated in the soil, but having stolons or rhizomes protected with shortened leaf plates, such as *C. dactylon*, *B. mutica* and *D. decumbens*; and (3) those with deep seated rhizomes, such as *P. purpureum* and *Sorghum halepense*.

The optimum cutting or grazing height needs to be related to plant structure, and 15 cm height may promote satisfactory use for the bunch grass *P. maximum* where 3 cm may be appropriate for *P. clandestinum*.

Genetic variation within the species *D. intortum* is evident in the structural response to defoliation of different genotypes. These mainly initiate growth from one or two axillary buds immediately below the cut surface, but Imrie (1971) noted that one accession (CPI 23189) regrew vigorously from buds close to the basal crown.

There are large differences in patterns of regrowth tillering amongst summer fodder crops. *Sorghum bicolor* × *S. sudanense* and *S. sudanense* develop strong basal tillering in response to grazing or cutting (Clapp & Chamblee, 1970), and this type of regrowth increases as the height of cut is lowered. By contrast, regrowth

of *P. americanum* cv. Gahi-1 is more dependent upon continued production from terminal buds, and their removal restricts regrowth and makes the management of the crop more exigent; the hybrid *P. americanum* × *P. purpureum* is capable of tillering following decapitation (Muldoon & Pearson, 1979a).

(viii)   Morphogenic effects of light quality also control tillering through the reversible red/far-red phytochrome system. Far-red light (*c.* 730 nm) is inhibitory, whilst red light (*c.* 660 nm) is promotive. The ratio of red/far-red light is reduced when sunlight is filtered through the green surfaces of the sward profile (Cumming, 1963) and this inhibits tillering in grasses such as *P. dilatatum* (Deregibus *et al.*, 1985).

Severe defoliation or burning exposes the soil surface to full sunlight, which promotes the germination of light-sensitive seeds (Section 5.2.5). It also exposes the crown of the plant to full sunlight with a high red/far-red ratio. The rate of tillering of *B. decumbens* is stimulated by cutting to ground level and sweeping the organic residues from the surface (Fig. 4.14; Stür &

Fig. 4.12. The tall bunch grass *Andropogon gayanus* in Colombia.

Humphreys, 1988*a*); this produced more tillers than cutting at 10 cm, and since in the study reported there were no buds in the 0–10 cm stem zone, a red 660 nm promotive effect on tillering provides an attractive hypothesis. Burning immediately after cutting with a low fuel load also promotes tillering, but other burning treatments which apparently cause injury to the crown give lower tiller densities, as discussed in Section 10.3.

4.4.3
Homeostasis and constraints other than light

Variations in the management of the leaf surface by defoliation have only transitory effects on growth in environments in which the primary constraints to growth are water or nutrients rather than light.

An attempt was made to relate residual L to the growth of swards of *P. maximum* var. *trichoglume* on a fertile basaltic soil at Gayndah, Queensland (lat. 26 °S, 730 mm rainfall). Swards were grazed by cattle and then slashed at two cutting heights or not slashed, to generate treatments differing in residual L, designated heavy, medium or lenient defoliation (Humphreys, 1966*a*).

Growth and light interception were measured in the subsequent six weeks, and 12 grazing cycles were imposed over the summer seasons of three years, using a balanced design to avoid persistent effects of previous treatments. Residual L averaged 0.1, 0.3 and 0.7 in the severe, medium and lenient defoliation treatments respectively.

If accumulated net shoot growth over the three summers in the medium defoliation treatment (116000 kg-ha^{-1}) was designated as 100, the severe and lenient treatments yielded 95 and 99 respectively; the differences were not significant. Compensatory effects which created homeostasis were evident. Treatment effects were inconsistent between particular defoliation cycles, but some general trends were present. For the first three weeks after defoliation the initial advantage of the lenient treatment in L present was maintained in 10 of the 12 periods, although the rate of increase in L was significantly less in the lenient treatment for 7 of the 12 periods. The net assimilation rate of the well-illuminated swards of the severe and medium defoliation treatments, which also had an advantage in the production of young leaves, tended to be greater than for the lenient treatment. Light values at ground level did not

Fig. 4.13. Well grazed, sod-forming *Paspalum dilatatum* sward.

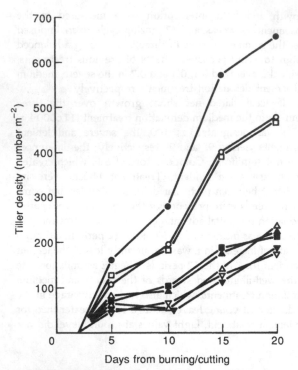

Fig. 4.14. Effects of different residue disposal treatments on tiller density of *Brachiaria decumbens* cut at 10 cm (open circles), 0 cm (closed circles); burnt with low (open symbols) or high (closed symbols) fuel load immediately (boxes), 4 days (triangles) or 8 days (inverted triangles) after cutting. (From Stür & Humphreys, 1988a.)

reach the critical 0.05 level, and the maximum L recorded in the experiment was 3.4. At some stage in each of the 12 periods of six weeks the pastures wilted; the onset of wilting was delayed by up to 10 days in the severe defoliation treatment, which commenced each period with lowest L.

These pastures were grown with a modest input of 68 kg N ha^{-1} yr^{-1}, and it appeared that constraints to pasture growth were primarily due to intermittent water stress or to nitrogen availability, and not to the capacity of the leaf surface to intercept radiation. This study illustrates the lack of relevance to subhumid areas of the leaf area index concept as a management tool which might increase shoot growth. On the other hand, in subhumid and semi-arid regions grazing and cutting management affects pasture persistence and botanical composition (Chapter 5); the distribution of assimilate to plant organs is also affected.

4.5
Distribution of assimilate

4.5.1
Roots, stems and leaves

Defoliation alters the balance between root, stem, leaf and inflorescence growth. Defoliation which greatly reduces leaf area decreases photosynthetic rate, which decreases the distribution of assimilate to roots; the ratio of leaf to stem increases, partly as a reflection of differences in shoot ontogeny.

In the last example (Humphreys, 1966a) net shoot growth of *P. maximum* var. *trichoglume* was not consistently affected by variation in the intensity of defoliation. However, the relative proportions of leaf and stem were changed, and the net growth of green lamina during 12 defoliation cycles was in the proportion 59:79:100 for the lenient, medium and severe defoliation treatments respectively. Severe defoliation left a sward with a low residual leaf/stem ratio, since there was little leaf remaining; superior leaf growth then usually led to a higher leaf/stem ratio than occurred under more lenient management, where more of the shoots reached flowering stage. The N concentration of green lamina was greater than that of the remaining attached plant material by an average factor of 2.9, so that lenient management gave a penalty to nutritive value, as discussed in Chapter 6.

Effects of defoliation on flowering and seed production are described in Section 5.2.3.

Reduced root growth following defoliation is a long-known and universal phenomenon in herbage plants (Ruby & Young, 1953 for *P. plicatulum*; Troughton, 1957; Barnes & Hava, 1963 for *P. maximum*; Humphreys & Robinson, 1966 for *C. ciliaris* and *P. maximum* var. *trichoglume*; Taerum, 1970 for *P. maximum*). Moderate systems of defoliation which do not impair shoot growth may still reduce net photosynthetic rate and cause unrecognised reductions in the movement of assimilate to the roots and in their growth. For example, similar rates of shoot growth occurred in a study where *M. atropurpureum* was (1) undefoliated or had (2) 40% or (3) 80% of its shoots removed every 32 days but the percentage of root growth to total growth was 45%, 26% and 20% respectively (N. D. Young, unpublished).

Severe defoliation causes cessation of root growth. The delay in the initiation of new roots varies according to species and conditions. New adventitious roots appeared in *P. maximum* in Ivory Coast *c.* 10 days after cutting, but their development was interrupted if a further cut was applied (Picard, 1977). In *C. gayana* root respiration also decreases after cutting (Kipnis, Dovrat & Lavee, 1977). Richards (1984) regards reduced root growth following defoliation as a desirable adaptive mechanism which assists the more rapid re-establishment of the leaf surface and the eventual restoration of a normal balance between root and shoot growth.

The regulation of shoot/root ratio has been described (Wilson, 1988) in terms of allometric models, functional equilibrium models, hormone models, and a mechanistic model (Thornley, 1972) based on the supply, transport and utilisation of carbon and nitrogen. The effects of defoliation can be accommodated in Thornley's model.

4.5.2
Regrowth and non-structural carbohydrate

The translocation of mobile non-structural carbohydrate from roots or crown tissue for reassembly at the growing points influences growth in a minor and transitory fashion. It is expected that proteins and other compounds, as well as labile carbohydrate, contribute to tissue maintenance under conditions of energy starvation.

(i) Considerable interest used to be evinced in management which affected the accumulation of total non-structural carbohydrate (TNC) in the roots and crown of pasture plants, since it was believed this controlled the rate of regrowth after defoliation and the persistence of plants (Weinmann, 1961). Grassland scientists accept the latter view that the survival of plants requires a minimum level of energy residual for tissue maintenance, but have now, in the main, discarded the former hypothesis. The principal evidence with respect to tropical species is as follows.

(ii) C_4 grasses accumulate sucrose with or without starch in roots and crown, in contrast to C_3 grasses, in which fructosan is the main labile carbohydrate stored.

Concentration of TNC in C_4 grasses is low relative to that of C_3 grasses. Adjei, Mislevy & West, (1988) report levels in roots/crown of 3.7–6.4 % in *Cynodon aethiopicus*, *C. nlemfuensis* and *D. decumbens*, whilst 12.2 % occurred in *P. notatum*. It is usual for TNC to accumulate more in N-deficient plants (Adegbola & McKell, 1966; Gallaher & Brown, 1977 for *P. notatum*) than in N-adequate plants where assimilate can be more readily used in the production of new shoots; Wilson (1975) believes this generalization has more applicability to C_3 than to C_4 grasses. Frequent defoliation reduces TNC levels in storage organs (Reis *et al.*, 1985 for *B. decumbens*), although these are not impaired at high SR in some cultivars such as *C. aethiopicus* cv. McCaleb (Adjei *et al.*, 1988).

A simple technique for monitoring energy 'reserve' is to record the dry weight of etiolated growth produced by swards or plugs in the dark. This has been correlated with TNC content in roots and crown (Burton & Jackson, 1962; Adegbola, 1966; Humphreys & Robinson, 1966) and with the previous rates of pasture growth (Akinola, Mackenzie & Chheda, 1971 for *Cynodon* spp.)

(iii) Steinke & Booysen (1968) recorded a positive relationship in *E. curvula* between regrowth in the light and TNC in roots and crown, but the evidence from numerous studies (Adegbola, 1966, Humphreys & Robinson, 1966; Jones, 1974; Christiansen, Ruelke & Lynch, 1981; Kipnis *et al.*, 1977; Jones & Carabaly, 1981; Richards & Caldwell, 1985; Hodgkinson, Mott & Ludlow, 1985) against a close causative relationship is strong.

Table 4.3. ^{14}C *content and leaf regrowth of* Paspalum notatum *defoliated after exposure to* $^{14}CO_2$ *and subsequently defoliated at different frequencies*

	Frequency of defoliation		
	3 days	6 days	9 days
^{14}C in plant (counts min^{-1} × 10^{-4})	1040	1198	913
^{14}C in leaf regrowth (counts min^{-1} × 10^{-4})	246	208	168
% of total ^{14}C in leaf regrowth at:			
3 days	12.9		
6 days	4.5	13.6	
9 days	2.6		16.6
12 days	1.4	2.6	
15 days	1.2		
18 days	0.9	1.2	1.7
Total leaf regrowth (g)	0.71	0.91	1.93

Source: Beaty *et al.* (1974).

Carbon-14 studies have illuminated the processes of shoot regrowth after defoliation. When $^{14}CO_2$ was fed to *P. notatum* swards before defoliation, translocation of ^{14}C to subsequent regrowth was virtually complete three to six days after defoliation (Ehara, Maeno & Yamada, 1966; Beaty *et al.*, 1974). When swards were repeatedly defoliated to ground level at intervals of three, six and nine days (Table 4.3), the total ^{14}C movement into leaf growth was 24%, 18% and 18% of the 'reserves'. This indicates that a large proportion of TNC was not available for regrowth. Severe defoliation resulted in an immediate loss of weight from stolons, root and leaf sheaths.

In *P. americanum* × *P. purpureum* movement of ^{14}C into new growth from stem bases and leaf sheaths (but not roots) occurred for up to eight days and involved protein and other compounds (Muldoon & Pearson, 1979b). In *E. curvula* respiratory demands account for a large fraction of the TNC lost (Steinke, 1975), but some export from the stubble to young leaves and growing shoots takes place.

The size of and the losses from the labile pool are insufficient to account for much regrowth. Hodgkinson *et al.* (1985) suggested that TNC recorded in their study of *T. triandra* would only last for two days if it were all used in new growth; TNC may be regarded only as a buffer in the system and not a large 'reserve' pool. Shoots usually become photosynthetically independent within three days of their exsertion and commence exporting to other organs. Management is better directed to the maintenance of tiller density and seed reserves than to the excessive and wasteful accumulation of carbohydrate below ground or in the surface crown.

Table 4.4 *Effects of cutting height and regrowth period on rate of phosphorus absorption ($\mu mol\ P\ g^{-1}$ fresh roots d^{-1}) of Centrosema pubescens*

Cutting height	Leaf area index	Time (days)		
		0–7	7–14	14–28
Undefoliated	1.1	17.8	14.8	6.8
12.5 cm	0.7	14.1	12.7	6.9
7.5 cm	0.5	12.8	12.4	6.9
2.5 cm	0.1	6.7	9.5	6.0

Source: Chantkam (1978).

4.5.3
Uptake of minerals and water

The absorption of minerals and water by pasture plants has an energetic cost; severe defoliation which restricts root activity temporarily decreases the uptake of minerals and water. The fertiliser needs of the pasture may also be modified by effects of defoliation on growth rate and on shoot ontogeny.

The requirements of pastures for maintenance applications of fertilisers are often estimated from small-plot cutting experiments in which the cut material is removed from the plots. This leads to misleading conclusions if the results are extrapolated to grazed conditions where the cycling of nutrients through the animal modifies the fertiliser requirement (Section 3.1.3). We need to consider additionally the effects of defoliation on the capacity of the plant to absorb nutrients and on its growth responses to fertiliser application.

(i) Defoliation which restricts root activity reduces mineral uptake, as illustrated for *C. pubescens* in Table 4.4 (Chantkam, 1978). In this solution culture experiment, progressive reduction in cutting height, which decreased both residual leaf area and residual bud number, gave a progressive reduction in the rate of P uptake. The decrease was especially marked in the plants cut at 2.5 cm height where little residual leaf area ($L = 0.1$) occurred, and in the first seven days after defoliation rate of P uptake was 38% of that of the undefoliated control. The rate of P uptake for all treatments decreased with time, and treatment differences became insignificant in the period 14–28 days after defoliation. Regrowth over 28 days was positively associated with cutting height.

The size of the active root system will also affect the degree to which the soil mass is explored, both in intensity and depth. Intense pasture use which restricts root growth may therefore reduce the exploitation of the nutrients present; effects on water supply are mentioned in Section 5.2.2.

This question has received little research attention in the tropics. Mears & Humphreys (1974a) measured in field exclosures the growth and N uptake of shoots of *P. clandestinum* which had been grazed continuously at varying SR (see Fig. 4.11). When an additional ammonium nitrate application of 350 kg N ha^{-1} was applied in the exclosures, N uptake in the succeeding 14–28 days was unaffected by previous SR, suggesting that root activity in these *P. clandestinum* swards was well maintained, irrespective of defoliation management.

(ii) The labile pool of nutrients in the sward may influence the rate of new shoot growth, through the export of minerals from senescing tissue to young shoots. Obviously the size of the labile pool will be restricted in heavily used pastures; Davidson & Milthorpe (1966) suggest the detrimental effect of defoliation on the rate of leaf extension in *Dactylis glomerata* was exacerbated under conditions of low nutrient supply because of the lack of internal plant nutrient sources.

(iii) The 'demand' for nutrients is related to the rate of pasture growth. Intense pasture use which reduces the rate of pasture growth through reductions in L or in shoot density may therefore be expected to decrease nutrient demand.

(iv) The last factor may be offset by the differences in stage of shoot ontogeny which reflect differences in the intensity of use. Young shoot tissue containing a high ratio of green lamina to stem, leaf sheath and old lamina, as is more prevalent in frequently defoliated pastures, has higher concentrations of nutrients than older shoots. The critical P concentration, which may be defined as the P concentration in apical tissue at which 90% of the asymptotic maximum plant yield occurs, decreased in *S. hamata* cv. Verano from 9 to 21 weeks after germination from 0.30% to 0.14% P in plants which were cut to 7 cm every three weeks (Wilaipon *et al.*, 1981). Critical P concentration was lower in undefoliated plants and decreased from 0.25% to 0.08% over the same period.

This example illustrates the difficulty of using critical concentration as a diagnostic tool for nutrient deficiency in the field grazing situation, where the ontogenetic stage and previous defoliation history of the shoots sampled are not easily defined in precise terms.

(v) In mixed pastures the effect of defoliation management on botanical composition (Chapter 5) will indirectly affect fertiliser needs, especially if the balance of grasses and legumes, with their differing nutrient requirements, is altered.

(vi) The net balance of these factors will produce differing outcomes in the field situation. The effects of defoliation in reducing nutrient uptake, decreasing the internal plant pool of nutrients and maintaining young tissue will increase fertiliser needs, whilst its effect in decreasing growth rate will decrease fertiliser needs.

Chantkam (1982) found that in a flowing solution culture experiment, maximum yield of tops and roots of *C. pubescens* occurred at the same nutrient concentration (3 μM P) independently of defoliation treatment; maximum yield of defoliated plants in a still solution culture experiment in which P was depleted with time required a higher initial solution concentration than occurred with undefoliated plants. The two systems are analogous to strongly or poorly buffered soils with a high or low

P 'capacity' respectively. Robertson, Humphreys & Edwards, (1976) observed in north-east Thailand a greater response to P in *S. humilis* with infrequent cutting, but the P application rate giving maximum yield was independent of cutting frequency.

In Zimbabwe Rodel & Boultwood (1981*a*) noted maximum yields of *C. aethiopicus* of 12 200 kg ha^{-1} where swards were grazed, and of 15 700 kg ha^{-1} where swards had been rested for one to four years. Maximum yield occurred in the treatments receiving 340 and 510 kg N ha^{-1} respectively, so reduced grass yield under grazing was accompanied by a reduced N requirement. On the other hand Mears & Humphreys (1974*a*) at Lismore, New South Wales, were unable to relate growth rate of *P. clandestinum* consistently to SR, and the slope of the response to added N was independent of SR, which facilitates management decision.

This whole question of the effect of defoliation on fertiliser requirement needs to be distinguished from the levels of animal response to different levels of forage availability and quality, as induced by variation in fertiliser input (Section 7.2).

4.6
Nitrogen fixation in legumes

4.6.1
Nodule demography

The nodules on the roots of legumes, in which atmospheric dinitrogen is fixed in forms which plants are able to use, are the plant organs most sensitive to defoliation. Bowen (1959) observed that cutting *C. pubescens* caused the premature senescence and sloughing of nodules, as well as a loss of roots. The severity of defoliation influences the amount of nodule tissue present. In a field experiment by Whiteman (1970) the mean nodule dry weight of *D. intortum* and *M. atropurpureum* was 405, 142 and 135 mg respectively in plants which were undefoliated, cut to 7.5 cm, or had all leaves removed. Nodule weight was 214 mg if young leaves were removed to halve the leaf density, or 340 mg if old leaves, which contributed less assimilate, were removed.

This study recorded the time changes in nodule weight and number. The severity of defoliation determines first the proportion of the original nodule population which senesces, and subsequently the initiation of new nodules and the size to which these grow.

4.6.2
Assimilate supply

The amount of N fixed by effectively nodulated legumes is determined by the supply of assimilate to the nodules, which is directly related to growth.

Othman & Asher (1987) found after defoliating *Macroptilium lathyroides* at different heights and frequencies that N fixation was positively and linearly related to the dry weight of new shoots. Pate, Layzell & McNeil, (1979)

have modelled the transport and utilisation of carbon and nitrogen in nodulated legumes, based on the interrelations of plant parts with respect to their consumption of C and N and the C:N weight ratios of the xylem and phloem fluids. The source–sink interaction is a convenient conceptual framework within which to consider the effect of defoliation.

Monitoring of ^{14}C can be used to assess defoliation effects on the assimilate supply to the nodules. Othman, Asher & Wilson, (1988) cut *M. lathyroides* at *c.* 14 cm (8th node) when plants were at an early podding stage; this removed *c.* 70% of the leaf area, which was 230 cm^2 plant^{-1} in the undefoliated plants. The stubble leaves were retained or not. Cutting reduced ^{14}C-activity in the nodules to 25–40% 2–3 hours following defoliation (Fig. 4.15*a*). Subsequently assimilate supply to nodules in the cut plants which also had stubble leaves removed decreased to 5%. Assimilate movement to nodules was restored to the levels measured in the control undefoliated plants 25 days after cutting if stubble leaves had been retained, but this did not occur if these had been removed.

These differences are reflected elegantly in the similar patterns of assimilate supply and nitrogen fixation (Fig. 4.15*b*). Nitrogenase activity was reduced to *c.* 5% in the most severely defoliated treatment and only recovered to *c.* 30% 25 days later; if stubble leaves had been retained N fixation reached 80% of the level of the undefoliated control after 25 days.

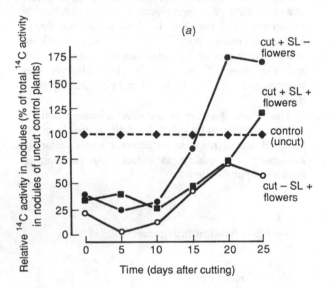

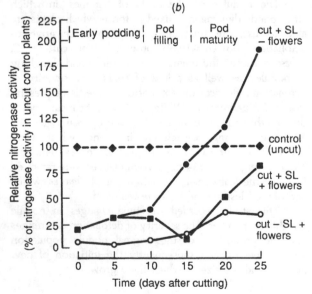

Fig. 4.15. Effects of cutting, stubble leaf (SL) removal and flower removal in *Macroptilium lathyroides* on (*a*) supply of ^{14}C-labelled assimilate to nodules and (*b*) nitrogenase activity. (From Othman *et al.*, 1988.)

4.6.3
Competition with inflorescences

Developing inflorescences act as a competitive sink with nodules for assimilate, and appear to compete with axillary buds for nitrogen.

The continual removal of flowers (Fig. 4.16) of *M. lathyroides* (Fig. 4.15*a*) was beneficial to the supply of assimilate to the nodules, which recovered to the level of the undefoliated control *c.* 16 days after defoliation, and subsequently exceeded it by 75%. This was associated with a similar increase in N fixation (Fig. 4.15*b*). Nitrogenase activity, averaged over six harvests, was increased 120% by flower removal in the defoliated treatments, and this increase was especially pronounced from the pod filling phase. In other work Othman (1983) suggested that the inhibition of axillary bud expansion by flowers and developing pods, which was released by their removal, was associated with internal competition for nitrogen. The source–sink balances are obviously altered both by defoliation and by the removal of reproductive structures, which are strong competitors with nodules for assimilate.

It is difficult to find field studies of tropical pastures in which the effects of SR or cutting management on the N

Fig. 4.16. Regrowth of *Macroptilium lathyroides*; LHS, cut at 8th node and flowers retained; centre, cut at 8th node and flowers removed; RHS, cut at 4th node, removing flowers and stubble leaves.

output from the legume have been monitored. In the absence of these studies it is best to assume that legume growth is the best index of N fixation. For 1000 kg legume shoot yield, 20–40 kg N fixation is expected. Opportunities to increase legume growth, even at the expense of short-term stock acceptance, may be viewed as positive for N accretion, as mentioned in Section 5.3.3.

4.7
Conclusion

The maximisation of the intake of nutrients by the animals from the pasture on a sustained basis requires a sophistication of knowledge which is lacking for many tropical situations and which in any event is difficult to apply in unpredictable climates. The pasture manager who discards maximum pasture growth as an objective in favour of minimising senescence and increasing utilised pasture growth of high nutritive value encounters problems associated with discontinuity of forage supply, except in favoured environments or in intensive production systems which employ irrigation and fertilisers to control growth in relation to animal feed requirements. The alternative is to manage the base pastures at high levels of use, and to arrange alternative sources of feed to buffer the discontinuities in the base pasture supply (Chapters 8 and 9).

Management in humid zones will take into account the need to maintain sufficient leaf surface to avoid gross wastage of sunlight. This chapter has described the compensating factors which influence net photosynthetic output and the restoration of the leaf surface after grazing or cutting.

Management in subhumid and semi-arid zones cannot be directed to increasing pasture growth through more efficient use of sunlight, but the maintenance of at least a periodic minimum canopy is needed to ensure pasture persistence, and this, together with litter accumulation, is required to minimise runoff and erosion. Botanical composition (Chapter 5) can also be modified.

The management of pasture legumes will be directed to maximising growth and hence N accretion; the special role of many legumes in meeting animals dietary needs in the dry season will usually be compatible with this objective, as mentioned in the next chapter.

5

The modification of botanical composition by grazing: plant replacement and interference

5.1
Objectives of management

Grazing modifies the proportion of different plant species which occupy the sward. Systems of grazing may be directed to producing particular combinations of plants for specific purposes.

(i)   One objective assumes that the dominance of the most productive plants leads to higher plant yields. Considerable activity is directed to the eradication or reduction of less productive species. However, the loss of yield associated with the occurrence of less productive plants in the sward may have been overstated, since vigorous plants growing next to less vigorous plants grown even more than if next to other vigorous plants (Smith & Alcock, 1985).

(ii)   Management may be more legitimately directed to the promotion of plants of high nutritive value, to the control of unpalatable plants, which do not contribute to the feed ingested, or of toxic plants, which damage animal health.

(iii)   A frequent objective is the control of the balance between grasses and legumes, so that growth, nitrogen accretion and the maintenance of nutritive value may be optimised.

(iv)   Species composition, especially with respect to the content of sod-forming plants, bears on the resistance of the landscape to soil erosion and on the runoff characteristics of the sward.

(v)   Finally, some grassland managers place a premium upon species diversity. It is hoped that species having some complementarity of seasonal growth rhythms will provide better continuity of forage supply than

monospecific swards. Mixtures provide more insurance against the occurrence of natural disasters, such as flood or fire, against the possible occurrence of nutritional imbalances, against the evolution or incidence of new pests or diseases, or against changes in management policy related to grazing pressure or fertiliser use. Grassland managers look for botanical stability; it is perhaps more realistic to seek pasture mixtures which display resilience, or the capacity to recover after disturbance (Grime, 1979). I quote the late J. F. Kennedy out of context by suggesting the grassland world might be made 'safe for diversity'.

Botanical composition is controlled by grazing or cutting when these affect (1) the processes of individual plant persistence and replacement, and/or (2) the capacity of plants to interfere with the availability of environmental growth factors to their neighbours. An understanding of these processes may subsequently be used in the design of grazing systems, which is discussed later in Sections 8, 9 and 10.

5.2
Plant replacement

5.2.1
Pathways of plant persistence

The persistence of plant yield may arise from (1) longevity of the original plants; (2) plant replacement through the cycle of flowering, seed formation, accretion to soil seed reserves, seedling regeneration, and seedling survival to flowering; or (3) plant replacement from perennating vegetative buds.

The relative significance of each of these three pathways to persistence conditions the choice of management strategies. Longevity is firstly a character under genetic control. For example, many shrub legumes have great longevity. In south-east Queensland *Leucaena leucocephala* showed 87% survival after 16–20 years at two sites, and most fatalities occurred in poorly drained areas (Jones & Harrison, 1980). More recently the introduced psyllid *Heteropsylla cubana* has shortened the life cycle of this plant, but clearly plant replacement is less of a management issue in woody plants than in herbaceous species.

Many 'perennial' tropical herbage legumes prove upon examination to be quite short-lived as individuals. *Stylosanthes hamata* cv. Verano has a half-life of *c.* 3 months at Lansdown, Queensland (lat. 90° S, 870 mm rainfall) (Gardener, 1980). At Mt Cotton, Queensland (lat. 28° S, 1430 mm rainfall), the persistence of various legumes in mixture with either *Paspalum plicatulum* or *Brachiaria decumbens* was studied under continuous grazing. The half-life survival (Fig. 5.1) was 6 months for *S. guianensis* cv. Graham, 15 months for *Macrotyloma axillare* cv. Archer, 18 months for *Macroptilium atropurpureum* cv. Siratro and 24 months for *Desmodium intortum* cv. Greenleaf. The shorter lived plants depend for their replacement upon seedling regeneration, whilst *D. intortum* and *D. uncinatum* seed late in the growing season when frost or drought may occur, exhibit poor seedling survival, and rely upon stolon rooting to provide new centres of plant growth (Jones, 1989). The sterile aneuploid *Digitaria decumbens* cv. Pangola depends wholly upon vegetative buds for perennation.

The annual cycle of plant replacement from seed is illustrated (Fig. 5.2) for the twining perennial legume *M. atropurpureum* cv. Siratro grazed continuously at 1.7 beasts ha^{-1} at Samford, Queensland (lat. 27° S, 1100 mm rainfall) (Jones & Bunch, 1987b). This diagram represents the average situation over a 12-year period. Seed production was 100 seeds m^{-2} of which 3% were ingested by the grazing animal. Some 43 seeds m^{-2} of the seed shed were lost by predation and deterioration before incorporation in the soil seed store of 255 seeds m^{-2}. Losses from seed reserves were in equilibrium with seed inputs at this stocking rate, but of the 55 seeds m^{-2} lost from the pool, only 14 seedlings emerged. Seedling survival was 14%. Fig. 5.2 indicates the wasteful character of natural reproductive processes, and, indeed, this example illustrates more efficient plant replacement than occurs in many other pasture species. Grazing management and seasonal conditions permitting the production of 100 seeds m^{-2} were necessary to provide 2 plants m^{-2} as replacements for the plants which died in the current year. This was sufficient to maintain a minimum of *c.* 5–6 plants m^{-2} in the sward.

Climatic and site conditions modify the pathways to persistence under grazing. Table 5.1 illustrates the demography of *M. atropurpureum* cv. Siratro which was grazed with cattle for the 6-month wet season (early June–early December) at different stocking rates at Khon Kaen, Thailand (lat. 16° N, 1260 mm rainfall) (Gutteridge,

1985). In this table the figures on any one line show the survivors of that seedling cohort against time. About 40% of plants perennated into the second year, and about 2% survived into the fourth year. Each year at the beginning of the rains in June, seedling regeneration (indicated by the italicised figures) provided plant replacement. Some scientists regard 6 plant crowns m^{-2} as a critical figure for the maintenance of Siratro yield, and after 4 years this was only attained at the lightest stocking rate of 2.5 beasts ha^{-1}. At this site stolon rooting was not a source of new plants; plant survival of the 6-month dry season, when only 15% of annual rainfall occurs, apparently depended upon the tap root of the seedling exploiting moisture at depth. By contrast, stolon rooting of Siratro is a significant aspect of perennation at Samford, Queensland, where 30% of rainfall occurs in the six winter months (Jones & Bunch, 1987a).

Stocking rate alters the balance between replacement by seedlings and replacement by vegetative buds. Fig. 5.3 (Walker, 1980) shows the density of crowns, runners and seedlings of *M. atropurpureum* cv. Siratro after 4 years of continuous grazing at different stocking rates at a site near Mackay, Queensland (lat. 22° S, 1490 mm rainfall). The density of crowns and of runners (shaded) was greatest at a lenient stocking rate of 1.7 beasts ha^{-1}. It was slightly less at the lowest stocking rate of 1.2 beasts ha^{-1}, where the incidence of the fungal disease *Rhizoctonia solani* was highest (as mentioned in Section 3.2.2), and was reduced at the higher stocking rates. Plant crowns were smaller and stolons not present at the higher stocking rates of 2.5 and 3.3 beasts ha^{-1}, where seedlings constituted the majority of plants in the population. Similarly, Jones & Bunch (1987a) reported that 100% of Siratro crowns arose from seedlings in a heavily stocked pasture, grazed at 3.0 beasts ha^{-1} for 4 years and 2.0 beasts thereafter. By contrast, 7% arose from adventitious roots on stolons in a moderately grazed pasture stocked at 1.7 beasts ha^{-1}. The half-life of seedlings (Fig. 5.4) averaged 8 months in the heavily stocked and 22 months in the moderately stocked pasture, indicating the strong dependence of yield in the former upon regular seedling replacement. The population in the heavily stocked pasture showed a characteristic linear relationship between the log of density and time; at the moderate stocking rate the survival of the few remaining old plants was greater than expectation from the linear log model.

5.2.2
Plant longevity

The processes causing plant death are not well understood, and defoliation may have positive as well as the more usual negative effects on survival. Defoliation delays ontogeny, and in annual plants, which die because they have flowered, the prevention of flowering promotes persistence. This phenomenon is also observed in longer-lived shrub legumes. In Belize (lat. 17° N, 1800 mm) undefoliated *Codariocalyx gyroides* flowered, seeded and died, whereas plants subjected to regular cutting survived for more than 2 years (Lazier, 1981). Plants cut at 5, 25 and 50 cm height showed 0%, 33% and 51% survival after 17 months treatment.

Defoliation has negative effects on plant survival if (1) regenerative buds are destroyed, (2) energy substrate in the roots and crown is exhausted, or (3) a reduced root system increases the susceptibility of plants to drought, disease or insect predation.

(i) Plant habit influences the susceptibility of plants to mortality following defoliation, according to the height of the growing points and the density of basal buds. Plant improvement programmes may be directed to increased basal branching, as has occurred with the shrub legume *Leucaena leucocephala* (Gray, 1966). Ecotypes of *Stylosanthes guianensis* also vary in their degree of basal branching and a shortage of regenerative tissue has been regarded as contributing to plant mortality under the types of grazing system which have led to elevated stems.

The failure of the twining tropical legumes to persist at high stocking rates arises from a number of factors;

those related to seed reserves, seedling regeneration and competitive relations are mentioned subsequently in this section. Additionally the elevated apices and the sparse density of basal buds contribute to mortality under grazing. Clements (1986) contrasted at Samford, Queensland, the rates of destruction of the growing points of legumes of differing habit and acceptability. Fig. 4.2 indicated the high proportion of terminal growing points of *M. atropurpureum* cv. Siratro removed when grazed at a stocking rate (SR) of 1.3 beasts ha^{-1}. Concurrently, *Trifolium repens* was grazed at 2–4 beasts ha^{-1}; 21–50% of the runners present (according to stocking rate) were grazed during 3 weeks. In this case only 1–5% of terminal growing points of *T. repens*, which were low set to the ground, were removed. It is postulated that the abundance of basal regenerative buds in *T. repens* contributes to its longevity under grazing, and in part explains its contrasting response to that of Siratro.

Grazing modifies plant habit. The promotive effect of defoliation on tillering and basal branching was discussed in Section 4.4.2. For example, the height of first branching in seedlings of *S. hamata* cv. Verano is influenced by SR. Verano has a comparatively erect habit when grown with the tall bamboo grass

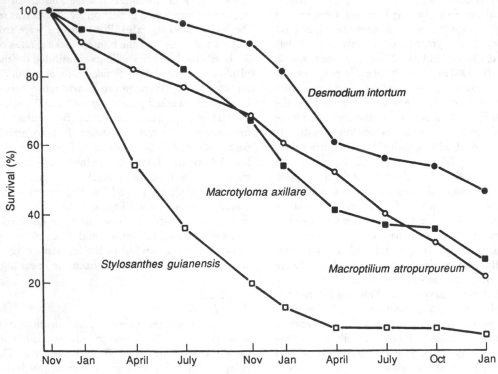

Fig. 5.1. Survival (%) of individuals of four legumes grown at Mt Cotton, Queensland and continuously grazed. (From F. Von Sury *et al.*, unpublished data.)

Arundinaria ciliata at Khon Kaen, Thailand. First branching occurred at 9.3 cm mean height when grazed seasonally at 2.5 beasts ha^{-1}; this decreased in a curvilinear fashion with increasing SR to 4.2 cm height at 6.5 beasts ha^{-1} (Gutteridge, 1982). Systems of deferred grazing in which elevated apices and little basal regenerative tissues occur can create conditions in which plants are subsequently more vulnerable to mortality from grazing than where regular grazing develops branching near to ground level.

Grazing also causes physical damage to basal tissue through treading (Section 3.1.1) and prehension.

(ii) Labile energy substrate in the roots and crown is used in respiration and maintenance as discussed in Section 4.5. Defoliation interrupts the flow of assimilate from the green surfaces of the plant to these organs, whose survival requires a minimum level of residual energy. Mineral and water uptake also have an energetic cost (Naidoo & Steinke, 1979).

The capacity to accumulate non-structural carbohydrate in the crown and below-ground organs varies between species. Under conditions where identical shoot growth occurs, *Cenchrus ciliaris* may accumulate more than double the non-structural carbohydrate accrued in the roots and crown of *Panicum maximum* var. *trichoglume* (Humphreys & Robinson, 1966); the latter plant is less persistent under heavy grazing. Kobayashi & Nishimura (1978) ranked the autumn carbohydrate accumulation of grasses in southern Japan as *Paspalum dilatatum* > *Panicum coloratum* var. *makarikariense* > *Setaria sphacelata* > *C. ciliaris* > *P. maximum* var. *trichoglume*. As mentioned in Section 4.5 the C$_4$ grasses generally have lower concentrations of non-structural carbohydrate in storage organs relative to C$_3$ grasses (Wilson, 1975). Dovrat, Deinum & Dirven (1972) report levels of 0.8–3.4% in roots and 1.6–3.7% in stubble of *Chloris gayana* whilst many C$_3$ grasses have levels in excess of 20% (Davies, 1965). Despite these differences, many C$_4$ grasses persist equally well under heavy grazing; the lower concentration of non-structural carbohydrate is compensated in part by the greater below-ground biomass, or other factors predominate in determining plant survival.

Levels of defoliation which are moderate in terms of both frequency and intensity will result in the diversion of assimilate to new shoot growth and restrict the accumulation of labile carbohydrate in below-ground

Table 5.1. *Changes in* Macroptilium atropurpureum cv. Siratro *plant density (plant m^{-2}) over four years at three stocking rates at Khon Kaen, Thailand*

Stocking rate (beasts ha^{-1})	1977 June	1977 Dec	1978 June	1978 Dec	1979 June	1979 Dec	1980 June	1980 Dec
2.5	18.1[a]	9.4	7.2	7.0	3.8	2.6	0.6	0.6
			79.6	3.8	0.8	0.8	0.4	0.4
					161.2	11.2	4.2	2.8
							51.6	2.0
Total density Dec		9.4		10.8		14.6		5.8
4.5	14.0	12.4	7.6	6.2	2.2	1.4	0.6	0
			73.8	5.0	1.6	0.4	0	0
					41.4	3.6	1.8	1.0
							19.6	1.4
Total density Dec		12.4		11.2		5.4		2.4
6.5	16.4	10.6	8.4	3.0	0.2	0.2	0.2	0
			90.4	8.8	0.6	0.6	0.2	0
					100.2	6.6	0.6	0.6
							18.4	0.6
Total density Dec		10.6		11.8		7.4		1.2

The figures on any one line show the survivors of that seedling cohort against time.
[a] Figures in italic indicate seedling regeneration.
Source: Gutteridge (1985).

organs. This need not occasion mortality; grazing-resistant plants only die after frequent and severe defoliation. Stolons of *Paspalum notatum* required complete leaf removal at weekly intervals for 13 weeks before death resulted (Sampaio *et al.*, 1976).

(iii)   An indirect effect of defoliation upon plant mortality may be mediated through the increased susceptibility of plants to drought, due to reduced root growth. Taerum (1970) has shown in Kenya that repeated defoliation of *Panicum maximum* not only reduced root mass, but restricted rooting depth. It is postulated that heavily grazed pasture may exhibit poorer exploitation of soil moisture at depth. The tap roots of legume explore deeper soil levels than do adventitious roots. The morphological change from a root mass based primarily on seedling tap roots to one based on stolon roots

(Jones, 1980) is believed to increase drought susceptibility; grazing which accelerates the death of tap roots may accentuate this effect. It is also expected that larval feeding on root systems, whose effects are most evident in dry seasons (Mears, 1967 for *Amnemus quadrituberculatus*), would be most damaging to the survival of susceptible legumes such as *Desmodium intortum* if the root system had been reduced by heavy grazing.

5.2.3
Flowering and seed production

The effects of grazing and cutting on seed production are considered here primarily in the context of plant replacement. A secondary consideration is the dual use of pastures for seed crops and for grazing, which is reviewed by Humphreys & Riveros (1986). The objectives of diversification and maximisation of income lead many seed growers into mixed farming with grazing animals, or graziers may take opportunist seed crops. Seed growers are obliged in any event to dispose of crop residues, and skilled managers will risk the damage of accidental grazing or seed contamination arising from animal vectors. Judicious management increases seed production in circumstances where convenience of harvesting or the manipulation of botanical composition is favoured by defoliation.

Grazing animals affect the accretion of seed to soil reserves directly by prehension of flowers or of seeds in

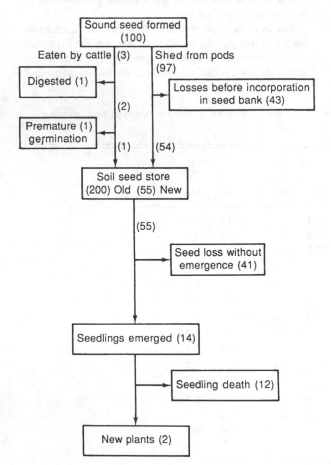

Fig. 5.2 The annual flow of seeds and seedlings (no. m^{-2}) of *Macroptilium atropurpureum* cv. Siratro in pastures grazed at 1.7 beasts ha^{-1} at Samford, Queensland. (From Jones & Bunch, 1987*b*.)

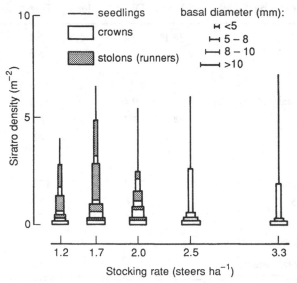

Fig. 5.3. Effects of stocking rate (steers ha^{-1}) on density of plant crowns, stolons and seedlings and on basal diameter of crowns of *Macroptiluim atropurpureum* cv. Siratro. (From Walker, 1980.)

different stages of development. The fate of these seeds is considered in Section 5.2.4. The indirect effects of grazing arise from the manner in which defoliation (1) modifies inflorescence density, (2) alters the supply of assimilate to inflorescences, (3) delays ontogeny, thereby altering the environmental conditions which occur during seed formation, and (4) changes the competitive relations of the constituent species in the sward.

These direct and indirect effects of the grazing animal are often difficult to separate in the field, but the net effects are substantial. *Lotononis bainesii* pastures at Mt Cotton, Queensland, grazed with sheep in a system of 3 days grazing and 15 days rest showed a negative effect of stocking rate on inflorescence density (Fig. 5.5). Sheep appeared to consume inflorescences selectively, despite restlessness in the presence of bees. At Khon Kaen, Thailand, Gutteridge (1985) observed that cattle consumed almost all the flowers of *M. atropurpureum* cv. Siratro present until late in the growing season when flower density exceeded 1–2 m^{-2}. At Samford, Queensland, Jones & Bunch (1987b) found that over 14 years of observation the peak flowering of Siratro in autumn averaged 33, 17, and 3 flowers + pods m^{-2} at stocking rates of 1.1, 1.7 and 2.3 beasts ha^{-1} respectively. Thus for each unit increase in

stocking rate there was a linear and negative response of *c.* 25 flowers + pods m^{-2} and this represented a total reduced input for the year of *c.* 140 seeds m^{-2}.

Let us now consider the factors involved in these responses.

(i)   Inflorescence density is increased by defoliation in some grasses by unexplained processes. The arid-zone bunch grass *Astrebla lappacea* seeds poorly in the absence of grazing; this phenomenon incidentally casts doubt on the value of long-term grazing enclosures for assessing range condition. Orr (1986) clipped *A. lappacea* tussocks; seed production increased from 131 to 334 seeds m^{-2} as the level of utilisation increased from 10% to 90% respectively. Defoliation did not affect spikelet number/inflorescence, or seed number/spikelet, but increased inflorescence density, mainly through an increase in the number of inflorescences formed on axillary tillers. Reproduction in the axillary buds of *Pennisetum clandestinum* is also favoured by decapitation of the main shoot (Carr & Eng Kok Ng, 1956).

The loss of apical dominance may affect a hormonal balance modifying flowering, as may be postulated for the above two examples; it also may generate more

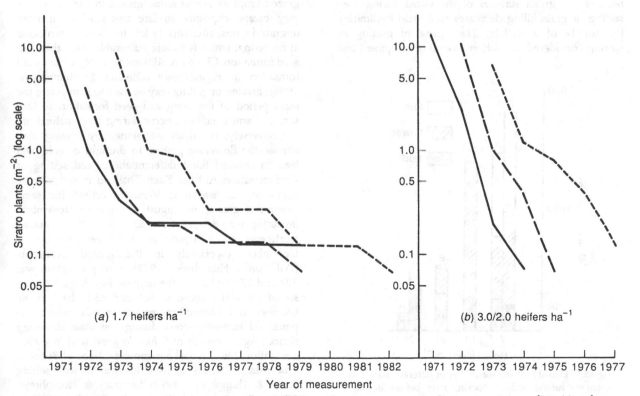

Fig. 5.4. Density of successive cohorts of seedlings of *Macroptilium atropurpureum* cv. Siratro at an intermediate (*a*) and a heavy (*b*) stocking rate at Samford, Queensland. (From Jones & Bunch, 1987b).

sites for inflorescences, since branching and tillering are stimulated, as was discussed in Section 4.4. Seed yield of the annual legume *Stylosanthes humilis* may be increased by defoliation early in crop development, which is associated with better branching (Fisher, 1973). Similarly, removal of 60% of shoot apices, laminae or whole shoots of *S. hamata* cv. Verano results in an immediate increase in the rate of leaf differentiation (Wilaipon, Gigir & Humphreys, 1979). This leads to a rapid restoration of the leaf canopy, so that seed yield is independent of early defoliation.

The first formed inflorescences contribute the highest seed yield per inflorescence (for example, Chadhokar & Humphreys, 1973 for *Paspalum plicatulum*). There are exceptions to this generalisation; first appearing inflorescences of *Brachiaria decumbens* cv. Basilisk have a lower proportion of florets setting seed (Stür & Humphreys, 1987), but normally grazing or cutting which removes first-formed inflorescences is expected to reduce seed yield. Haggar (1965) emphasised the contribution of early-formed tillers to seed production in *Andropogon gayanus*.

(ii)   The rate of seed production depends primarily upon current photosynthesis; grazing or cutting which reduces the green surfaces of the sward during seed setting or grain filling decreases seed yield by limiting the supply of assimilate. The timing of grazing or cutting considered in relation to crop development and

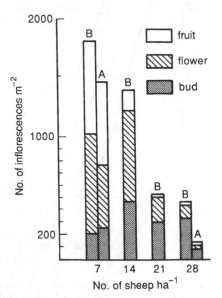

Fig. 5.5. Density of inflorescences at different stages of *Lotononis bainesii* at four stocking rates, before (B) and after (A) a 3-day grazing event with sheep. (From Pott & Humphreys, 1983.)

to subsequent environmental conditions determines whether seed yield is reduced. Loch, Hopkinson & English (1976) found that when the obligate short-day legume *S. guianensis* cv. Cook was slashed to a leaf area index of 2.5 40 days before floral initiation, it restored its canopy to the same level as undefoliated swards by the time floral initiation occurred, and that both swards gave the same seed yield. Slashing at later dates, when autumn conditions were less favourable for crop development, gave reduced seed yields.

In a similar type of study at Planaltina, Brazil (lat. 16° S, 1500 mm rainfall) the short-day bunch grass *Andropogon gayanus* gave increased seed yield if cut or grazed in mid-January relative to undefoliated swards, due mainly to increased seed yield per inflorescence (CIAT, 1983). Defoliation up to mid-March greatly reduced seed yield. Again, seed yield of *Galactia striata* near Belo Horizonte, Brazil, was highest if grazing deferment commenced no later than 21 February (Vera *et al.*, 1983).

(iii)   The pronounced delay in ontogeny occasioned by grazing or cutting is a feature of the above examples. Some circumstances arise where delayed flowering is advantageous to seed production. Seed development of grazed crops of *Avena sativa* grown in the subtropics may escape exposure to late frosts which damage ungrazed crops; alternatively late frosts may then occur at flowering, which is a more vulnerable stage then late seed formation. Cloudy conditions probably reduce seed formation in *P. maximum* (Oliveira & Humphreys, 1986); grazing or cutting may be used to manipulate the main period of flowering and seed formation so that sunny, warm conditions occur during crop maturation.

Conversely, the delay occasioned by grazing may expose the flowering pasture to drought or cold, and lead to reduced floret differentiation, seed setting or seed formation. At Khon Kaen, Thailand, mixed pastures containing *S. hamata* c. Verano received the same amount of grazing in August, October or November; flowering was delayed from 25 June in the ungrazed treatments to 9 September, 8 November and 10 December respectively in the grazed treatments (Wilaipon & Humphreys, 1981). Seed production was 374 and 379 kg ha^{-1} in the ungrazed or August grazed swards; it was reduced to 165 and 85 kg ha^{-1} in the October and November grazed swards, which experienced moisture stress during the main flowering period. Again, swards of *S. humilis* grew well following defoliation, but the cool autumn conditions of St Lucia, Queensland (lat. 28° S) were inimical to seed setting (Loch & Humphreys, 1970; Skerman & Humphreys, 1973) and seed production was reduced relative to undefoliated swards.

(iv) Grazing or cutting which modifies the competitive relations of species in mixed swards also indirectly affects the seed production of the component species. The phenomenon is most evident when plants of short stature are shaded by tall plants in underutilised pastures. This principle is illustrated from a study at Mt Cotton, Queensland (Wilaipon & Humphreys, 1976) in which the tall bunch grasses *Setaria sphacelata* and *P. maximum* var. *trichoglume* were grown with the low-growing legume *Stylosanthes hamata* cv. Verano. Various grazing and mowing treatments were applied. Pastures crash-grazed by sheep and mown early in the growing season on 27 December produced grass growth of 3350 kg ha^{-1} by 25 February, relative to 5570 kg ha^{-1} in ungrazed pastures. Verano seed production was 355 and 221 kg ha^{-1} respectively in these treatments. The advantage to Verano seed production was attributed to the more favourable conditions for the growth of Verano, which had light values at the surface of its canopy on 4 February of 0.73 full sunlight in the grazed and 0.48 in the ungrazed swards. Similarly, grazing management directed to increasing seed production of *Trifolium repens* emphasises the control of competing tall species.

5.2.4
Seed reserves in soil

The optimum seed bank for plant replacement contains (1) sufficient germinable, soft seed to give emergence upon any rainfall event favourable for successful seedling regeneration, and (2) sufficient density of long-lived seed to maintain seed reserves at a satisfactory level if replenishment is interrupted by adverse climatic or other factors.

Standards need to be developed to define adequate levels of seed reserves for particular species and districts. These depend upon (1) the level of seed inputs from the sward, (2) the rate of seed losses by predation, deterioration (Priestley, 1986), and movement to inaccessible soil depths, (3) the incidence of germination, and (4) the probability of successful establishment. Rapid deterioration places a premium upon regular replenishment for ecological success. In Belize seed of the weed grass *Paspalum virgatum* lost 96% of its viability after one year and all viability after two years, whilst fire destroyed all seed on the soil surface (Kellman, 1980). However, high levels of hardseededness do not guarantee plant replacement. *L. bainesii* may have reserves of c. 6000 seeds m^{-2} (Pott & Humphreys 1983) of which 97% is hard; c. 63% of the seed bank may occur below 20 mm depth, which is the limit of emergence for seedlings arising from these small seeds (0.3 mg seed^{-1}) (H. Fujita and L. R. Humphreys, unpublished data), and in some seasons the pasture may be devoid of *L. bainesii*

plants, despite the apparently abundant seed reserves below.

Fig. 5.2 illustrated a seed bank of 260 seeds m^{-2} of *M. atropurpureum* cv. Siratro which was adequate for plant replacement at Samford, Queensland (lat. 27° S). The rate of softening of hard seed of legumes is temperature dependent; in the subtropical Samford environment 15% of Siratro seed was still in the soil seed bank after 6 years and 88% of this seed was viable (Jones, 1981). At hotter sites at lower latitudes most of the hard *Stylosanthes* spp. seed softens by the summer following its formation (Gardener, 1975), and the rate of softening may be predicted from the number of days during which maximum soil temperature exceeds 50–55 °C (Mott *et al.*, 1981). Gardener (1982) recorded an average seed bank of 6900 seeds m^{-2} of *S. hamata* cv. Verano at Lansdown, Queensland (lat. 19° S, 870 mm rainfall), and this was adequate for plant replacement.

How does grazing and cutting affect the seed bank?

(i) Stocking rate exerts the primary control of the level of seed reserves. *M. atropurpureum* cv. Siratro seed (number m^{-2}) in the 0–10 cm soil layer was estimated as the linear function $903 - 203x$, where x is stocking rate as beasts ha^{-1} (Walker, 1980, near Mackay, Queensland). Increased stocking rate similarly reduced soil seed reserves in *L. bainesii* (Pott & Humphreys, 1983) and *Desmodium uncinatum* (Jones & Evans, 1977), but in the latter study *T. repens* reserves benefited from a higher stocking rate, presumably due to a favourable shift in botanical composition. *S. hamata* cv. Verano had seed reserves of 4 kg ha^{-1} in pastures heavily grazed through the wet season, and 149 kg ha^{-1} when protected from wet season grazing at Khon Kaen, Thailand (Shelton & Wilaipon, 1984); at a more leniently grazed site presence of ruminants did not affect seed reserves.

(ii) The digestion and excretion of seed influences animal performance (as mentioned in Chapters 6 and 8), the spread of seed to new locations, and the accretion of seed to soil reserves. Seed may be consumed from the pasture sward, and may be prehended from the soil surface. Playne (1974) estimated that cattle might consume up to 450 kg ha^{-1} of the pods of the annual *Stylosanthes humilis* in a single dry season at Lansdown, Queensland, and maximum seed content of the faeces reached 38 g^{-1} dry weight. At the same site C. J. Gardener (pers. comm.) found that cattle voided 0.9 million seeds ha^{-1} yr^{-1} of *S. scabra* grazing at 1 beast ha^{-1}; at Narayen, Queensland (lat. 26° S, 710 mm rainfall) c. 4 million seeds of *S. scabra* cv. Fitzroy were excreted by each animal grazing from January to June (Jones, Kerridge & McLean, 1987). These examples

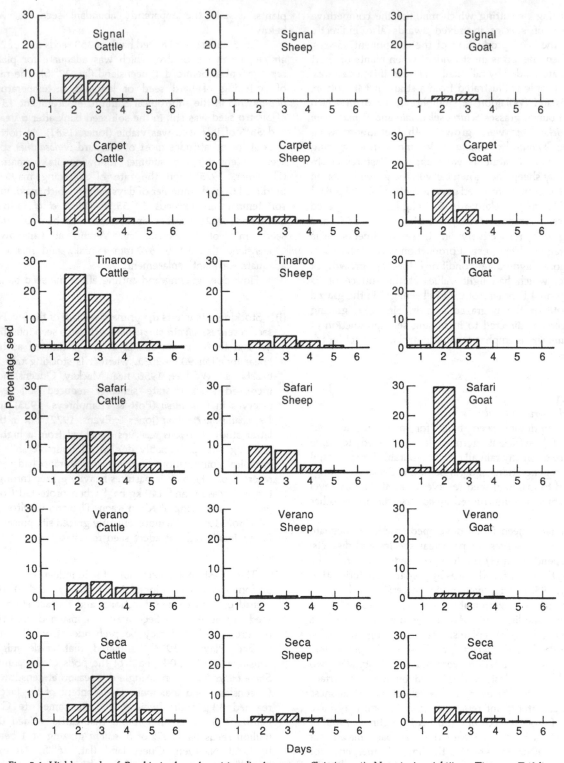

Fig. 5.6. Viable seeds of *Brachiaria decumbens* (signal), *Axonopus affinis* (carpet), *Neonotonia wightii* cv. Tinaroo, *Trifolium semipilosum* cv. Safari, *Stylosanthes hamata* cv. Verano, and *S. scabra* cv. Seca voided each day after feeding to three animal species, expressed as % total seed ingested. (From Simao Neto, 1985.)

indicate that seed passing through the gut may have a major ecological influence; grazing animals also act as a vector through adherence of seed to the animal coat, as for *D. intortum* and *D. uncinatum*.

Most of the seeds eaten by ruminants are excreted within 4 days of ingestion; Fig. 5.6 shows that on average cattle, sheep and goats excreted respectively 92%, 94% and 99% of the total viable pasture seeds by day 4 after ingestion. Most seeds passed through the gut on day 2 or day 3, and an interval of 7 days would be necessary to complete the processes of digestion and excretion. Passage of seed through goats was faster than through sheep and cattle; 50% seed was excreted after 48, 62 and 69 hours respectively. The rate of passage is influenced by diet quality. Seed fed in diets of high digestibility pass through the gut more quickly than seed mixed with feeds of low quality and high roughage content. These conclusions are helpful if animals are being used to disseminate seed, or alternatively if contamination of seed crops is to be prevented.

The recovery of viable seed is clearly much higher from cattle than from sheep or goats (Table 5.2). *S.*

hamata was digested more than *Neonotonia wightii*, *S. scabra* or *Trifolium semi-pilosum*. The soft legume seeds were digested, and many hard seeds were passed intact, but were voided as soft, germinable seed. The proportion of damaged seed passed in the later days of voiding (days 4 to 6 after ingestion) increased, and these were seeds which had been in the digestive tract for a longer period. The aspect of seed morphology most influential in digestion appears to be the length of the seed; this is apparent for sheep and goats where long seeds such as *Brachiaria decumbens* or *S. hamata* (pods) exhibited lesser recovery in dung. Survival of seed in cattle dung is negatively related to the duration of fermentation of the dung deposited and to the temperature of fermentation (Simao Neto & Jones, 1986). Small, hard legume seed best survives the digestion and excretion processes.

(iii) The modification of soil cover by grazing or cutting alters seed survival and seed movement. As mentioned earlier, the softening of hard seed is temperature dependent, and light grazing which provides continual

Table 5.2. *Viability (%) and hardseededness (%, in parentheses) of legume seeds excreted by cattle, sheep and goats*

	Cattle	Sheep	Goats
Neonotonia wightii cv. Tinaroo	56 (41)	11 (8)	24 (18)
Stylosanthes hamata cv. Verano	18 (10)	2 (1)	4 (2)
Stylosanthes scabra cv. Seca	39 (26)	8 (5)	10 (7)
Trifolium semi-pilosum cv. Safari	39 (31)	23 (17)	36 (24)

Source: Simao Neto (1985).

Fig. 5.7. Seedling regeneration of *Calopogonium mucunoides* in a dung pat in Brazil.

litter and plant cover may be expected to reduce the rate of breakdown of hardseededness. It will also alter the microclimate and feed supply for the soil fauna and micro-organisms which affect the levels of seed predation and seed deterioration. The movement of seed in runoff water will be impeded by soil cover; *S. humilis* spreads downhill faster than it spreads uphill (Graham & Mayer, 1972), implying that water is a dispersal agent. Finally, the degree of trampling will affect seed burial.

5.2.5
Seedling regeneration

Regeneration from seed which provides replacement plants which survive to flower follows (1) germination and emergence, and (2) seedling growth and survival. Grazing influences these separate processes in different ways.

(i)   Removal of cover promotes the germination of light-sensitive seeds. For example, seedling emergence was 40 m^{-2} in swards of *B. decumbens* cut to 10 cm at Mt Cotton, Queensland, and 290 m^{-2} in swards clipped to ground level and swept to remove litter. This effect is attributed to the high red/far red ratio of sunlight unfiltered by a pasture canopy, as mentioned in Section 4.4.2 (Cumming, 1963); red light about 630 nm is promotive whilst the far-red irradiation of 730 nm is inhibitory to germination. Germination of many other tropical grasses is favoured by light, including *Chloris gayana*, *Melinis minutiflora*, *Paspalum dilatatum* and *Setaria sphacelata* var. *sericea*. The germination of light-sensitive seeds of weed species is also promoted by the removal of cover and by disturbance.

Dung pats, especially when deposited on short pasture, provide disturbed sites favourable for seedling regeneration (Fig. 5.7). Wilson & Hennessy (1977) fed seed of *Pennisetum clandestinum* to cattle and noted seedling regeneration in every dung pat within 18 weeks of excretion.

Heavy grazing may be used to create a 'gap' in the sward, which is favourable for seedling regeneration. As mentioned previously, the adverse effects of treading on plant growth and survival may be counterbalanced by favourable effects on seedling emergence, as in *L. bainesii* (Pott, Humphreys & Hales, 1983; Section 3.1.1). Hoof cultivation has been used to establish 'fodder banks' of *Stylosanthes* spp. in Nigeria. The converse effect applies where cover reduces temperature and increases soil moisture availability to favour seedling emergence (Mott, McKeon & Moore, 1976).

(ii)   Seedling growth and survival is sensitive to seasonal conditions and a knowledge of the most favourable times for seedling regeneration may be used to modify grazing management suitably. *T. repens* is often treated

as an annual or weak perennial in subtropical environments; grazing is directed to favouring seed production and seedling regeneration. Fig. 5.8 shows the relative seedling survival of seedling cohorts of *T. repens* emerging in successive months near Grafton, New South Wales (lat. 30° S). The increasing slope of the lines showing survival as emergence moves from February through to October indicates the importance of seedling regeneration in February, March and April to the maintenance of plant density; plants emerging in winter or spring were short lived.

The net effect of grazing on seedling growth and survival will depend on whether favourable influences on the seedling environment are counterbalanced by the negative effects of treading and defoliation. *M. atropurpureum* cv. Siratro seedlings showed 0–1% survival in pastures grazed at 1.1 beasts ha^{-1}, and *c.* 14% survival at 1.7 beasts ha^{-1} (Fig. 5.2; Jones & Bunch 1987*b*). Seedlings emerging in lightly grazed pastures were tall and etiolated. Light values at 5 cm height during the growing season when most seedlings emerged averaged 0.04, 0.20 and 0.25 full sunlight at stocking rates of 1.1, 1.7 and 2.3 beasts ha^{-1} respectively. Light values exceeding 0.4 full sunlight did not occur at the lightest SR, and it is considered that seedling death was caused mainly by low radiation level. Successful seedling regeneration is also favoured by medium to heavy grazing if there is negative selection by the animal for the species concerned, as often occurs for

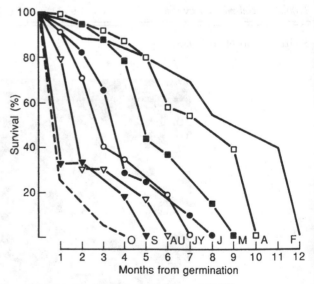

Fig. 5.8. Survival (%) of seedlings of *Trifolium repens* emerging in successive months (F February, M March, A April, J June, JY July, AU August, S September, O October) near Grafton, New South Wales. (From D. L. Garden, unpublished data.)

tropical pasture legumes when green grass is addition-ally available (Bohnert, Lascano & Weniger, 1985).

5.2.6
Vegetative plant replacement
There are several studies of tillering in seed crops of C_4 grasses (Boonman, 1971; Chadhokar & Humphreys, 1973; Bahnisch & Humphreys, 1977; Stür & Humphreys, 1987), where synchrony of tillering is sought, but the effects of grazing on the persistence of planted C_4 grasses have not been of sufficient agricultural moment to generate much research interest in tiller dynamics under grazing.

We know little of the life-history of C_4 grass tillers and even less about the effects of grazing on these life histories. Field observation indicates that the central portion of the circular bunch grass crown eventually dies, and daughter plants develop from the younger tillers on the periphery of the old crown. Sod-forming grasses develop a reticulate network of stolons or rhizomes, in which the first formed stems gradually die and disappear.

More is known about vegetative plant replacement in tropical and subtropical pasture legumes, since their persistence in field practice has been more uncertain. Reference was made in Section 5.2.1 to legumes such as *D. intortum* and *D. uncinatum* whose main pathway of persistence is vegetative, and to the lesser dependence of *M. atropurpureum* cv. Siratro upon plant replacement by seedlings if pastures were leniently grazed.

Grazing promotes stolon rooting if circumstances are favourable for this process (Pott & Humphreys, 1983 for *L. bainesii*): it will not occur in marginal climatic environments. Stolon growth and rooting of *T. semi-pilosum* cv. Safari is strongly influenced by seasonal grazing pressure. Summer deferment of grazing was deleterious to Safari when growing with the grasses *Setaria sphacelata* var. *sericea*, *Chloris gayana* and *Digitaria didactyla* at Samford, Queensland; in April stolon length was 14 m m^{-2} if summer grazed, and 3 m m^{-2} if rested, whilst density of stolon rooted points was 380 m^{-2} and 130 m^{-2} in the same treatments. Winter deferment of grazing promoted stolon growth but decreased the amount of stolon rooting and the density of growing points below 5 cm height (Sproule, Shelton & Jones, 1983).

5.2.7
Seasonal variation in grazing pressure
Variation in the seasonal density of animals on the pasture or the timing of cutting may be employed to favour vegetative replacement, as in the last example, and/or to increase plant replacement from seedlings. Table 5.1 (Gutteridge, 1985) was drawn from a study where *M. atropurpureum* cv. Siratro was grazed during the wet season only. This was consistent with the forage needs of the local mixed farming system in north-east Thailand where abundant cropping land is available for grazing during the dry season, but less forage is available close to settlement during the wet season. Withdrawal of cattle in early December was compatible with maximising the quantitative short-day flowering response of Siratro; flowering is also favoured by mild water stress (Kowithayakorn & Humphreys, 1987). Replenishment of soil seed reserves were promoted by the seasonal grazing applied. Similarly, R. M. Jones (1988) found that Siratro pastures at Samford, Queensland, which produced 70 seeds m^{-2} under continuous grazing produced 250 m^{-2} if spelled from January to May, or 340 seeds m^{-2} if spelled from September to May. Deferment of grazing late in the growing season led to substantially increased Sirato yield in subsequent years.

Seasonal variation in grazing pressure requires alternative feed sources on the farm, or a policy regarding animal purchase and sales adapted to pasture needs. It is a happy situation if pastures with complementary seasonal grazing needs can be identified. This occurs with mixed pasture in south east Queensland based on *M. atropurpureum* cv. Siratro and on *T. repens*. Additional summer grazing of *T. repens*, often grown in the valley bottoms, controls grass dominance and promotes the legume (Jones, 1984). Concurrently, late summer deferment of grazing of Siratro based pasture, often grown on adjacent hillsides, favours replenishment of legume seed reserves. The feed

Table 5.3. *Density (plants m^{-2}) of* Lotononis bainesii *plants following the application of different seasonal stocking rates*

Spring–early summer stocking rate (sheep ha^{-1})	Late summer–autumn stocking rate (sheep ha^{-1})		
	9	18	27
5	3	8	13
18	3	3	4
27	1	2	6

Source: H. Fujita and L. R. Humphreys (unpublished data).

available on the more frost-susceptible valley bottoms is thereby utilised by the late autumn, when animals are returned to the Siratro pastures on the frost-free hillsides.

A further illustration comes from studies of the creeping legume *L. bainesii* cv. Miles. This legume has a high nutritive value, is selectively eaten, and has better cold tolerance than other tropical legumes. Its performance in grazed pastures is erratic in farm practice, and seasons in which *L. bainesii* makes a significant contribution to pasture yield are often followed by its virtual disappearance. Demographic studies revealed that this apparently perennial plant was short lived, despite the presence of a deep tap root and stolons which root at the nodes (Pott & Humphreys, 1983). Variation in seasonal grazing pressure was therefore applied to devise a basis for a grazing system favouring plant replacement. Established pastures of *L. bainesii* grown with the rhizomatous *Digitaria decumbens* at Mt Cotton, Queensland, were heavily grazed with sheep or goats in mid-summer to create a 'gap' for seedling regeneration, and were lightly grazed by sheep in the winter months. Flowering occurs in late summer and autumn, but the main flowering period is usually October–November. Table 5.3 shows that plant density of *L. bainesii* was increased by lenient grazing in the spring–early summer period. Seedling regeneration (Fig. 5.9) occurred mainly in the summer months, and late summer–autumn heavy grazing increased both plant survival and *L. bainesii* recruitment. *L. bainesii* is most susceptible to trampling damage in the young seedling stage (Pott, Humphreys & Hales, 1983), but the contractile

growth of the hypocotyl which produces a more grazing resistant buried crown is impeded in low radiation environments. Reduction of the shade offered by the companion *D. decumbens*, which would also disadvantage *L. bainesii* in other ways, appeared to outweigh the trampling damage associated with an increased stocking rate during late summer–autumn. The combination of lenient spring–early summer grazing to favour soil seed reserves and heavy late summer–autumn grazing to favour seedling regeneration and to reduce competition from the grass during its most rapid growth phase led to a density of *L. bainesii* plants capable of making a substantial contribution to yield.

5.2.8
Catastrophic events
Unusual events create opportunities for substantial changes in botanical composition. An unexpected flood which kills susceptible species in the sward, a fire or a severe drought may create a 'gap' which will be occupied by resistant, surviving species or by the dominant species in the soil seed bank. A hay cut, or pasture renovation which kills some existing plants and exposes seed for regeneration may alter the balance of species. Similarly, a concatenation of conditions favourable for seedling emergence, growth and survival may have profound effects on the balance of species which persist for years. In a study of mixed pastures of *S. sphacelata* var. *sericea*, *M. atropurpureum* cv. Siratro and *Stylosanthes guianensis* near Mackay, Queensland, there was an unexplained artefact in

Fig. 5.9. Seedling regeneration of *Lotononis bainesii* following heavy late summer grazing.

Walker's (1980) data. A high density of *S. guianensis* in some grazing treatments was not simply explained by treatment, but appeared to arise from a single germination event which was more favourable for *S. guianensis* in some paddocks than in others.

These catastrophic events may or may not be influenced markedly by grazing conditions before or after their occurrence. In the semi-arid Warrego region of Queensland the density of *Astrebla* spp. grassland was monitored for 45 years. The only major seedling recruitment events occurred in 1941 and 1983 (Roe, 1987), and in the intervening period climatic conditions were insufficiently favourable for recruitment to arrest a decrease in plant density. The deliberate intervention of the pasture manager and the effects of the grazing animal are secondary in these circumstances.

5.3
Plant interference

5.3.1
Light relations in pastures

Relations between plants in a sward have been described traditionally in terms of competition (Donald, 1963), which commences when the supply of a limiting resource falls below the combined demands of the plants. There may be instantaneous competition for light, or sustained competition for water or nutrients. Plants are successful if they gain a greater share of these environmental resources for growth, thereby reducing the supply of these to their neighbours. It has become fashionable to discuss these relations in terms of interference (Hall, 1978) rather than of competition, since the latter does not embrace some aspects of plant relations: the modification of a neighbour's microenvironment which may alter the incidence of pests or disease, the accretion of soil nitrogen by the legume–*Rhizobium* symbiosis and its uptake by companion grasses, or allelopathy, where toxins are secreted.

When plants gain a greater share of one resource, such as light, there is usually a secondary and positive effect on their capacity to gain a greater share of the other resources, and the availability of water and nutrients to their companions is also reduced. We know little about the way in which defoliation modifies the supply of water and nutrients to companion plants. It is expected that selective grazing which reduces the root system of one plant more than that of a non-preferred plant would advantage the latter in terms of its access to water and nutrients, but I have not sighted good tropical studies of this question.

The amount of light intercepted by the leaf surface, the age of the leaves present and the light environment in which leaves are differentiated control growth significantly,

unless shortages of water and/or nutrients are acute, as discussed in Chapter 4. Two important assumptions are that competitive relations are modified by grazing when (1) high stocking rates ensure the access of plants of short stature to a favourable light environment, and (2) low stocking rates enable twining legumes or shrubs to elevate their leaf surface to the top of the canopy and to reduce the light supply to companion plants.

(i) Reference was made in Section 5.2.3 to the manner in which early defoliation increased the irradiance at the canopy surface of *S. hamata* cv. Verano growing with tall bunch grasses. The positive effect of SR on the light environment and the survival of *M. atropurpureum* cv. Siratro seedlings were mentioned in Section 5.2.5. D. A. Carrigan (personal communication) grew *S. humilis* and *Heteropogon contortus* in narrow soil panels which separated their roots but not their shoots. Defoliation which prevented the bunch grass *H. contortus* from shading the shade-sensitive *S. humilis* (Sillar, 1967) increased the contribution of *S. humilis* to the total yield of the two species. This confirmed the role of light in relations between the two species, and support grazing recommendations which have been directed to the control of the height of companion grasses in order to favour the low-growing legume *S. humilis* (Norman & Phillips, 1973).

S. guianensis var. *intermedia* cv. Oxley is an unusual member of this species, since it is a short-statured legume which develops a buried crown; it is resistant to heavy grazing and fire. A mixed sward of *H. contortus* and another bunch grass *Rhynchelytrum repens* was oversown with cv. Oxley at Gayndah, Queensland (lat. 26° S, 730 mm rainfall). Bowen & Rickert (1979) conducted a study in which the pasture was continuously grazed by cattle, or was rested from grazing for 2, 4 or 6 months, commencing in September, December, March or June. The six spring and summer months receive 70% of the annual rainfall, and temperatures are sufficiently cold in winter to preclude growth of the principal species. The legume density (Fig. 5.10) was decreased by resting, and this effect was greatest if deferment in spring or summer, during the main growing season, was of sufficient duration to permit the accumulation of standing forage. The deleterious effect of protection from grazing on the legume component may be interpreted as an effect of restricted light; alternatively it may be associated with possible differences between species in dietary preferences, as is discussed in Section 5.3.2.

(ii) Tropical pasture plants differ in their shade tolerance (Shelton, Humphreys & Batello, 1987), but adaptation to conditions of low radiation may be of less ecological significance for the outcome of competition in mixed pastures than the capacity of a plant to shade its neighbours (Ludlow, 1978). The first steps in studying this question are the determination of the profile of light interception in the different layers of the sward, and the measurement of the green surface area contributed by the component species to each stratum.

This is illustrated (Fig. 5.11) for mixed swards of *S. sphacelata* var. *sericea* and *D. intortum* which Riveros (1970) grew at Redland Bay, Queensland (lat. 28° S). *S. sphacelata* var. *sericea* has a well-developed basal crown with ascendant culms arising from short internodes, while *D. intortum* is a broad-leaved legume with long trailing stems which are carried upwards by a companion grass. Light interception is shown at intervals of 7.5 cm height, and leaf area index (*L*) (determined from stratified clips) is shown for the same vertical intervals. Fig. 5.11*a* illustrates a sward in which *D. intortum* dominates the grass; it has a greater *L* in all strata but one. There is a sharp cut-off in the light value at 15 cm due in that stratum to well-grown *D. intortum*, whose near-horizontal leaves are effective in restricting the light supply to lower layers. Fig. 5.11*b* shows a sward in which grass and legume are co-dominant. The slope of the curve indicating light interception is quite different, and light is attenuated more gradually where the grass leaves predominate in the two upper strata.

The second consideration is the photosynthetic activity of the component species in each layer of the profile, which will determine the level of interference between the plants. This is illustrated for the same pasture (Fig. 5.12), but in a manner which now takes into account the leaf net photosynthetic rate of the two species in each stratum, and the proportion of net canopy photosynthesis represented by each of these components.

In this study Ludlow & Wilson (1983) fed $^{14}CO_2$ about noon, when the photosynthetic rate was steady into a chamber enclosing the sward. A stratified harvest was made shortly afterwards, and the absorption of $^{14}CO_2$ was used to estimate photosynthesis in each component of each stratum. Sward A (Fig. 5.12) was legume dominant, and there was a relatively sharp cut-off in illuminance in the 0.2–0.3 m stratum where legume *L* exceeded grass *L*. In the layers above 0.2–0.3 m the leaf net photosynthetic rate was highest, and was greater for the C_4 grass than the C_3 legume, but these layers contributed little assimilate because of their low *L*. Most (55%) of the net canopy photosynthesis came from the 0.2–0.3 m layer; the 0.1–0.2 m stratum contributed little, despite an appreciable *L*. The legume contributed 63% of net canopy photosynthesis and clearly interfered with the light supply to the lower sward layers.

Sward B (Fig. 5.12) shows a contrasting situation. The legume had a slightly greater *L* than the grass, but the superior net photosynthetic rate of the C_4 grass in the upper layers compensated, so that its photosynthetic output was slightly greater than that of the legume. The upper three layers of the canopy each contributed about

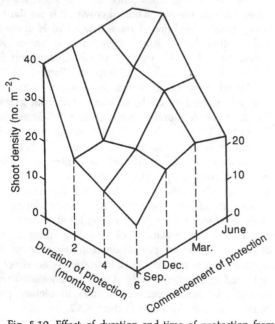

Fig. 5.10. Effect of duration and time of protection from grazing on the density (shoots m⁻²) of *Stylosanthes guianensis* var. *intermedia* cv. Oxley. (From Bowen & Rickert, 1979.)

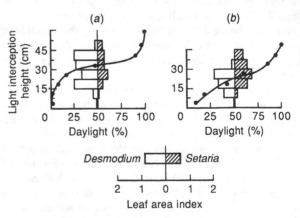

Fig. 5.11. Profiles of light values (% daylight) with depth and *L* in different strata of a mixed sward of *Desmodium intortum* and *Setaria sphacelata* var. *sericea* showing (a) legume dominance and (b) legume and grass co-dominance (From Riveros, 1970.)

25% of net canopy photosynthesis; decreasing irradiance and photosynthesis with depth was compensated by increasing L.

Light relations in tropical pastures and their effects on botanical composition are changed by the manner in which the sward is grazed or cut, but there are few critical studies of the type just described which can validate recommendations for management; recommendations are usually developed from empirical or anecdotal evidence.

5.3.2
Selective grazing

Defoliation by grazing is relatively uniform under management systems in which high animal densities are applied to the pasture for short periods. The more common situation is for selective grazing to occur, in which there is spatial heterogeneity (Section 4.2) of defoliation (patch grazing), and differences in animal acceptance between plant species, between plants of the same species, and between plant organs. The basis of these differences and

their effects on animal performance are discussed in Chapter 6, and mention is made here of the effects of selective grazing on competitive relations.

The growth of pasture plants is increased if the surrounding herbage is removed; the presence of companion herbage also increases the proportion of assimilate diverted to shoots rather than roots (Norman, 1960a). We expect that selectively grazed plants will receive lower levels of illuminance and have a lesser capacity to exploit the water and nutrients in the soil mass than ungrazed plants.

Seasonal variation in dietary preferences influences the degree of interference. Animal preference for green grass and negative selection for green legumes when green grass is present is a commonly reported occurrence (Stobbs, 1977a; Lascano, 1987); in extreme cases this leads to complete legume dominance and to a reduction in pasture productivity. This phenomenon is widespread in pastures of the Amazon basin, Brazil, where *Pueraria phaseoloides* dominates *P. maximum* (Spain, Pereira & Gualdron, 1985). *Calopogonium mucunoides* and *C. caeruleum*

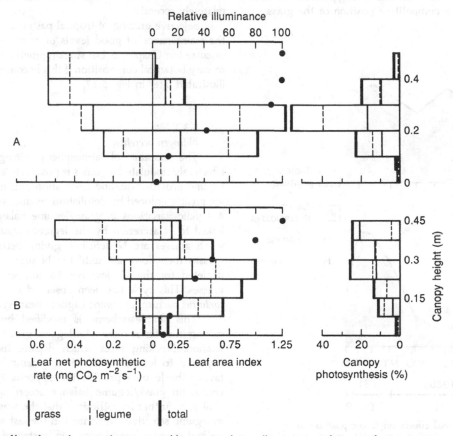

Fig. 5.12. Profiles of net photosynthetic rates of leaves, L, relative illuminance and % contribution to canopy net photosynthesis in mixed swards of *Setaria sphacelata* var. *sericea* (grass) and *Desmodium intortum* (legume). (From Ludlow & Wilson, 1983.)

(Middleton & Mellor, 1982) are less well accepted by ruminants than many legumes, and the disappearance of the grass component has occurred in many tropical areas where they have been planted. There is a growing use of sheep in the rubber plantations of Malaysia, since selective grazing of grasses and of edible weeds well maintains the legume cover crops, reduces herbicide costs, and diversifies income.

At Carimagua, Colombia (lat. 5° N, 2160 mm rainfall), mixed pastures containing the bunch grass *Andropogon gayanus* are dominated by the twining legume *P. phaseoloides* when grazed continuously (2 beasts ha^{-1} wet season, 1 beast ha^{-1} dry season; Bohnert, Lascano & Weniger, 1985). The legume component comprised *c.* 55–75 % of the pasture available (Fig. 5.13). However the dietary intake of legume, as measured by sampling the material ingested by steers fitted with oesophageal fistulas, varied from *c.* 10 % in July (in the middle of the wet season) to *c.* 90 % in January–February (in the middle of the dry season). Under this grazing regime the twining legume is able to maintain dominance; Santillan (1983) and Spain *et al.* (1985) believe the introduction of long periods free from grazing restores the competitive position of the grass.

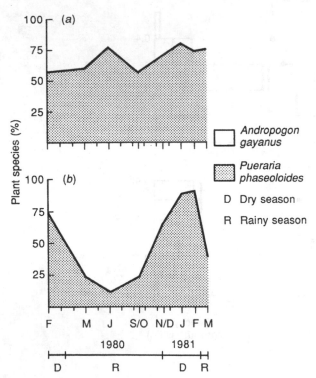

Fig. 5.13. Seasonal effects on % composition of *Andropogon gayanus* and *Pueraria phaseoloides* in (*a*) pasture and (*b*) animal diet. (From Böhnert, Lascano & Weniger, 1985.)

The dominance of weeds which are unacceptable to the grazing animal is a common phenomenon of heavily grazed pastures. In the highlands of northern Thailand cattle are housed at night in the village and taken out to pasture during the day. There is gradient of decreasing grazing pressure away from the village centre. Some of these originally forested lands have regressed to grassland under shifting 'slash and burn' cultivation as the interval between cultivation has narrowed with increasing population pressure. The grassland is dominated by the fire-resistant, tall grass *Imperata cylindrica*. Falvey & Hengmichai (1979) measured botanical composition at intervals on transects 2 km outwards from the centers of six villages (Fig. 5.14). There was a positive association between percentage of *I. cylindrica* and distance from settlement. Conversely, the incidence of the unpalatable weeds *Eupatorium adenophorum* and *E. odoratum* increased with closeness to settlement. The relationships were modified by the period which cattle had been grazed by villagers; villages such as Mae Tho where cattle had grazed for many years had less *Imperata* and more *Eupatorium* than villages such as La Pa Ta, where cattle had been introduced relatively recently.

Selective grazing of tropical pastures is necessary for the maintenance of good levels of animal production, as discussed in Chapter 6, but stocking methods are available to vary botanical composition if this becomes necessary, as illustrated later in Fig. 5.17.

5.3.3
Nitrogen accretion

The amount of atmospheric nitrogen fixed by effectively nodulated legumes is positively associated with legume growth rate, and nodulation and nodule activity are greatly reduced by defoliation, as discussed in Chapter 4. Cyclic variations in grass/legume balance have been linked to N accretion by the legume: legumes dominate when grasses are N deficient, giving better N fixation; grasses then dominate until the N supply is sufficiently depleted for the C$_3$ legumes to compete with the C$_4$ grasses. This cycle has been observed with *S. humilis/H. contortus* pastures at Swans Lagoon, near Ayr, Queensland.

This simple scheme is modified by grazing and cutting in so far as these alter legume growth and hence N fixation. Stocking rates which deplete the legume are expected to lessen N accretion; grazing systems which favour the legume provide the opportunity for a cyclic change in grass/legume balance according to N availability. A further hypothesis is that the seasonal variation in legume selectivity described in the last section may be utilised to maximise N accretion, sustained grass production, and dietary quality. High acceptability is a criterion of merit used in plant improvement programmes. An alternative view is that low acceptability of the legume during the main growing season provides the opportunity for high legume growth rates and N fixation; if this is combined with high legume acceptability during the dry season when deterioration of grass quality is greater, animal production benefits and the augmented soil N supply may promote a satisfactory grass/legume balance.

High levels of animal production were sustained for some years on pastures of the sod-forming grass *Brachiaria dictyoneura* and the trailing legume *Desmodium ovalifolium* at Quilichao, Colombia (lat. 3° N, 1840 mm rainfall) (CIAT, 1986). At high stocking rates the grass component was maintained at *c.* 50%, despite negative selection for the legume. Crude protein content of the diet selected averaged 11%. One hypothesis is that the negative

selection of the legume has provided sufficient N to maintain grass competitiveness in the long term, and that the two species exhibit a desired compatibility over a range of grazing pressures in this environment.

The effect of grazing on the N balance of tropical pastures is poorly documented and further research is needed to validate the assumptions made in this section.

5.4
Case studies

Most of the former examples have been chosen to illustrate particular facets of the manner in which grazing or cutting influence botanical composition. In the con-

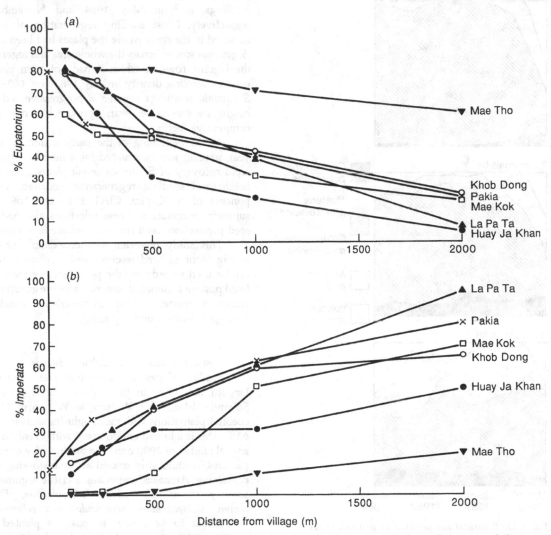

Fig. 5.14. Influence of distance from settlement for different highland Thai villages on (*a*) % *Eupatorium* spp. and (*b*) % *Imperata cylindrica*. (From: Falvey & Hengmichai, 1979.)

cluding part of this section some integrative field studies from different low latitude sites are presented to supplement earlier material.

5.4.1
Carimagua, Colombia
There has been little demographic study of pastures under grazing in Latin America, and work reported by Toledo, Giraldo & Spain (1987) is a notable exception. The experiment was conducted at Carimagua (lat. 5° N, 175 m altitude) on a well-drained oxisol in the Llanos. The site has an annual rainfall of 2160 mm with a dry season occurring from December to March. *S. capitata* cv. Capica and its five

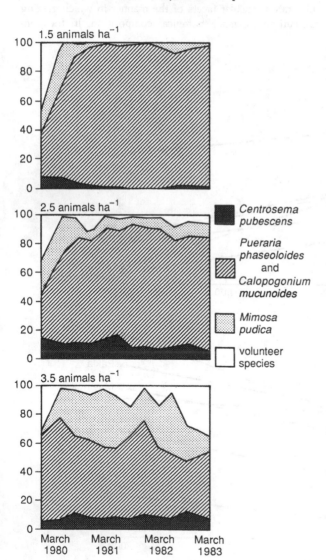

Fig. 5.15. Botanical composition of pastures under coconuts grazed at differing stocking rates. (From Smith & Whiteman, 1985a).

component selections were planted with the tall bunch grass *Andropogon gayanus* in July 1982. The pastures were continuously grazed at 2 beasts ha⁻¹.

S. capitata proved to be a short-lived perennial. Two of the components, CIAT 1693 and 1728, were longer lived than the other three selections, with a half-life of 16 and 10 months respectively. This indicated the importance of flowering, seed production and seedling regeneration for plant replacement and for the persistence of legume yield. All selections flowered for a similar period during the early dry season, when the legume is well eaten (Böhnert *et al.*, 1985). Mean legume seed reserves in the soil (to 5 cm depth) increased to 6.1 kg ha⁻¹ by December 1983, but progressively decreased under the grazing regime imposed to 0.4 kg ha⁻¹ by November 1985. Seedling regeneration occurred repeatedly; it was, for example 39 and 16 seedlings m⁻² in May 1984 and November 1985 respectively. Most seedling regeneration of *S. capitata* occurred in the rows where the plants had been sown, but *A. gayanus* spread across the whole area and regenerated in the legume rows in which it had not been planted; *A. gayanus* seedling density in May 1983 was 140–200 m⁻². *S. capitata* seedlings showed a progressive reduction in height in successive years, presumably due to some competitive factor.

At the beginning of the rainy season in the fourth year, grazing was interrupted for 4 months. This led to a rapid recovery of *S. capitata*; adult plants grew to 25 cm height and seedling regeneration occurred. Two components of cv. Capica, CIAT 1728 and 1693, showed superior recuperation. These selections also had superior seed production, seed reserves and seedling regeneration.

This study provides another example of how the management of seed reserves and seedling regeneration can be used to promote the persistence of yield in short-lived pasture legumes. It also indicates the genetic diversity within a species in the characteristics which favour ecological success under grazing.

5.4.2
Pastures in plantations, Solomon Islands
Studies of pasture production were undertaken at Lingatu (lat. 9° S) in the Russell Islands group of the Solomon Islands by Watson & Whiteman (1981b) in coconut plantations with mean light transmission (PAR) of 62%. This is a humid environment with evenly distributed annual rainfall of 2900 mm. The productivity of naturalised pastures, continuously grazed at three stocking rates and containing *Axonopus compressus* and the legumes *Centrosema pubescens*, *Desmodium heterophyllum*, *Desmodium canum*, *Calopogonium mucunoides* and *Mimosa pudica*, was found to be similar to pastures planted with *C. pubescens*, *Pueraria phaseoloides* and *Stylosanthes guianensis* and the grasses *Brachiaria mutica*, *B. decumbens* and *B.*

humidicola. Initially the planted pastures had higher presentation yields of forage, but these differences disappeared as the introduced grasses were replaced by *A. compressus* and *Mimosa pudica*.

A further study was undertaken (Smith & Whiteman, 1985a) to see if rotational grazing (28 days on, 28 days off) and the use of the more shade-tolerant *B. miliiformis* planted with *B. decumbens* would give better performance of the introduced grasses. Botanical composition in the planted pasture series (Fig. 5.15) showed an increased incidence of volunteer weeds and of *M. pudica* and a decreased contribution of *P. phaseoloides* as stocking rate increased. The planted grasses disappeared within 4 months of the commencement of grazing. Animal production was similar on the planted pasture and the pasture with naturalised legumes.

Two important generalisations arise from this work.

(i) Dominance of pasture legumes is more likely to occur in tropical plantation agriculture, since the growth potential of C_3 legumes and C_4 grasses is often similar under shaded conditions (Ludlow, 1978).

(ii) Grasses will be especially disadvantaged under grazing in humid environments lacking a pronounced dry season. Positive selection for grasses and negative selection for legumes may operate throughout the whole year to the advantage of the legume.

5.4.3
Kluang, Malaysia

A grazing experiment at Kluang (lat. 2° N) in Johore, Malaysia, used a mixed pasture of the tall bunch grass *Panicum maximum* with the legumes *C. pubescens*, *P. phaseoloides*, and *S. guianensis*. The annual rainfall of 1760 mm was relatively evenly distributed, with some short dry spells occasionally occurring between January and April. Eng, Kerridge & Mannetje (1978) varied SR from 2 to 6 beasts ha^{-1}, using continuously grazed Kedah-Kelantan bull calves averaging 80 kg hd^{-1} initial weight.

Botanical composition over the last 18 months of the experiment (Fig. 5.16) was expressed as % frequency of occurrence in 2×0.5 m quadrats. *P. maximum* was well maintained at 2 and 4 beasts ha^{-1} but decreased sharply in the final year at 6 beasts ha^{-1}. *P. phaseoloides* was the

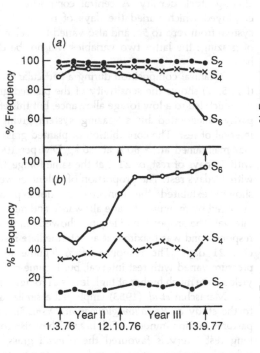

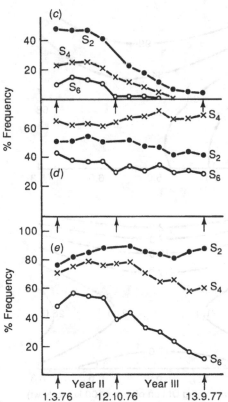

Fig. 5.16. Effects of stocking rate (S$_2$, S$_4$ and S$_6$ refer respectively to 2, 4 and 6 beasts ha^{-1}) on % frequence of occurrence in 2×0.5 m quadrats: (*a*) *Panicum maximum*, (*b*) volunteer spp., (*c*) *Pueraria phaseoloides*, (*d*) *Stylosanthes guianensis*, (*e*) *Centrosema pubescens*. (From Eng, Kerridge & Mannetje, 1978.)

legume most sensitive to grazing; this contrasts with many Latin American reports and with Fig. 5.13, but conforms with experience in northern Australia and the Pacific Islands. *C. pubescens* also decreased at high SR whilst *S. guianensis* was most evident at the intermediate SR. The incidence of volunteer species increased with SR and with time. The principal volunteer grasses and sedges at the highest SR in descending rank, were: *Paspalum conjugatum*, *Digitaria fuscescens*, *Eleusine indica*, *Cyperus* spp., *Axonopus compressus*, *Eragrostis malayana* and *Paspalum commersonii*. Broadleaf volunteer species were mainly *Borreria lactifolia*, *Erechthites valerianifolia*, *Sida acuta*, *Solanum turvum*, *Physalis minima*, *Emilia sonchifolia*, *Melastoma malabrathicum*, *Clidemia hirta* and *Lantana cinerea*.

At the highest SR, bare ground increased to 30%. This study illustrates the degradation of pasture and of the environment caused by overgrazing, and the varying susceptibility of improved species to damage.

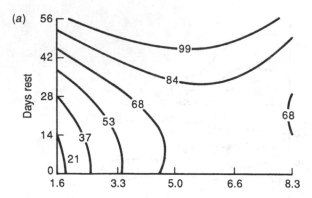

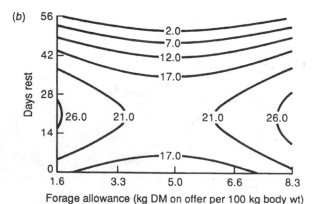

Fig. 5.17. Contours of (*a*) % grass and (*b*) % legume as influenced by duration of rest and level of forage allowance. (From Santillan, 1983.)

5.4.4
Pichilingue, Ecuador

The final case study was located at Pichilingue (lat. 1° S) in Provincia de los Rios, Ecuador. This is a low altitude site in the tropical moist forest zone with annual rainfall of 2150 mm; 82% occurs from December to June. The pasture (Santillan, 1983) comprised two twining legumes, *C. pubescens* and *Neonotonia wightii* cv. Malawi, and two tall grasses, *P. maximum* and *Pennisetum purpureum*.

Previous illustrations in this chapter have simplistically contrasted SR. Whilst they provide valid comparative effects within an experiment, they need to be qualified in terms of animal biomass. Any constant stocking rate implies a varying temporal grazing pressure, according to pasture growth and according also to changes in animal weight. Some workers therefore prefer to contrast differing forage allowances, maintained at particular levels of (usually green) dry matter on offer per 100 kg body weight. These studies are applicable in field practice if it is possible to vary stocking rate on a pasture seasonally, either through the integration of different pasture types, through the use of alternative feeds, or through stock mating and sales policies. This experiment maintained daily forage allowance at specific levels from 1.6 to 8.3 kg dry matter (DM) on offer per 100 kg body weight (BW), by altering stock density. A central composite design was employed which varied the days of rest in the grazing system from zero to 56, and also varied P level and period of grazing; the latter two variables will not be discussed here.

Botanical composition during a particular wet season (Fig. 5.17) shows the sensitivity of the planted grasses to SR which led to a low forage allowance, but this effect was partly ameliorated by a grazing system giving a long interval of rest. The contribution of planted grass to yield was maintained at > 80% at 1.6 kg DM per 100 kg BW with 56 days of rest, or 21% at the same forage allowance with 14 days rest. The proportion of volunteer weeds (not shown) exhibited the converse relationship and was favoured by minimum forage allowance and minimum rest interval. The proportion of legume showed a more complex response, and was maximal at an intermediate rest interval of *c.* 21 days. The response of % legume to grazing pressure varied with rest interval, but legume presentation yields were much higher at high levels of forage allowance.

Maraschin *et al.* (1983) employed a similar approach to the study of *Paspalum guenoarum/Desmodium intortum* pastures in Rio Grande do Sul, Brazil. They also found that long rest intervals favoured the planted grass, whereas reduced grazing interval and low forage allowance increased the incidence of weeds. However, long rest intervals also increase legume content, as occurred with *M. atropurpureum/S. sphacelata* var. *sericea* pastures at Samford, Queensland (Jones, 1979). The prescription that the grass/legume ratio may be increased by long duration of rest does not apply universally in tropical pastures; the reverse may apply for the more palatable legumes or for legumes especially dependent upon plant replacement by seed.

5.4.5
Conclusion

In this section an attempt has been made to unravel some of the complexities of the response of botanical composition to varying grazing and cutting practices by considering effects of the latter on the component processes of plant replacement and of plant interference. It is, however, difficult to model the total pasture response in a satisfactory predictive way for the general case. Integrative field studies are still necessary at the local (or at least regional) level to identify the predominant seasonal and edaphic influences on pasture combinations of different types, as these influences are modified by their interaction with utilisation. We also need to know the relative significance of different factors. For example, *L. bainesii* was identified as a legume with poor resistance to treading (Pott *et al.*, 1983); the most susceptible stage is the seedling before branching occurs. Subsequent work (Table 5.3) showed that the grazing management most favouring density of *L. bainesii* was heavy grazing during the late summer seedling regeneration phase, since the control of competing companion grass, which provided a better light environment for the seedling, was apparently of greater moment than the need to minimise treading damage to the seedling.

The justification for directed manipulation of botanical composition does not lie in the satisfaction of the pasture manager when the favourite species dominate. It relies on the hard evidence of improved animal performance, or alternatively upon the protection of the environment from erosion and degradation.

6

The response of grazing animals to tropical pastures

The availability of pasture, the structure of the sward and the nutritive value of its components reflect the characteristics of the species present and the environment which determines their growth and senescence. The last three chapters have described the influence of the environment in which the pasture grows on the sward system, the effects of defoliation on the rate of pasture growth and N fixation of legumes, and the manner in which grazing and cutting can modify the botanical composition of tropical pastures. The remainder of the book is focused on the animal response to the sward condition, and on the ways in which the feed on offer may be managed in synchrony with animal requirements.

Chapter 6 is concerned with the pasture attributes which determine the level of animal production, especially as these are modified by management practice. The effects of management on the rate of pasture growth and on progress to flowering affect mineral concentration and the organisation of carbohydrate into cell contents, cell wall constituents and cell wall type. The rate of plant removal decides what material is available for senescence, and the presence of senescing material is a major constraint on nutritive value. The stage and season of forage use exerts a powerful control on the relative proportions of plant organs present and their utility to the grazing animal. The level of feed availability, as modified by the intensity of previous use, determines the opportunity for diet selection.

The emphasis in this chapter is on the effects of management on pasture quality and some introductory material about the latter topic is needed to understand management effects. This is not a text on animal nutrition, but there are many comprehensive reviews of the feeding value of tropical forages which the specialist reader may consult (McDowell et al., 1977; Minson, 1980, 1990; Hacker, 1982; Preston, 1984; Butterworth, 1985; McDowell, 1985a; Leng, 1986; Hacker & Ternouth, 1987; Orskov, 1988). A conventional approach is to view pasture quality in terms of (1) digestibility and intake of energy, (2)

nitrogen nutrition, (3) availability of minerals and vitamins, and (4) the occurrence of toxicities and imbalances. The composition of the feed may then be considered in relation to the structure of the sward and the manner in which grazing behaviour and selective grazing modify the diet ingested.

6.1
Pasture quality

6.1.1
Digestibility and intake of energy

(i) A shortage of digestible energy (DE) is more commonly the primary constraint to individual animal performance on tropical pastures (Romero & Siebert, 1980) than other nutrient deficiencies. Animal requirements (expressed as metabolisable energy, ME, which is $c.\ 0.81 \times DE$) for different classes of stock are tabulated subsequently from British standards in Tables 8.1 and 8.2. Usually in farm practice the long term animal performance on tropical pastures alone is limited by pasture quality to $c.\ 0.7$ kg LWG $hd^{-1}\ d^{-1}$ for beef cattle, or $c.\ 12$ kg milk $hd^{-1}\ d^{-1}$ for dairy cows.

The attainment of production goals depends upon the feeding value of the pasture, which reflects the level and balance of nutrients which it contains. These may be considered (Fig. 6.1, Egan et al., 1986) in terms of the amount of intake and the nutrient yield per unit intake which is generated. The feeding system determines the availability of feed whose acceptability to and preference shown by the animal influence intake, which is also subject to metabolic limitations and the physical limitation of rumen fill. Low retention time and good rate of flow from the rumen, which are related to intake, may be balanced by reduced digestion, which is favoured by a long retention time, and which is a feature of fibrous tropical feeds. The rate, site and nature of digestion influence the nutrient yield. Both the physical and the chemical aspects of plant characteristics need to be considered in these connexions.

(ii) The potential digestibility of pasture material conveniently falls into four plant fractions (Minson, 1976):

(1) the readily hydrolysable carbohydrates, such as monosaccharides, disaccharides, fructosan and starch, proteins, minerals and some cutin, which are potentially fully digestible;

(2) higher molecular carbohydrates including cellulose and hemicellulose, which are not protected from digestion and which are potentially digestible;

(3) higher molecular carbohydrates, which are fully encrusted by lignin or cuticle and which are not digestible; and

(4) indigestible lignin, silica and cuticle.

These groups are complex and variable in character, and are each composed of many different compounds, so that their behaviour in nutrition is not fully predictable. The cellular contents belong to the first group, whilst the cell wall constituents and epidermis are spread over the other three groups.

The relative proportions of the four fractions vary in different plant organs according to ontogenetic stage and the proportions of anatomical tissues present. Leaf has higher organic matter digestibility (OMD) than stem (Wilson & Minson, 1980), although young elongating stem may have similar OMD to leaf (Haggar & Ahmed, 1970) and lamina is more digestible than leaf sheath. The midrib of lamina is poorly digested. Senescent material has low OMD and intake, as discussed later. Seed may be highly digestible (Playne, McLeod & Dekker, 1972).

The mesophyll and phloem are the most digestible constituents, whilst the epidermis and parenchyma

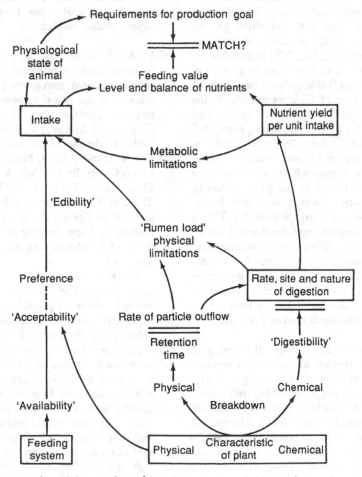

Fig. 6.1. Relationships between digestibility, intake and retention time. (From Egan *et al.*, 1986.)

bundle sheath are more degradable than sclerenchyma (Akin, Wilson & Windham, 1983). The lignified vascular tissues are mainly indigestible, but some variation in resistance may be associated with the retention of phenolic compounds in cell walls (Akin, Willemse & Barton, 1985). The content of silica has been discounted as a negative factor in digestion (Minson, 1971a), but recent work (Shimojo & Goto, 1989) prompts a reconsideration of this.

The tropical C_4 grasses are on average 13% units less digestible than the temperate C_3 grasses, whilst the tropical C_3 legumes are c. 4% units less digestible than temperate C_3 pasture legumes (Wilson & Minson, 1980). This mainly arises from the temperature of the environments contrasted, as discussed later; the differences between the grasses sometimes reflect differences in leaf/stem ratio (Minson & Wilson, 1980), but more importantly are due to the Kranz ('wreath') anatomy (Wilson & Hattersley, 1983) of C_4 grasses (see Fig. 6.2) The bundle sheaths of tropical grasses lead to efficient use of assimilate and efficient water transport; they also have thick, suberised outer walls which are resistant to mechanical breakdown (Fig. 6.3) and there are no intercellular spaces. Their characteristics depend on whether they belong to the C_3/C_4, PCK, NADP-ME, or NAD-ME C_4 groups (Fig. 6.4). The higher proportion of vascular bundle and sheath (Table 6.1) may be associated with longer retention time of C_4 grass material in the rumen. C_4 grasses contain less of the easily digested mesophyll tissue than C_3 grasses and the C_4 mesophyll is more densely packed (Wilson, Brown & Windham, 1983) both of which appear to delay access of rumen micro-organisms to the mesophyll surface area. These characteristics are accentuated as ontogeny proceeds, and leaves which are formed later in the season, higher up the stem, have lower dry matter digestibility (DMD) and higher cell wall content than first-formed leaves (Wilson, 1976a,b). Genetic differences in anatomy may be used in plant improvement programmes. In vitro dry matter digestibility (IVMD) of *Cenchrus* genotypes varied 9% for leaf and lignified tissues in cross-section, and total number of vascular bundles in the stem section also varied (Wilson, Anderson & Hacker, 1989; Fig. 6.5).

(iii)   Climatic conditions modify strongly the quality of pastures grown in the tropics, irrespective of whether the species are of temperate or of tropical origin. The C_4 grasses grow well in response to increasing temperature (see Fig. 8.3), but this is associated with the production of more supportive vascular tissue (Ford, Morrison & Wilson, 1979); this leads to an average decrease in DMD of 0.6 units for each 1 °C rise in growth temperature (Wilson & Minson, 1980). This effect is greater on DMD of C_4 stem (-0.86% per unit

temperature increase) than on DMD of C_4 leaf (-0.57% per unit temperature increase). The decrease in DMD is less in tropical legumes (-0.28% per unit temperature increase). This partially explains regional and seasonal differences in individual animal production, which reflect higher quality pastures in the highland tropics and subtropics than in the lowland tropics, and better digestibility in the cool season relative to the hot season, provided other factors (drought or frost) do not preclude cool season growth (Hacker & Minson, 1972).

Severe drought causes loss of pasture quality, since leaf growth is arrested and senescent material is less digestible and less acceptable to stock. Mild water stress increases pasture quality, since ontogeny is delayed, cell wall content is less than in fully watered plants, and DMD is better maintained (Wilson & Ng, 1975; Wilson, 1983). Pasture managers sometimes observe that a season with intermittent rains which is sufficient to maintain continuity of forage supply leads to higher individual animal gains than occur in a season of superabundant rainfall.

(iv)   A first principle is that young pasture growth leads to higher organic matter digestibility and intake than pasture reserved from grazing. The decrease in nutritive value with age is a phenomenon observed throughout tropical countries (for example, Soneji, Musangi & Olsen, 1971 in Uganda; Ademosun, 1973 in Nigeria; Veitia & Marquez, 1973, and Lamela, García-Trujillo & Cáceres, 1980 in Cuba; Nascimento & Pinheiro, 1975 in Minas Gerais, Brazil; Nada & Jones, 1982 in south-east Queensland; Wanapat & Topark-Ngarm, 1985 in Thailand; Peiris & Ibrahim, 1985 in Sri Lanka; Pond *et al.*, 1987 in Texas). The decrease in digestibility and N content and the increase in acid detergent fibre and lignin of *P. notatum* with age in Florida is illustrated in Table 9.1.

There are two main conclusions from this inescapable phenomenon. The first bears on management practice. Animals whenever possible should be given maximum access to young pasture growth. This explains the failure of schemes of rotational grazing which present more aged feed to the animal than occurs under continuous grazing, as discussed in Chapter 10. It also suggests that rotational grazing with long rest interval is more deleterious to animal production than rotational grazing with a short rest interval, and that cut-and-remove systems should avoid a long interval between cuts. In Perico, Cuba, milk production from *P. maximum* pastures decreased from 7.8 to 6.5 kg cow^{-1} d^{-1} as cutting interval increased from 38 to 50 days (Lamela *et al.*, 1980).

The second conclusion is that a premium is placed on slow rate of deterioration in pasture quality with age when choosing pasture species to plant. The rate of deterioration with age is greater for plants of high digestibility than for plants of low digestibility, but for any level of digestibility there is considerable variation available in the rate of deterioration. Norton (1982) recalculated a collation of data from the literature to contrast IVDMD of 28 day regrowths with the rate of decrease in digestibility. The plants tested fall into two groups, in which the tropical legumes and the grasses in the genera *Brachiaria*, *Digitaria* and *Setaria* exhibit low rates of decrease in digestibility relative to those in the genera *Chloris*, *Hyparrhenia* and *Panicum* (Fig. 6.6). This question is discussed further in Section 8.2.1.

(v) There is greater opportunity to choose species and cultivars with higher intrinsic intake characteristics than to select plants of higher digestibility (Minson, 1971*b*), and the two attributes are not necessarily closely associated. Aspects of sward structure (Section 6.2) and of grazing behaviour (Section 6.3) which bear on intake

are discussed later, and this section offers comment on the relevant physical and chemical attributes of the forage (Minson, 1982).

Leafiness increases pasture intake. This generalisation does not always hold for comparisons between plant species, but is usually true for comparisons within species (Minson & Laredo, 1972; Minson & Bray, 1986). More is known of this topic since scientists began separating leaf and stem fractions and feeding them separately. In these circumstances grass leaf having a similar digestibility as stem may have 46% greater intake than stem (Laredo & Minson, 1973). Again, the intake of the legume *L. purpureus* leaf fraction was 79% and 61% greater than the intake of stem by cattle and sheep respectively (Hendricksen, Poppi & Minson, 1981).

The greater intake of leaf results from its shorter retention time in the rumen, which leads to a greater rate of passage. Intake is controlled by bulk in the rumen, which is relatively inelastic (Thornton & Minson, 1973). The shorter retention time of leaf is associated with its greater surface area per unit weight, and the small

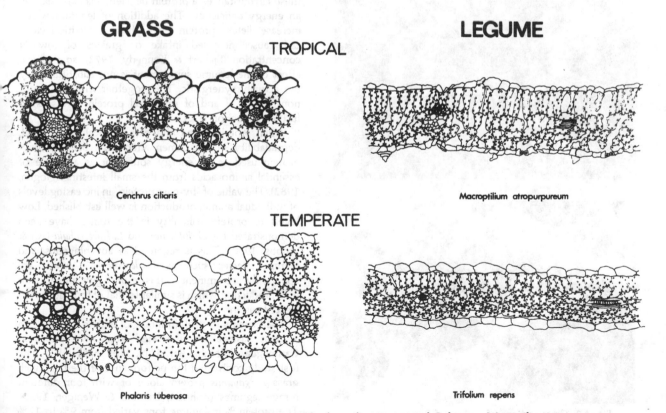

GRASS

TROPICAL

Cenchrus ciliaris

LEGUME

Macroptilium atropurpureum

TEMPERATE

Phalaris tuberosa

Trifolium repens

Fig. 6.2. Cross-sections of laminae of a tropical C$_4$ grass (*Cenchrus ciliaris*), a tropical C$_3$ legume *Macroptilium atropurpureum*, a temperate C$_3$ grass (*Phalaris aquatica* formerly *P. tuberosa*) and a temperate C$_3$ legume (*Trifolium repens*) showing different proportions of vascular tissue. (Photo: J. R. Wilson.)

particle sizes which result from chewing, detrition and digestion. There is a small outlet from the rumen which restricts the egress of particles greater than 1 mm in size. Rate of passage out of the rumen is related to the time taken for the particle size distribution of the ingested forage to move from particles in excess of 1 mm to particles less than 1 mm size. This is associated with the proportion of structural elements, which is greater in stem than in leaf, and with the relative degradability of these structural elements. Grinding forages before feeding reduces differences in rate of passage, which identifies mechanical factors as having significance.

Forage legumes usually have greater intake than grasses compared at the same level of digestibility, and this is associated with shorter retention time of legume material in the rumen and a greater density of packing legume forage in the rumen (Thornton & Minson, 1973).

The presence of thorns and the occurrence of secondary compounds (Barry & Blaney, 1987) may

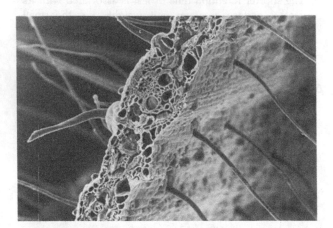

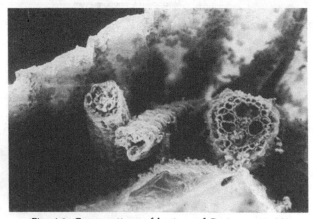

Fig. 6.3. Cross-sections of laminae of *Panicum* spp. (*a*) before digestion (*b*) after digestion, showing undigested bundle sheaths. (Photo: J. R. Wilson & D. E. Akin.)

reduce acceptability of forage to livestock. Protein content is discussed in Section 6.1.2, and mineral content may limit the rate of intake. Some suggested critical values below which forage intake is reduced are: P 0.12%, S 0.17%, Na 0.05%, Co 0.1 ppm and Se 0.05 ppm. There are many reports in the literature where the application of fertiliser or the provision of mineral supplements enhances intake (for example, Rees & Minson, 1976 for calcium; Gardener, Freire & Murray, 1988 and Murray, Freire & Gardener, 1988 for superphosphate; Rees, Minson & Smith, 1974 and Hunter and Siebert, 1985 for S).

6.1.2
Nitrogen

(i) The protein content of the forage also limits intake; the critical value quoted used to be 1.1% N but may extend to 1.3% N (Hennessy, 1980). Animals respond to higher levels of N in the forage, but intake *per se*, as controlled by the N metabolism of the rumen organisms, requires a minimum of 6–8 crude protein (N × 6.25); in these circumstances a protein deficiency is expressed as an energy deficiency. The addition of legumes which increase dietary protein levels above the critical value can cause increased intake of grasses of low N concentration (Siebert & Kennedy, 1972), so that the effect of the legume supplement is not substitutionary but actually synergistic. This, together with the use of non-protein N and of protected protein, is discussed further in Section 9.3.1.

When protein is fed above the critical 6–8% level, degradability in the rumen affects the balance of N protein used as an energy source or as a source of essential amino acids from the small intestine (Hogan, 1982). The value of 'by-pass protein' in increasing levels of individual animal production is well established. Low levels of protein solubility in the rumen have been demonstrated for *D. intortum* and *L. leucocephala* (Aii & Stobbs, 1980). This is regarded as advantageous, and legume species such as *L. leucocephala* which are apparently pre-eminent in this respect give an effect on milk production which is similar to that of formaldehyde-treated casein.

The benefits to protein intake of adding a legume to a grass pasture are illustrated from two studies at Carimagua, Colombia (lat. 5 °N, 2160 mm rainfall). The first study contrasted the performance of the tall bunch grass *A. gayanus* grown alone or with four different forage legumes (Böhnert, Lascano & Weniger, 1986). The protein % in legume tops varied from 9% to 12% in the dry season, and 11% to 15% in the wet season, being highest in *P. phaseoloides*. The presence of a companion legume increased the protein content of the companion grass (Fig. 6.7a). This effect was greater with leaf than stem and with *P. phaseoloides* than with *S. capitata* or *Zornia latifolia*; the differences were greater in the wet season. This was reflected in a higher protein % in the forage ingested by steers (Fig. 6.7b).

In the second study, cattle grazed *B. decumbens* alone or *B. decumbens* with *P. phaseoloides* (CIAT, 1985). After five years of continuous grazing, marked differences in the N% of the pasture on offer became apparent, and this was reflected in greater differences in LWG in favour of the legume-based pasture. Protein intake of cattle (Fig. 6.8) was estimated to be substantially greater and at an acceptable level on mixed pasture, partly due to the effect of the legume on the level of N concentration in the associated grass. A low value occurred during a dry February.

Rate of LWG has been linearly related to the N content of the diet over a wide range (Fig. 6.9). These data derive from cattle grazing native pastures and legume/grass pastures at Lansdown, Queensland (Siebert & Hunter, 1977).

(ii) Management options for overcoming N deficiency are (1) the planting of legumes, (2) the feeding of protein and N supplements, (3) the maintenance of pasture on offer at a young growth stage, and (4) the application of fertiliser N.

Nitrogen concentration of available pasture decreases with age, depending on soil N supply and the N nutrition characteristic of the species; for example, *P. clandestinum* consistently has higher N concentration in its leaves than does *P. maximum* var. *trichoglume* (Wilson & Sandland, 1976). Legumes maintain their N% relatively well with age, but grass N% decreases markedly, together with digestibility and intake (Table 9.1). The comments made earlier in the chapter

Table 6.1. *Percentage of different tissues in cross-sectional area of grass lamina*

Tissue	Tropical (C_4)	Temperate (C_3)
Epidermis	33.0 ± 2.1	28.7 ± 1.5
Mesophyll	34.6 ± 2.4	61.0^a
Bundle sheath	19.3 ± 2.1	–
Vascular bundle	7.6 ± 0.4	5.5 ± 0.4
Sclerenchyma	5.0 ± 1.1	3.4 ± 0.4

Source: Wilson & Minson (1980). ^{a}includes 6% parenchyma sheath.

concerning the advantages of offering young pasture to animals were made in the context of maintaining available energy levels, but this applies equally to N concentration. C_4 grasses exhibit efficiency of N use in DM production which leads to a rapid reduction in N% unless soil N supply is continually augmented (Wilson & Minson, 1980).

Synchrony between growth rate and defoliation ameliorates this effect. High SR which delays ontogeny of the grass shoots present reduces forage allowance but consistently increases the nitrogen concentration of the forage on offer. This was demonstrated for a range of SR and N application levels by a study of *P. clandestinum* pastures at Lismore, New South Wales (Mears & Humphreys, 1974a). Alternatively, low SR when N is applied may lead to advanced ontogeny and the average N% of the forage available may actually be reduced by N fertiliser application.

The use of N fertiliser increases pasture intake in circumstances where N supply is so limited that N% falls below the critical 1.1–1.3 %. Nitrogen fertiliser has little effect on pasture quality above this level, and increases in animal production result from carrying more animals on the pasture to consume the additional feed supply, as discussed in Section 7.2. The effects of fertiliser N level are illustrated for 28 day regrowths of *C. gayana* under irrigated conditions at Gatton, Queensland (lat. 28 °S), which received 125 (low N) or 500 (high N) kg N ha^{-1} as urea (Minson, 1973). The high N treatment produced forage of slightly higher (2.4%) organic matter digestibility, and intake was increased by 3.3 units on average (Table 6.2). This result occurred where N% of the low N treatment did not fall to the critical level for intake.

6.1.3
Minerals

(i) The soils of the tropics are highly leached and weathered, and are mainly of low base exchange capacity. Cropping is concentrated on the more fertile soils, and the residual areas of lower fertility are more commonly devoted to pasture. In consequence the mineral content of tropical forages is relatively low, and the widespread mineral deficiencies in the stock which graze them reflect this situation.

Mineral deficiencies in tropical ruminants merit a book in their own right, and the cursory mention they receive in this text arises from constraints of space; the mineral needs of grazing animals require universal recognition by pasture managers if good levels of animal performance are to be achieved. Available reviews include Gartner *et al.* (1980), Little (1982),

McDowell, Ellis & Conrad (1984) and Minson (1990). The pattern of deficiency is local in character and the manager requires good diagnostic techniques based on observable syndromes in the animal and on analytical information from the pastures.

A survey of 2615 Latin American forages (McDowell *et al.*, 1983) suggested that borderline or deficient elements were present in a high proportion of entries: Zn 75%, P 73%, Na 60%, Cu 47%, Co 43%, and Mg 35%. In the Mato Grosso of Brazil Sousa *et al.* (1979, 1981, 1982) recorded widespread mineral defi-

ciencies of P, Na, Zn, Co, Mn, whilst in Bolivia a survey (Peducassé *et al.*, 1983) suggested P, Na, Cu, Zn, Ca and Se were probably deficient. Salih *et al.* (1983) drew attention to deficiencies of P, Cu and Fe in Florida. In central Thailand Cu, Na, P, Ca and Zn were the most likely elements to be deficient on the basis of forage and animal tissue analysis (Vijchulata, Chipadpanich & McDowell, 1983). In northern Australia Gartner *et al.* (1980) emphasise the field occurrence of animal responses to P, Na, S, Co and Cu. Many elements interact in the expression of deficiencies (Little, 1982)

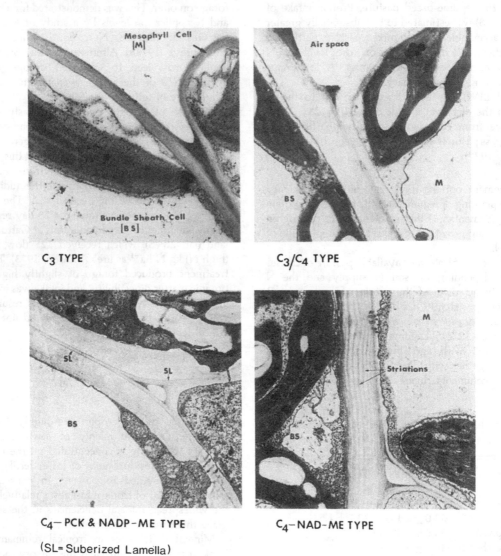

Fig. 6.4. Differences in cell wall structure of bundle sheath (BS)-mesophyll (M) interface in the laminae of *Panicum* spp. C$_3$ type *P. trichanthum*. No suberised lamella in BS cell wall. C$_3$/C$_4$ type *P. milioides*. No suberised lamella in BS cell wall. Intercellular air space. C$_4$ PCK type (*P. maximum*). Suberised lamella (SL) present in outer walls; also in C$_4$ NADP-ME type. C$_4$ NAD-ME type (*P. decompositum*). Striations (STR) but no suberised lamella in cell wall. (Photo: J. R. Wilson & P. W. Hattersley.)

and this adds to the complexity of diagnosing local problems. It is unwise to assume the local presence of a particular deficiency because of its widespread occurrence; for example, animals responding positively to a sodium phosphate supplement may be benefiting from the sodium and not the phosphate (Falvey, 1983).

(ii) The mineral nutrients essential to animal growth and reproduction strongly overlap those which are needed for maximum pasture growth, but there are some significant differences. Na is a widespread animal deficiency and has been shown to be essential for plant growth but plant responses are insignificant. Co has been implicated in legume N fixation but is not used in field practice for this purpose; it is a widespread

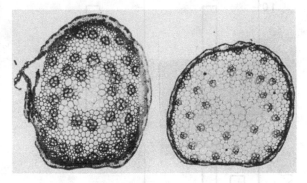

Fig. 6.5. Variation in density of vascular bundles in *Cenchrus* spp, which is negatively associated with IVDMD. (Photo: J. R. Wilson.)

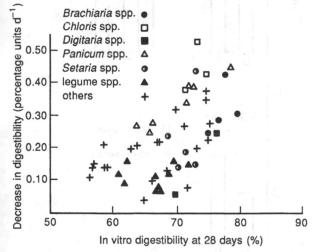

Fig. 6.6. Relationships between IVDMD of 28 day regrowths and rate of decrease in digestibility. (From Norton, 1982.)

deficiency of cattle. Again, I is not found as a plant deficiency, but goitre is a common animal problem. Many animal deficiencies are overcome by fertilising pastures, and the application of single superphosphate increases levels of P, S and Ca in the forage. However, some planted pasture species have been selected for their adaptation to low fertility soils; critical P levels in *S. scabra* may be as low as 0.10% and animals grazing fertilised pastures containing *S. scabra* may still require a P supplement.

In animal nutrition the macro-elements Ca, P and S are mainly used for structural purposes (Little, 1982), whilst Na, K and Cl are involved in the maintenance of the acid-base balance; K, Ca and Mg contribute to energy transfer, transmission of nerve impulses and the activation of enzymes. The micro-elements Mn and Cu act as enzyme co-factors; Zn, Mo and Se are involved in the structural or functional activities of enzymes whilst I and Co are required in hormones and vitamins respectively. Primary deficiencies of energy and protein limit the expression of most mineral deficiencies; a Cu supplement may raise Cu level in the liver but the animal remains unthrifty unless it is well fed.

Suggested mineral requirements for young cattle growing at a moderate rate or for cows lactating to give 10 kg milk $cow^{-1} d^{-1}$ (Table 6.3, Little, 1982) may be expressed as the net requirement based on animal tissue concentrations adjusted for endogenous losses; this figure when further adjusted for the absorption co-efficient suggests the required concentration in the animal diet. The latter is more readily related to plant mineral analysis, especially if this is based on the plant material being selectively grazed. These figures are indicative only; 0.12% P may be a satisfactory dietary level for growing stock in some situations.

The more common mineral interactions include the following. The balance between Ca and P has attracted a good deal of research interest, and the wide ratio which commonly occurs in the tropics may be deleterious. High levels of K exacerbate the condition of hypermagnesaemia. The ratio of N:S is desirably c. 14:1, and low levels of S inhibit protein formation. There is an antagonism between dietary Cu, Mo and S, and excess Mo (as occurs with liberal use of Mo fertiliser on pasture) reduces the availability of Cu if adequate S is present. Cu and Zn exhibit a mutual antagonism.

These examples indicate the requirement for specialist treatment of the problems of mineral availability. Case studies of particular mineral responses include: P (Winks, Lamberth & O'Rourke 1977; Awad, Edwards & Huett, 1979; Little, 1980, Loxton, Murphy & Toleman, 1983), Ca (Rees & Minson, 1976; Awad *et al.*, 1979; Arias *et al.*, 1984), S (Kennedy & Siebert, 1973; Rees *et al.*, 1974; Hunter, Siebert & Webb, 1979), Na (Winks & O'Rourke,

1977; Davison et al., 1980; Gutteridge et al., 1983; Falvey, 1983; Little, 1987), Co (Norton & Hales, 1976; Mannetje, Singh Sidu & Murugaiah, 1976; Winter, Siebert & Kuchel, 1977; Nicol & Smith, 1981; Barry, 1984; Norton & Deery, 1985) and Se (Wilson et al., 1981).

Vitamin supplementation (McDowell, 1985b) is rarely necessary for grazing ruminants exposed to sunlight and consuming green forage. It has been suggested that animals eating dry feed may need vitamin A supplement, but field responses are rare (Gartner & Anson, 1966).

(iii)   Two relevant aspects of pasture management are mentioned in conclusion. Animals grazing young pasture are less likely to show mineral deficiencies, since almost all elements are present in higher concentration than in aged forage (Little, 1982). Calcium concentration is exceptional in that it shows little change with advancing season, and the rate of change for other elements in tropical grasses in Florida is tabulated by Rojas et al. (1987).

The mineral concentration present varies with the pasture species planted; deficiencies appear (or not) on

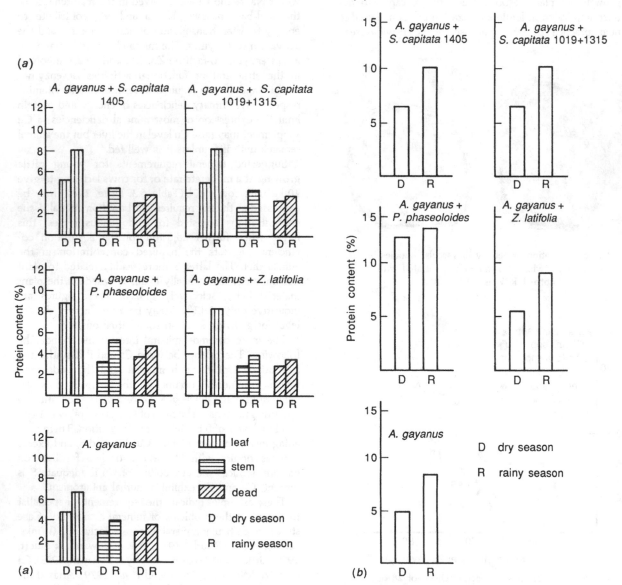

Fig. 6.7. Protein concentration (a) in the organs of *Andropogon gayanus* and (b) in the diet of cattle grazing pastures in Colombia with and without legumes. (From Böhnert et al., 1986.)

a particular soil according to the species present. In coastal south-east Queensland animals grazing pastures based on *A. affinis* show Cu deficiency where *P. dilatatum/T. repens* pastures on the same soil are Cu adequate. Na deficiency might be expected on a solodic soil growing *M. atropurpureum/H. contortus*, where animals grazing pastures of *C. gayana* on the same soil would be Na sufficient (Playne, 1970). These differences extend to the intraspecific level (Hacker, Strickland & Basford, 1985). Problems of mineral deficiency may therefore be mitigated through plant improvement programmes directed to maximising mineral uptake and the maintenance of nutrient concentrations in plant tissue.

6.1.4
Toxicities

Excess mineral content causes toxicities, and the most common occurrences in the field are with Se, F, and Mn (McDowell *et al.*, 1983); Cu and Mo excess also cause problems. Contamination of pastures with heavy metals may arise from impurities in fertiliser (Williams, 1977) or from the use of sewage waste.

Table 6.2. *Effects of fertiliser N on attributes of 28 day regrowths of* Chloris gayana *in south-east Queensland*

	N (%)		OMD (%)		Intake (g kg $W^{-90.75}$)	
Month	Low N	High N	Low N	High N	Low N	High N
Dec	2.04	2.27	63.8	64.6	65.4	68.1
Jan	1.54	2.27	61.2	66.3	63.9	64.7
Feb	1.26	1.89	62.0	61.8	59.1	58.6
Mar	1.66	2.61	65.2	66.1	69.3	72.5
Apr	2.04	2.27	58.0	63.3	57.4	67.7
Mean	1.71	2.26	62.0	64.4	63.0	66.3

Source: Minson (1973).

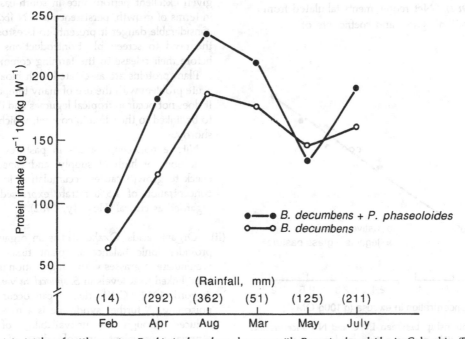

Fig. 6.8. Protein intake of cattle grazing *Brachiaria decumbens* alone or with *Pueraria phaseoloides* in Colombia. (From CIAT, 1985.)

Compounds which are toxic to animals, or which are regarded as anti-quality factors because they reduce forage intake or animal performance, are classed as secondary plant metabolites (Rosenthal & Janzen, 1979; Hegarty, 1982; Barry & Blaney, 1987), since they are not essential for the growth and development of plants. Some of these

Table 6.3. *Net and dietary mineral requirements of cattle for growth (200 kg liveweight; 0.5 kg d^{-1}) and lactation (500 kg liveweight; 10 kg d^{-1})*

	Growth		Lactation	
Mineral	Net[a]	Dietary	Net[a]	Dietary
% in dry matter				
Calcium	0.19	0.43	0.22	0.32
Phosphorus	0.19	0.24	0.17	0.30
Magnesium	0.03	0.15	0.03	0.18
Sodium	0.06	0.07	0.09	0.10
ppm in dry matter				
Zinc	3.6–6.0	12–20	5.4–7.5	18–25
Copper	0.3–0.6	8–14	0.3–0.6	10–14
Manganese	0.1–0.2	10–20	0.1–0.2	10–20
Colbalt		0.11		0.11
Selenium		0.03–0.05		0.03–0.05
Iodine		0.4		0.5

Source: Little (1982). [a]Net requirements tabulated from dietary concentration figures and coefficients of absorption.

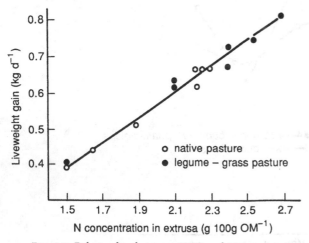

Fig. 6.9. Relationship between LWG and N concentration in extrusa of cattle grazing native and legume–grass pasture. (From Siebert & Hunter, 1977.)

compounds may be regarded as a plant defence against insect or fungal attack or consumption by the larger herbivores. The main groups of these compounds identified as causing problems in the consumption of tropical pasture plants are now listed.

(i)   Nitrogenous compounds. Mimosine is one amino acid which accumulates in *Leucanea leucocephala* to the detriment of its use as a monogastric feed; in Australia and Papua New Guinea (but not other countries) ruminants showed hair loss, goitre and low rates of liveweight gain (Jones, Blunt & Holmes, 1976). This problem arose from the degradation of mimosine in the rumen to 3-hydroxy-4(IH)-pyridone, (DHP) which is a potent inhibitor of thyroid peroxidase, which is needed in the synthesis of thyroid hormones (Hegarty *et al.*, 1979). An alternative pathway of mimosine degradation was achieved by introducing bacteria from Hawaiian goats to Australian ruminants (Jones & Megarrity, 1986); these bacteria may be spread in stock drinking water and their presence in the cattle rumen lead to excellent performance from *Leucaena* pastures (Quirk *et al.*, 1988). This considerable scientific achievement suggests that it may not be sufficient to import the native *Bradyrhizobium* with a legume introduction to a new country; it may also be desirable to import the rumen organisms which have adapted to its utilisation.

Another amino acid, indospicine, is a potent liver toxin which occurs in *Indigofera spicata*. This legume has given excellent performance in south-east Queensland in terms of growth, persistence and N fixation, and the considerable danger it presents to livestock emphasises the need to screen plant introductions for toxicities before their release to the farming community.

Plant proteins are associated with bloat, which is an acute problem with the use of many temperate legumes. It does not occur in tropical legumes and this is believed to be linked to their tannin content, which is mentioned shortly.

Nitrate poisoning occurs in pastures grown under conditions of high N supply, and especially if some check to growth causes accumulation in leaf tissue. A concentration of 1.5% nitrate expressed as KNO_3 is regarded as critical (Hegarty, 1982).

(ii)   Organic acids. Oxalic acid is an organic acid which provides ionic balance in plant tissue; it tends to accumulate in grasses with a high cation uptake and has been linked to K levels in *S. sphacelata* var. *sericea* (Jones & Ford, 1972). Cattle deaths can occur from oxalate poisoning. A further syndrome is a mineral imbalance induced through the unavailability of Ca; this is especially prevalent on sole diets of *C. ciliaris*, *P. clandestinum* and *S. sphacelata* var. *sericea* (Gartner *et al.*,

1980). Prospects for plant improvement in this respect are not good, since both yield and digestibility are highly correlated with oxalic acid content in *S. sphacelata* var. *sericea* (Hacker, 1974).

(iii)   Heterocyclic compounds. Oestrogenic isoflavones cause widespread infertility problems in sheep grazing *T. subterraneum* pastures. Oestrogenic potency does not arise in tropical pasture legumes (Little, 1976).

(iv)   Alkaloids. Pyrrolizidine group alkaloids are hepatotoxins common in some legume genera such as *Crotalaria*. Lines of *C. juncea* have been selected for their low alkaloid content; the highest concentration occurs in the seeds, followed by the stems, whilst the young leaves have the lowest concentrations (Jobin & Shelton, 1980). This plant is successfully used in south-east Asia as a high quality hay or green manure crop, and toxicity problems are not evident in dried material cut at the bloom stage.

(v)   Cyanoglycosides. These occur in many herbage plants, and disorders are more prevalent amongst the domesticated species *C. aethiopicus* and *Sorghum* spp. The popularity of *P. americanum* and *Echinochloa crusgalli* as fodder crops arises in some districts from the fear farmers have that *Sorghum* spp. will kill their cattle. The production of hydrocyanic acid can have devastating effects on cattle, but the incidence on a farm of stock death from *Sorghum* grazing may be separated by periods of several years during which good production occurs. These events are associated with unusual weather conditions, such as a heat wave checking the growth of a young crop, or with the access of hungry or stressed stock to *Sorghum*. High N supply favours the production of cyanoglycosides.

(vi)   Tannins. The phenolic compounds designated as tannins were mentioned as protectants against bloat; they also have a role in reducing insect and fungal attack which has a secondary effect on the rapidity of organic matter turnover in the soil (Barry & Blaney, 1987). Their presence may also reduce forage intake. A positive effect is the protection of proteins, and the occurrence of condensed tannins in the range up to 11% dry matter increases the duodenal flow of non-ammonia N; digestion of protein in the rumen is reduced and the 'bypass' protein which escapes may be used effectively to increase animal production. Shrub legumes fed to ruminants differ markedly in their acceptability to stock and in the efficiency of N digestibility in the post-ruminal tract. The correct balance and concentration of condensed tannins is sought which minimises the negative effect on acceptability and optimises N

nutrition; this is found in genotypes of *L. leucocephala* and *Sesbania sesban*.

(vii)   Other compounds. These include toxic sulphur compounds, which sometimes occur in *Brassica* spp., and disorders caused by mycotoxins, or metabolites which result from fungal infection. The occurrence of ergot (*Claviceps paspoli*) in *P. dilatatum* may lead to abortion and to neuromuscular disorders. Photosensitisation periodically affects young stock grazing *B. decumbens*.

The occurrence of these disorders is sporadic. *T. repens* is regarded as the most nutritious of all the herbage plants, in that no other forage has produced such high rates of lamb growth. Cyanoglycosides are found in *T. repens*, which can also induce bloat. The pasture manager has to guard against the development of an obsession about natural poisons, since these are widely prevalent and cause infrequent problems on well-managed farms.

6.2
Sward structure

(i)   Effects of variation in sward structure on canopy photosynthesis and on the interference exerted by one species on another in the supply of environmental growth factors were discussed in Section 5.3.1. Attention is now drawn to the ways in which animal intake and production are influenced by the disposition and density of plant organs in successive strata above the ground. The arrangement of plant parts in both horizontal and vertical dimensions influences the capacity of the grazing animal to prehend and select the material on offer.

The bulk density, and particularly the leaf bulk density of tropical pasture plants, is less than that of temperate pastures. This has the effect of reducing the average amount of forage material ingested with each prehending bite (the bite size) relative to that ingested from good quality *L. perenne* pastures. Some scientists believe this represents a constraint to animal performance.

The sward structure of well fertilised *C. gayana* pastures (Table 6.4) grown at Samford, Queensland (lat. 27 °S) shows increasing height with time since cutting; after eight weeks 1.8 t DM ha^{-1} occurred above 60 cm height in these rapidly growing pastures (Stobbs, 1973b). The proportion of the available forage which was leaf decreased with time since cutting. The leaf bulk density decreased with height above ground and the maximum value observed was 91 kg ha^{-1} cm^{-1} in the layer 0–15 cm above ground level. Sward total bulk density increased with time and was also greater in the lower layers with a maximum value of 148 kg ha⁻¹ cm⁻¹. Bite size of Jersey cows grazing these pasture increased from 0.26 to 0.32 g OM bite⁻¹ at two and four weeks respectively, and then decreased to 0.17–0.15 g subsequently.

Stobbs (1973a) varied the canopy structure of *D. decumbens* and *C. gayana* by applying the growth regulators GA, which produced stemmy plants with long internodes, and CCC, which produced denser, leafier swards. Bite size was greater from the latter canopies, and Stobbs suggested that sward bulk density and leafiness influenced bite size. He contrasted the values for bulk density of temperate pasture swards (159–413 kg ha⁻¹cm⁻¹) with the lower values for *C. gayana* in Table 6.4.

Other workers have found similarly low values for sward density in tropical forages. The trailing legume *Lablab purpureus* cv. Rongai, which has a structure with contrasting features, was also grown at Samford (Hendricksen & Minson, 1985), and produced a yield of 6.3 t ha⁻¹ 121 days after sowing. The canopy structure (Fig. 6.10) showed a reduced proportion of leaf in the

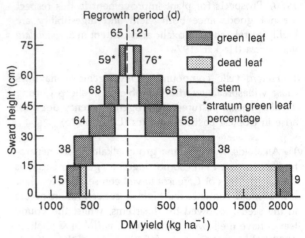

Fig. 6.10. Vertical distribution of dry matter of plant organs of *Lablab purpureus* 65 and 121 days after planting. (From Hendricksen & Minson, 1985.)

Table 6.4. *Changes in sward characteristics of* Chloris gayana *with age of regrowth*

Characteristic and vertical layer (cm)	Age of regrowth (weeks)			
	2	4	6	8
Shoot yield (kg ha⁻¹)				
> 60			1010	1810
45–60		50	740	1080
30–45		60	1190	1420
15–30		510	1640	1890
0–15	220	1420	2050	2230
Total	220	2040	6630	8430
Sward leaf bulk density (kg ha⁻¹ cm⁻¹)				
> 60			5	6
45–60			25	33
30–45		1	46	50
15–30		29	69	72
0–15	12	77	91	78
Total	12	27	43	42
Sward bulk density (kg ha⁻¹ cm⁻¹)				
> 60			42	69
45–60		3	49	72
30–45		4	79	95
15–30		34	109	126
0–15	14	95	136	
Total	14	34	79	98

Source: Stobbs (1973b).

lowest strata at 121 days relative to that at 64 days, and at least half of the leaf was in the 30–60 cm layer; the total bulk density in that layer was 35 and 51 kg ha^{-1} cm^{-1} at 65 and 121 days respectively. The sprawling *D. intortum* and *Desmodium sandwicense* averaged 32 and 53 kg ha^{-1} cm^{-1} for leaf and total bulk density respectively (Stobbs & Imrie, 1976). At Narayen, Queensland (lat. 26 °S, 710 mm), *C. ciliaris* pastures averaged 8.8, 11.7 and 30.5 kg ha^{-1} cm^{-1} for green leaf, green stem and dead material respectively, giving a total bulk density of 53 kg ha^{-1} cm^{-1} (Vieira, 1985). Bulk density was highest in the autumn and decreased at low stocking rate.

In Fiji pastures of *Pennisetum polystachyon/M. atropurpureum* under lax grazing showed leaf bulk density varying from 1 to 110 kg ha^{-1} cm^{-1} (Partridge, 1979b). On the other hand heavily grazed *D. heterophyllum* pastures exhibited leaf and total bulk density of 225 and 268 kg ha^{-1} cm^{-1} respectively, indicating that values within the range found for temperate pastures can occur in the tropics.

(ii) The nutritive value of forage decreases with depth in the canopy; the upper layers from which the animal selects the major part of its food is also of higher digestibility. Near La Habana, Cuba (lat. 23 °N) digestibility of *C. dactylon* (Fig. 6.11) varied by as much as 22 units between the uppermost and lowest stratum, and this effect of position applied to both leaf and stem material (Herrera *et al.*, 1986). Digestibility was also substantially higher in the cool dry season than in the rainy season. These differences in digestibility were negatively associated with cell wall content and lignin concentration; hemicellulose concentration was also slightly higher in the lower strata. In a similar study Ruiz *et al.* (1981) found decreased N and leaf % in the lowest stratum of *C. dactylon* swards.

The significance of these findings for selective grazing

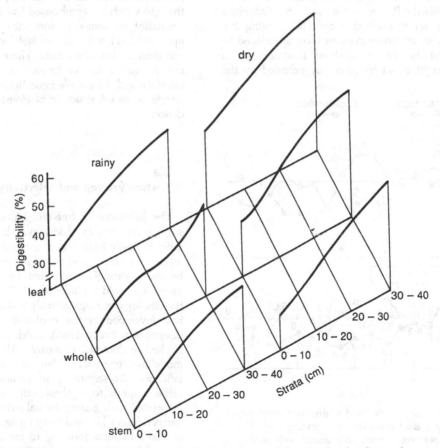

Fig. 6.11. Vertical distribution of IVDMD of leaf, stem and whole shoots of *Cynodon dactylon* during the rainy and dry seasons in Cuba. (From Herrera *et al.*, 1986.)

is clear, and will be further developed shortly. The significance of leaf and sward density for animal production is more controversial. Cows are loth to take more than 36 000 bites d^{-1} (Stobbs, 1973*b*); a cow of 400 kg LW ingesting 3% of body weight requires 10.8 kg OM d^{-1}, which is only achieved if bite size averages at least 0.3 g OM. Lower figures for bite size occur in field practice, suggesting that bite size constrains performance. A long term study of steers grazing *S. sphacelata* var. *sericea* and *B. decumbens* at differing stocking rates found a positive correlation between bite size and liveweight gain (Chacon, Stobbs & Dale, 1978); bite size was also correlated with yield of leaf and leaf bulk density. Many pasture attributes are well correlated with each other, so that it is difficult to establish causative relations. At the same herbage yield, steer growth was influenced by leaf %, leaf bulk density, leaf yield, IVOMD, and % N.

A study at Atherton, Queensland (lat. °S, 1300 mm rainfall), which attempted to manipulate sward structure to increase leafiness and to reduce sward height, is instructive (Davison, Cowan & O'Rourke, 1981). Nitrogen-fertilised *P. maximum* and *B. decumbens* pastures were set stocked at 5 cows ha^{-1} during the summer months, or two variations were introduced to control sward height: a slashing treatment at a decreasing height of 41 to 10 cm, as indicated by the

rate of pasture growth, or a variable SR in which additional cows were added whenever it appeared that growth was exceeding consumption and the pastures were becoming stemmy. Both these treatments were successful in increasing leaf % of the pasture on offer; the average increase due to slashing or variable stocking was 1–6%. Nitrogen % and IVDMD of diet were also increased by slashing or by variable SR.

The effects on milk yield are shown in Fig. 6.12. Milk yield decreased each time the pastures were slashed or extra cows were added. Some recovery then ensued as the pasture regrew, but milk production per cow in the set stocked treatment was superior overall to the slashed or variable SR treatments, which were especially disastrous for milk production per cow on the taller growing *P. maximum* cv. Gatton pastures.

This result is explicable in terms of treatment effects on leaf yield (as distinct from leaf %); slashing or variable SR reduced the yield of leaf on offer. This result is in sympathy with later discussion of the positive effect of forage allowance (Section 7.1.2) on milk production per head, but does not provide support for the school of thought which has emphasised leaf density as the major constraint on animal performance. Much will depend upon the level of forage available and whether extreme variation in structure occurs. There are few continuing critical studies in the tropics of the pasture–animal interface, and the complex questions associated with the effects of sward structure obviously need more elucidation.

6.3
Pasture grazing and selectivity

(i)   The behaviour of free-range grazing animals which survive on pastures of low availability at low SR may differ radically from that of more domesticated animals whose movements and access to particular pastures may be more strongly controlled by management intervention. Arnold & Dudzinski (1978) have written about the ethology of range animals, and considerable interest has developed in the evolution of foraging theory (Stephens & Krebs, 1986), which has only in part been applied to domestic ruminants. The principal theme is modelling optimality; 'an animal's energy budget influences its sensitivity to variance in reward', and cattle appear to optimise the energy expenditure associated with grazing by balancing added feed intake derived from additional foraging against the energy cost of that additional foraging to maximise the long term rate of net energy intake (Squires, 1982). Foraging theory is also directed to the 'prediction of environ-

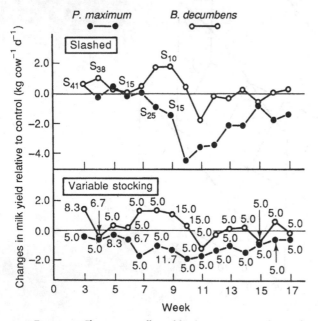

Fig. 6.12. Change in milk yield relative to control set of continuously grazed cows for cows grazing slashed pastures or pastures continuously grazed with varying stocking rate (Height of slashing (cm) and SR (cows ha^{-1}) are shown). (From Davison *et al.*, 1981.)

mental conditions under which a foraging animal should or should not show exploratory or sampling behaviour' (Stephens & Krebs, 1986).

Forage intake may be regarded as the product of (1) duration of grazing, (2) rate of biting, and (3) bite size.

(ii) Duration of grazing is related to forage availability. Grazing time is less at high levels of forage availability, especially if readily detachable leaf is abundant, than under conditions of medium forage availability, where the animal appears to compensate for the lesser forage supply by spending a greater time foraging, in order to fulfil appetite. However, physical factors, such as rumen distension, and chemostatic factors operate on appetite. Grazing time is reduced as forage availability further decreases, and usually in these circumstances the animal is obtaining less material with each bite; there are definite limits to the compensation which is available from the extension of grazing time.

This is illustrated for the behaviour of Jersey cows at Samford, Queensland, grazing a crop of *L. purpureus* cv. Rongai. Animals were selecting leaf, and the best relationship between grazing time and forage availability was with available green leaf (Fig. 6.13; Hendricksen & Minson, 1980). This was a quadratic relationship in which

$$Y = 265.1 + 0.67X - 0.0003X^2$$

where X is green leaf yield (kg ha^{-1}) and Y is grazing time (min d^{-1}). Maximum grazing time recorded was 685 min d^{-1}. Cowan, Davison & Shephard (1986) also found a quadratic relationship between grazing time of Friesian cows and availability of green leaf; grazing time was reduced at low levels of forage availability. The

response of grazing time to forage availability is dependent upon local circumstances and the range of forage availability tested; Vieira (1985) observed greater duration of grazing at high SR on *C. ciliaris* pastures at Narayen, Queensland.

The average time spent grazing appears to be generally greater on tropical than on temperate pastures. Some estimates for cattle (min d^{-1}) are: 561, 677 and 719 according to pasture type for Jersey cows (Stobbs, 1974a); 593 and 646 in spring and autumn for Jersey cows (Chacon & Stobbs, 1976); 600 for Friesian cows (Cowan 1975); 619 for Hereford steers (Chacon et al., 1978); 652 for Simmental × Hereford steers (Ebersohn, Evans & Limpus, 1983); and 681 for steers at Narayen (Vieira, 1985). In the humid lowlands of Nigeria grazing duration of Holstein-Friesian cows had a low value of 474 min d^{-1} (Breinholt, Gowen & Nwosu, 1981).

Duration of grazing is influenced by climatic conditions. The fear that European breeds of dairy cattle were not adapted to hot grazing conditions in the tropics has had disastrous consequences for the efficient development of dairying. The housing of dairy cattle in the tropics and the adoption of cut-and-carry feeding systems have raised the incidence of disease, caused a transfer of nutrients to the shed, with attendant pollution and additional fertiliser cost, reduced the opportunity of the animal to improve dietary intake by selective grazing, and greatly increased labour costs. Cut-and-remove forage systems are convenient for many farm situations, as discussed in Chapter 2, but are the least-preferred option if grazing is feasible. European breeds of cattle suffer heat stress, but the infusion of some zebu or other topical blood lines, and the provision of shade and of adequate length of night grazing mitigate this problem.

This is illustrated by a study at Muaklek, Thailand (16 °N, 220 m a.s.l.) in which 63–75 % Holstein-Friesian or Red Dane cows on *P. maximum/S. hamata* pastures either grazed for 24 hours or were fed with a cut-and-remove system plus night grazing (Hongyantarachai et al., 1989). The pastures were grazed at c. 25 day intervals from a maximum of 50 cm height down to c. 15 cm. The experiment was conducted for a period which included the hottest time of the year (April to May), when daily maximum temperature averaged 34 °C. Milk production was independent of treatment at 14.1–14.2 kg cow^{-1}d^{-1}, conception rate was unaffected, % butter fat was higher in the 24 hour grazing treatment, liveweight gain slightly greater in the cut-and-remove system, but the incidence of mastitis was higher in the latter.

Ruminants adapt to high day temperatures by increasing the proportion of grazing conducted at night, often with no reduction in total grazing time. There are

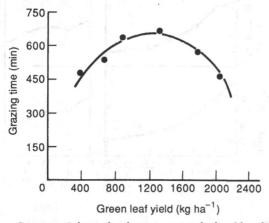

Fig. 6.13. Relationship between green leaf yield and time cows spent grazing *Lablab purpureus*. (From Hendricksen & Minson, 1980.)

reports of seasonal differences in grazing duration (Vieira, 1985), but these differences are confounded with differences in forage availability, composition and structure. At Atherton, Queensland, the proportion of the total grazing time spent grazing the pasture between the morning milking (0530 to 0630 h) and the afternoon milking (1430 to 1530 h) was negatively related to daily maximum temperature by a quadratic relationship (Fig. 6.14; Cowan, 1975).

Some animals are corralled at night to reduce theft and predation, but this is not without cost to production. In Nigeria Bayer & Otchere (1985) regarded the 5 hours per day grazing allocation as a serious constraint to animal performance. At Borabu, Thailand, steers with 24 hours of grazing access gained 74 kg hd^{-1} in 5 months compared with 39 kg hd^{-1} for animals in yards at night (McLeod, 1974). Near La Habana, Cuba, liveweight gain of Holstein-Friesian bulls varied from 0.56 to 0.88 kg hd^{-1} d^{-1} according to the timing and duration of access to pasture (Martin & Ruiz, 1986). These examples indicate the desirability of giving animals 24 hours of daily access to pasture wherever possible.

(iii) Rate of biting is also conditioned by forage availability and quality. There is a compensatory effect of increased rate of biting on mature swards or on swards containing more dead material (Stobbs, 1974a). The rate of biting decreases with time from the commencement of grazing, but this decrease is less on mature swards. Bites which prehend pasture need to be distinguished from masticating and ruminating jaw movements; prehending bites may be distinguished by the head being in a downward position (except when browsing) (Chacon, Stobbs & Sandland, 1976). Cattle also make manipulative jaw movements when selectively grazing in order to prehend the desired material. Thus bites per 100 jaw movements varied from 28 to 59 according to the degree of selection when cattle were selecting *A. americana* from the upper layers of *H. altissima* pastures; the rate of prehending bites decreased but bite size did not change (Moore *et al.*, 1985).

The grazing behaviour of cattle in response to changing diet is illustrated for *S. sphacelata* var. *sericea* pastures to which Jersey cows were introduced for 14

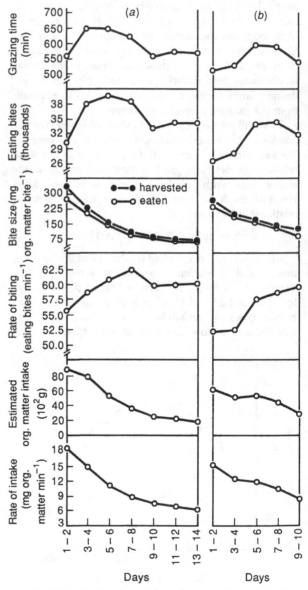

Fig. 6.15. Grazing time, biting characteristics and intake of cows grazing *Setaria sphacelata* var. *sericea* in (a) autumn and (b) spring. (From Chacon & Stobbs, 1976.)

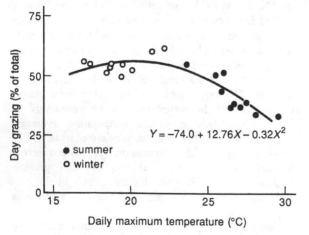

Fig. 6.14. Relationship between day grazing and daily maximum temperature. (From Cowan, 1975.)

day periods in autumn and for 10 day periods in spring (Fig. 6.15; Chacon & Stobbs, 1976). Forage allowance, leaf yield and leaf % decreased over these periods. Duration of grazing increased and then decreased as forage availability was further reduced; rate of biting increased to a plateau by days 7–8 in autumn and continued to increase to days 9–10 in the spring. However, bite size decreased throughout the grazing period, and this resulted in diminished OM intake over the grazing period.

(iv) The above example shows bite size to be below the suggested critical 0.3 g OM. Bite size is obviously controlled by the ease of detachment of material from stems and by the density of the material prehended, and is varied by the method of grazing of different animal species, as discussed in Chapter 2. The depth and density of leaf material in the upper layers of sward are modifying factors for bite size, and in many pasture situations these need to be related to the leaf or green pasture yield, as discussed in Section 7.1.2.

(v) Selective grazing as a factor influencing botanical composition was considered in Section 5.3.2. A perennial theme in this book is the necessity for maximum opportunity for selective grazing to operate in order that the quality of diet may be at a level giving acceptable animal performance. The avoidance of

selective grazing through forced high levels of utilisation decreases animal output per head, and is a main factor in the failure of the short-duration grazing schemes which are discounted in Chapter 10.

Animals selectively consume leaf, as discussed earlier in the chapter. In Florida *A. americana* was selectively eaten until the seeding stage, when cattle rejected the legume in preference to the grass *H. altissima* (Moore *et al.*, 1989). The superior diet selectively consumed relative to the forage on offer was illustrated in Fig. 6.7 for N%. The performance of leader and follower cows provides a good example of the effects of selective grazing; this is a system in which animals which the manager desires to favour are given first access to a pasture being rotationally grazed and less fortunate animals graze the residual pasture.

This example (Stobbs, 1978) used Jersey cows in milk for both groups; in farm practice a more common system would be one where leader cows in milk were followed by dry cows and young growing stock. *P. maximum* and *C. gayana* regrowths about three weeks old were grazed by leaders, and SR was adjusted to give followers a similar total forage allowance, although the leaf availability had been reduced by the leaders (Table 6.5). Digestibility and N% of herbage on offer was reduced, especially via the higher content of stem in the pasture offered to followers. This led to a reduced digestibility of the diet ingested by about three units,

Table 6.5. *Pasture and production attributes in leader–follower grazing systems on* Panicum maximum *and* Chloris gayana *pastures*

Attribute	Panicum maximum		Chloris gayana	
	Leader	Follower	Leader	Follower
Herbage allowance (kg DM cow^{-1} d^{-1})				
Leaf	24.8	20.3	23.7	18.3
Total	39.8	38.3	40.9	38.2
Herbage digestibility (%)				
Leaf OM	60.7	56.3	56.9	54.8
Stem OM	52.8	51.0	51.6	50.7
Herbage N (%)				
Leaf	2.7	2.1	2.6	2.4
Stem	1.6	1.3	1.9	1.0
Dietary intake				
DM digestibility (%)	58.6	55.2	56.1	53.1
N %	2.4	2.2	2.6	2.4
Grazing time (min d^{-1})	595	620	602	651
Bite size (mg OM $bite^{-1}$)	403	187	283	150
Milk yield (kg cow^{-1} d^{-1})	8.7	6.3	7.2	5.2

Source: Stobbs (1978).

and a slightly reduced nitrogen concentration. Followers increased their grazing time per day in compensation, but bite size was greatly reduced. The net effect on milk production was substantial, and leaders averaged 8.0 kg cow^{-1} d^{-1} relative to 5.8 kg cow^{-1} d^{-1} for followers. Milk from leader cows also contained higher solids-not-fat and protein %, but a lower butter fat %.

'Creep' grazing in which young, small animals are given first access to pasture or supplements is a further option.

Selective grazing poses some difficulty for the pasture manager in estimating animal requirements, since the diet ingested differs from the forage on offer.

The concluding emphasis in this chapter is on the relatively intransigent constraints on animal production which are currently imposed by the structural character-istics of the C$_4$ grasses. The high ecological success of these plants in tropical conditions, which reflects their efficient use of water and nutrients in maintaining photosynthesis and their persistence under adverse con-ditions, is achieved through mechanisms which lower or erode their nutritive value. Management therefore needs to be more exigent than in the case of temperate pastures if target levels of individual animal production are to be achieved. The management options usually revolve about (1) the provision of leaf, (2) the avoidance of aged forage, (3) the maximum opportunity for selectivity, (4) the absence of limitations to the duration of and free movement in grazing, and (5) the incorporation of legumes to improve the nutritive value of the diet, both directly and through the effects of the legume on grass nitrogen concentration. These options also require due attention to the con-centrations of essential minerals in the diet.

7

Stocking rate and animal production

The choice of stocking rate (SR) determines the level of animal performance, the sustainability of pasture production, and the profitability of the farm enterprise. It is the key management decision whose importance overrides that of other decisions, and schemes of pasture management fail if they are not based on the manager's appreciation of the relationship between the number of stock carried on the property and their individual performance.

Animals compete for the available supply of forage. At low SR animals have ample opportunity for selection, and their effect on each other's performance is minimal. As SR increases, the opportunity for selectivity decreases, and the reduced nutritive value of the diet ingested decreases individual animal performance. As SR increases further the availability of forage of maintenance quality to the individual animal may so decrease that the body condition

of the animal falls and a higher incidence of mortality results.

The rates of pasture growth and senescence (Chapter 4) and the botanical composition of pastures (Chapter 5) are sensitive to SR, and the degradation of the environment, as exemplified by high rates of runoff and soil erosion (Chapter 3), result from high SR. The objective is to synchronise animal demand with the level and type of forage supply which will result in the attainment of defined animal production targets. Managers who seek to raise production may increase the forage supply by the removal of competing trees, the replacement of natural grasses by improved, selected grasses, the planting of legumes, the rectification of soil mineral deficiencies and the application of irrigation. The cost of these inputs is only recovered if the additional forage grown is converted to animal product, through the adjustment of SR. The forage resource is destroyed and the environment damaged if the SR is set too high and above the 'crash point' for the particular pasture situation.

7.1
The shape of the animal response to pasture availability

7.1.1
Animal production per head and per unit area

(i) The base requirement is for the manager to have an estimate of the animal production from a pasture at a particular SR and the change in production which an alteration of SR, plus or minus, will cause. Fortunately the simplest mathematical function (the equation of a straight line) provides the form of that relationship for year-round grazing, and this linear function has proved remarkably robust over a wide range of tropical

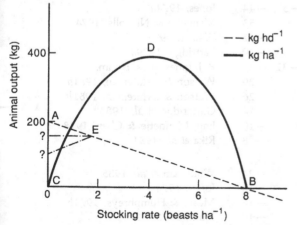

Fig. 7.1. Diagrammatic relationship between stocking rate (beasts ha^{-1}) and animal output per head and per unit area.

environments when tested for beef (Jones & Sandland, 1974) and, with limitations, for milk production (Cowan & Stobbs, 1976). Some qualifications are expressed later, but there has now been developed a great body of knowledge from SR experiments which are used in extrapolating to specific farm conditions.

The relationship between output per head (Y) and SR, expressed as animals per unit area (X), is given by the equation

$$Y = a - bX$$

where a represents the intercept on the y-axis and b is the slope of the line AB, or the regression coefficient giving change in Y for unit change in X (Fig. 7.1). The value of a may be regarded simplistically as reflecting the quality of the pasture, which sets a limit to individual animal performance when quantity of pasture is not limiting, as modified by the genetic capacity of animals to express that potential. In tropical conditions the value of a varies between 50 and 250 kg hd^{-1} yr^{-1} for LWG. The value of Y becomes negative at very high SR where losses exceed gains.

The manager is concerned with individual animal performance, but the gross margin of his farm output depends upon the sum of individual animal outputs, expressed as output per unit area. This is calculated as the product of average animal performance (Y) by the number of animals (X), which leads to a quadratic function (line CDB on Fig. 7.1), and

$$YX = aX - bX^2$$

where YX is the output per unit area. At low SR where individual animal performance is high the output per unit area is low, and the latter increases to a maximum at D. The SR giving maximum output per unit area occurs where SR = $a/2b$, and where Y has half its maximum value. This introduces the fundamental question of what level of individual animal performance the pasture manager is prepared to sacrifice in order to optimise output per unit area; considerations of risk, market premiums for quality, animal costs and environmental protection are all involved. As the SR at D is exceeded both animal and pasture productivity decrease sharply.

Fig. 7.1 has been given notional values of $a = 200$ and $b = 25$ for kg LWG yr^{-1}, and SR for yearling cattle of 0–8 beasts ha^{-1}. From these figures the SR giving maximum LWG ha^{-1} is

$$\frac{200}{2 \times 25} = 4 \text{ beasts ha}^{-1}$$

Table 7.1. *Selected regression coefficient b-values[a] for different pastures and sites*

Site and pasture type	Latitude	Annual rainfall (mm)	b-value[a]	Source
Native pasture				
Matopos, Zimbabwe	20° S	610	−180, −105	Carew, 1976
Grass/legume				
Samford, Australia	27° S	1100	−57, −44	Jones, 1974a
Narayen, Australia	26° S	710	−53	Mannetje & Nicholls, 1974
Mackay, Australia	21° S	1500	−42	Walker, 1977
Sigatoka, Fiji	18° S	1730	−32, −30	Partridge, 1979a
Lansdown, Australia	19° S	870	−32, −30	R. J. Jones, pers. comm.
Guadalcanal, Solomon Is.	10° S	2160	−29	Watson & Whiteman, 1981a
Lingatu, Solomon Is.	9° S	2900	−26	Watson & Whiteman, 1981b
Khon Kaen, Thailand	16° N	1260	−14	Gutteridge et al., 1983
Kluang, Malaysia	2° N	1760	−10	Eng, Mannetje & Chen, 1978
Sangiang, Bali	8° S	1710	−8	Rika et al., 1981
Grass + N				
Mt Cotton, Australia	28° S	1430	−24	Whiteman et al., 1985
Samford, Australia	27° S	1100	−16	Jones, 1974
Lismore, Australia	29° S	1660	−18	Mears & Humphreys, 1974b
Kunnunurra, Australia	16° S	irrigated	−10	Blunt, 1978

[a] (Kg liveweight gain head^{-1}yr^{-1}) decrease for unit increase in stocking rate (beasts ha^{-1}).

at which point $Y = 100$ kg hd^{-1} yr^{-1} and maximum output is 400 kg ha^{-1} yr^{-1}. It is mathematically preferable to express SR as beasts ha^{-1} and not as ha beast^{-1}.

(ii) The regression coefficient, b, in the above equation expresses the resilience of the pasture to change in SR; a low value of b is desirable, provided a is sufficiently great, since it indicates that SR can be varied without great detriment to individual animal performance. In these circumstances the manager has more flexibility and a greater margin of safety. Some selected b-values (Table 7.1) from grazing experiments indicate the great range in pasture LWG performance of cattle. This table is weakened by the absence of uniformity in the size of animals used in the experiments, and most of the experiments are based on only three stocking rates. It is dangerous to extrapolate beyond the SRs tested, since the rigid extrapolation of the quadratic formula may at times suggest a SR giving maximum gain per ha beyond the actual crash point of the pasture, especially at lower values of b. High values of b are expected in low rainfall areas, and most of the experiments used twining legumes which are less tolerant of high SR than other legumes discussed later in the chapter. The grass pastures fertilised with nitrogen have lower b-values (range 10–24) than the grass/legume pastures (range 8–57).

(iii) There are a few dairy experiments which take into account the whole year performance of the lactating cows, and experiments have not been sighted amongst these in which a sufficiently high SR has been included which reduces milk yield per ha below that attained at lower SR. The experiments reported are therefore within the line CD of Fig. 7.1 and do not extend to the line DB. The linearity of the SR/milk yield per cow response has still proved a useful approximation in field practice over a middle range of SR, with some departures from linearity, which will be discussed.

At Atherton, Queensland (lat. 17 °S, 1300 mm rainfall, 700 m altitude), annual milk production (kg cow^{-1}, Y) of Friesian cows grazing *P. maximum* var. *trichoglume/N. wightii* pastures was significantly related to SR (X) of 1.3, 1.6, 1.9 and 2.5 cows ha^{-1} by the following equation (Cowan & Stobbs, 1976):

$$Y = 4164 - 661X$$

In earlier work (Cowan, Byford & Stobbs, 1975) milk yield per cow decreased from 3810 kg at 1.3 beasts to 3290 kg at 2.5 beasts ha^{-1}, but the response to SR was flatter at high SR than from 1.3 to 1.6 beasts ha^{-1}. In the two years of the experiment the decrease in milk yield as SR increased became greater as the season progressed

from mid-summer into winter. Thus in year 1 of the experiment (Fig. 7.2) b-values expressed as kg milk cow^{-1} d^{-1} were about -1 in December–February, and increased to -5 in July, whilst in year 2 a similar seasonal difference in b-value applied. This was associated with decreased quality and quantity of pasture on offer.

In a later study at Atherton, Queensland, on *P. maximum* pastures fertilised with 200 or 400 kg N, Davison, Cowan & Shepherd (1985) found a significant relationship for the pooled data, as follows:

$$Y = 3476 - 276X$$

Their results (Fig. 7.3) suggest a non-significant flattening of response to increasing SR at 200 kg N, and a lack of response to reduced SR at 400 kg N at the highest SR, where feed was abundant. Their data are shown with results from *P. clandestinum* pastures at Lismore, New South Wales; milk production per cow in the study by Colman & Holder (1968) was similar at the two lightest SRs, whilst a slight flattening of response between the two highest SRs is suggested in the Colman & Kaiser (1974) data. Davison *et al.* (1985)

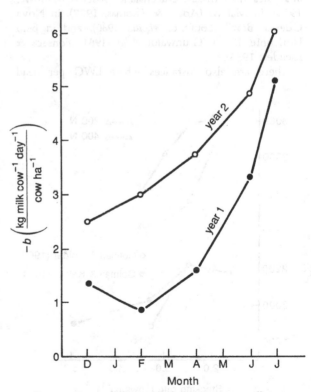

Fig. 7.2. Seasonal change in the regression coefficient relating milk production per cow to stocking rate. (From Cowan *et al.*, 1975.)

speculate that over a very wide range of SRs the yield per cow may show a sigmoid form, in which at low SR limitations of pasture quality lead to a plateau and at high SR the capacity of animals to use body reserves for lactation cushions the effect of feed shortage.

An annual fixed SR leads to considerable wastage of herbage, and better synchrony of forage availability and animal requirements becomes feasible if other sources of feed are introduced into the system, as described in Chapter 8. For example, at Coronel Pacheco in Brazil, the optimum use of pasture was attained with SR of 1.2 cows ha^{-1} during the October–March growing season and SR of 0.2 cows ha^{-1} from April to September; the alternative year-round fixed SR was only 0.4 cows ha^{-1} (Goncalves de Assis, 1984).

(iv) Liveweight gain (LWG) per head in some beef production experiments is independent of SR at low SRs where the feed supply does not limit individual animal performance, and this might be anticipated from the relationship of instantaneous forage allowance to LWG, as discussed later. At lighter SR than point E in Fig. 7.1 a plateau of LWG per head may be observed, and it is expected that understocking accounted for this response in studies in south-east Queensland (Bisset & Marlowe, 1974), in Malawi (Addy & Thomas, 1977), in Nova Odessa, Brazil (Lourenco *et al.*, 1980), and in Belo Horizonte, Brazil (Grunwaldt *et al.*, 1981; Fonseca & Escuder, 1983).

There are also instances where LWG per head

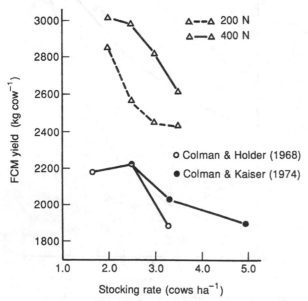

Fig. 7.3. Effects of SR on fat-corrected milk (FCM) yield per cow. (From Davison, Cowan & Shepherd, 1985.)

decreases at lighter SR than point E in Fig. 7.1. The botanical composition of the pasture may change with SR (Chapter 5), and in these circumstances the comparison of different SRs is in effect made upon different pastures. In Section 3.2.2 the greater incidence of the disease *Rhizoctonia solani* in *M. atropurpureum* at the lightest SR tested was mentioned, and this led to reduced legume content and was associated with decreased LWG per head (Walker, 1980). Similarly, reduced LWG per head occurred at a light SR on irrigated *P. dilatatum*/*T. repens* pastures on a tropical tableland, and this was linked to the shading and loss of the legume component in the under-utilised pasture. A further variation occurs if a more productive species appears at high SR. In Ankole, Uganda, better LWG per head occurred than expected at high SR and Harrington & Pratchett (1974) attributed this to an invasion by *B. decumbens* and the displacement of the less acceptable *T. triandra* and *Hyparrhenia filipendula*.

Another possibility is that the undefoliated pasture may develop a higher content of structural components and lesser accessibility of green leaf material. At Narayen, Queensland (lat. 26 °S, 710 mm rainfall), LWG per head was positively associated with pasture availability in spring and autumn, but in the summer growing season there was a highly significant negative relationship between LWG per head per day and availability of both dry matter (DM) and green dry matter (GDM) (Vieira, 1985). At Heathlands, Queensland (lat. 11 °S, 1710 mm rainfall), there is a strongly wet/dry monsoonal climate. *B. decumbens* or *P. maximum* were grown with *S. guianensis* and *M. atropurpureum* and grazed by cattle over a wide range of SR which did not materially alter botanical composition. In the summer the grass dominant pastures grew to 3 m tall in the ungrazed patches. Annual LWG per head was linearly and negatively related to SR in the first year (Winter, Edye & Williams, 1977). Subsequently opposing effects in summer and winter minimised the SR effect. During the wet season LWG per head increased with increasing SR to 1.4 beasts ha^{-1}and with increasing total GDM to 600 kg ha^{-1} to give a maximum LWG of 0.82 kg hd^{-1}d^{-1} (Fig. 7.4*a*; Edye, Williams & Winter, 1978). However, at the same stocking rate LWG per head decreased with increasing GDM yield; LWG per head also decreased at very high (2.3 beasts ha^{-1}) and very low SR (0.5 beasts ha^{-1}).

During the dry season LWG decreased linearly with decreasing green matter availability and with increasing SR (Fig. 7.4*b*); cattle LWG was positive at 0.8 beasts ha^{-1} if green matter was greater than 800 kg ha^{-1}.

The summer negative response of decreasing SR on LWG per head may in part be due to the greater compensatory gains occurring in animals which did poorly during the dry season (Section 9.2.2). It appears also to be associated with the amount of rank, ungrazed pasture at low SR. Pastures were relatively uniformly grazed at high SR, but at lower SR mosaics of closely grazed and of ungrazed tall pasture developed. From aerial photographs it was estimated that in May of one year the proportion of grazed pasture was only 29% at 0.7 beasts ha^{-1}, and 45% at 1.2 beasts ha^{-1}; the nominal, imposed SR was quite different from the 'effective' SR. This study shows how the SR/LWG per head relationship in fast growing C$_4$-dominant pastures which show a rapid decline in nutritive value with age can be changed from the conventional linear function previously described.

(v) There have been attempts to develop generalised mathematical functions to describe SR/animal pro-

duction relationships, and these result in more elaborate formulae than the linear function emphasised to this point. Harlan (1958) proposed a double exponential equation in which stocking rate was expressed as a relative pressure. Mott (1961) collated a number of SR experiments and suggested a function $Y = k - ab^x$, where k, a and b are constants. He suggested values for the constants associated with an 'optimum grazing pressure', and in which production became zero when SR/SR at optimum grazing pressure was 1.5. At low SR product per animal was maximal at 1.12 × product per animal at optimum grazing pressure. Mott's paper influenced many scientists to think seriously about the implications of varying SR on output at both the individual animal and the unit area levels. Petersen, Lucas & Mott (1965) developed a further model in which production per animal showed a plateau when feed was plentiful, and a concave function when it

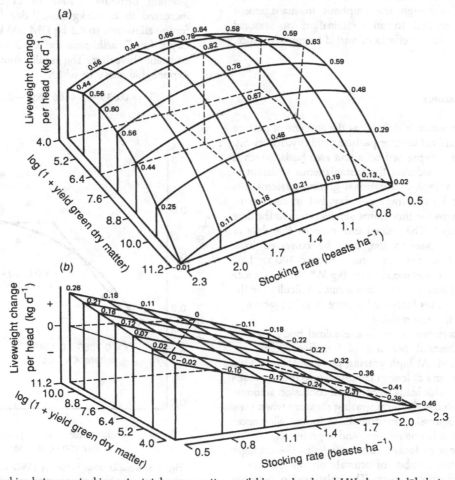

Fig. 7.4. Relationships between stocking rate, total green matter available per head, and LW change hd^{-1} during (a) wet and (b) dry seasons (From Edye *et al.*, 1978.)

became scarce; this led to a sharp peak if output were expressed on a unit area basis. This paper assumed that the amount and type of forage were independent of SR and that the type of forage consumed was the same as that available. Since both these assumptions are incorrect in farm practice the utility of the model was low; Owen & Ridgman's (1968) function had similar weaknesses and Coniffe, Browne & Walshe (1970) fitted a quadratic function to the SR/output per head relationship.

A linear model had also been in use (Riewe, 1961; Riewe *et al.*, 1963). Jones & Sandland (1974) surveyed many grazing trials to emerge with the linear model. Their mathematical technique has been criticised (Connolly, 1976; Walker, 1980); if the tested SRs are low, the SR giving maximum LWG per ha tends to be overestimated and the production per head at this point to be underestimated. However, the Jones and Sandland (1974) conclusions are valid for the middle range of SRs applicable in farm practice. None of the other models has withstood the test of time in the breadth of their applicability, although the emphasis in management studies has moved to an examination of seasonal relationships and the effects of varied forage allowance.

7.1.2
Forage allowance

(i)   Forage allowance is defined as the amount of pasture on offer per animal unit, or pasture yield divided by SR. It is sometimes expressed on a unit area basis, which is useful when considering the instantaneous situation in front of the animal. In rotational grazing systems it is expressed as kg of forage per day, but in continuous grazing systems the time frame simply refers to the date of measurement. The animal unit may be the class of animal under consideration, may be corrected to a standard AU, or may be expressed per 100 kg LW. Conversion to metabolic size (kg $W^{0.75}$) is a more sophisticated measure but adds a small difficulty for the manager who uses forage allowance in field practice in day-to-day decision making.

Grazing pressure refers to the animal body mass or animal requirement per unit of forage available, e.g. kg LW/kg DM. At high grazing pressure there is little feed on offer, and at low grazing pressure there is much feed on offer. This leads to continual confusion amongst scientists who use the term grazing pressure when they are in fact expressing treatments as forage allowance; the latter term is unequivocal and is preferred.

The concept of forage allowance helps the manager manipulate the number of animals on the pasture to meet a production target. The relationship between forage allowance and animal production needs to be

understood if this end is to be achieved. The problem centres firstly about what is to be measured as forage. Is yield measured from ground level or from above some notional grazing height? Does it include all plant material or only accessible green material?

(ii)   Mannetje & Ebersohn (1980) identify two contrasting situations:

(1) 'Herbage consumed is a constant proportion of herbage present or herbage allowance.' This applies under high SR in humid regions, under irrigation, or for short periods of the main growing season when most forage is consumed as green material. LWG, milk secretion, wool production and pasture intake per head under these conditions are related asymptotically to herbage allowance on a DM basis. For example, at Ona, Florida (lat. 28° N) nitrogen-fertilised grass pastures were grazed (14 days on, 28 days off) during the main growing season at SR varying from 7.5 to 15 yearling cattle ha^{-1}. Rate of LWG (Y) (Fig. 7.5) increased to *c.* 0.6 kg head^{-1} day^{-1} with increasing forage allowance to 6.3 kg DM 100 kg^{-1} body weight day^{-1} for *C. aethiopicus* cv. UF-5 and was relatively constant thereafter. The results could be fitted to an exponential curve in which

$$Y = 0.639 - 1.4466 \exp(-0.3445X)$$

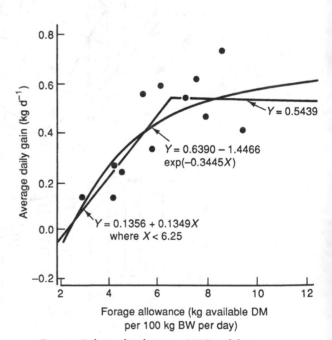

Fig. 7.5. Relationship between LWG and forage allowance of *Cynodon aethiopicus* cv. UF-5 at Ona, Florida. (From Adjei, Mislevy & Ward, 1980.)

where X is forage allowance. Alternatively, a 2-stage regression function could be fitted with the linear equation $Y = 0.1365 + 0.3449X$, accounting for the segment of the data up to 0.625 kg hd^{-1} d^{-1}, and little response occurred thereafter. For the same experiment *C. aethiopicus* cv. McCaleb and *C. nlemfuensis* cv. UF-4 reached maximum LWG of 0.6 and 0.5 kg hd^{-1} d^{-1} at forage allowances of 7.7 and 6.7 kg DM 100 kg^{-1} body weight d^{-1} respectively. This suggested that for these pastures maximum LWG ha^{-1} would be attained with forage allowances of 6–8 kg DM 100 kg^{-1} d^{-1}. At Kluang, Malaysia (lat. 2° N, 1760 mm rainfall), on grass–legume pastures Eng, Mannetje & Chen (1978) found an exponential relationship for bulls of

LWG (g hd^{-1} d^{-1}) = 282.6 − 460.2 exp(0.0179FA)

where *FA* is forage allowance (kg DM 100 kg^{-1} LW × 100 per 28 days). The asymptote occurred at *c.* 14 kg DM 100 kg^{-1} LW d^{-1} for 0.28 kg LW hd^{-1} d^{-1}.
(2) 'Herbage consumed is not a fixed proportion of herbage present or herbage allowance.' This is the more common situation in most tropical situations, where a seasonally variable amount of dead material is present. In these circumstances animal output per head is better related to some measure of GDM (green dry matter or green leaf yield) than to DM. Watson & Whiteman (1981*a*) in the Solomon Islands grazed mixed pastures continuously at four stocking rates and recorded asymptotic relationships (Fig. 7.6) between

LWG (as kg hd^{-1} d^{-1}, Y) and GDM (kg hd^{-1}, X) as follows:

> *B. mutica*
> $Y = 0.516 − 0.527 \exp(−0.002X)$
> *B. decumbens*
> $Y = 0.465 − 0.379 \exp(−0.003X)$
> *P. maximum*
> $Y = 0.394 − 0.229 \exp(−0.001X)$

There are many other attempts in the literature to develop these relationships, and looking at the diversity of results it is clear that local equations need to be established. At Lingatu, Solomon Islands, Smith & Whiteman (1985*a*) also found an exponential relationship between LWG hd^{-1} and GDM hd^{-1}, but the asymptote occurred at a lower level of forage allowance than in Fig. 7.6. Mears & Humphreys (1974*b*) reported a quadratic relationship between LWG hd^{-1} and GDM hd^{-1} for *P. clandestinum* pastures at Lismore, New South Wales, for autumn–winter, but no clear relationships were evident in spring and summer.

Milk production is also related to forage allowance. At Atherton, Queensland, milk yield per cow on *P. maximum* pastures increased from 15 to 55 kg DM cow^{-1} d^{-1} on *P. maximum* var. *trichoglume*/*N. wightii* pastures. A subsequent study (Cowan, Davison & O'Grady, 1977) developed this relationship further for Friesian cows receiving different levels of supplementation (Fig. 7.7). Jersey Cows at Samford, Queensland (Stobbs, 1977*b*),

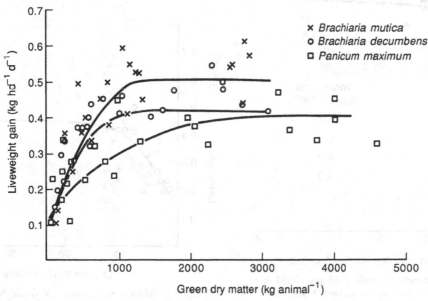

Fig. 7.6. Relationship between LWG and GDM available per animal for *Brachiaria mutica*, *B. decumbens*, and *Panicum maximum* pastures. (From Watson & Whiteman, 1981*a*.)

yielded 8.9–10.6 kg milk cow^{-1} d^{-1} as herbage allowance of *P. maximum* pastures increased from 15 to 55 kg DM cow^{-1} d^{-1}. Molar percentage of short-chain fatty acids in milk fat (C$_6$, C$_8$, C$_{10}$, C$_{14}$) increased and oleic acid (C$_{18:1}$) decreased as herbage allowance decreased, indicating a greater dependence of milk production upon cow body reserves. In Perico, Cuba, cows grazing *C. dactylon* cv. Coastcross produced 9.5, 10.4 and 11.5 kg cow^{-1} d^{-1} at forage allowances of 15, 32 and 50 kg DM cow^{-1} d^{-1} (Milera *et al.*, 1986).

(iii)   It is obvious from the scatter of these points that the relationships developed are crude approximations. Whiteman *et al.* (1985) found a better relationship between LWG per head and % green leaf than with yield of green leaf on offer, suggesting that accessibility of green leaf was a significant factor. Experience to the contrary has been described in Section 6.2. Some workers have found a relationship between animal production and legume composition of the sward as at Guadalcanal, Solomon Islands (Fig. 7.8), which is drawn from the same data as were used to produce Fig. 7.6. Levels of legume content below 10–12% would clearly modify the production relationship with GDM.

Guerrero *et al.* (1984) assessed the forage allowance according to the IVDMD of the forage on offer. A predictive production function for different *C. dactylon* cultivars could be developed by classifying pasture on offer as falling above 60%, 53–60% and less than 53% IVDMD; forage allowance for these respective classes needed to be 6.8, 8.3 or 8.9 kg 100 kg^{-1} LW to maintain 0.94 kg LWG hd^{-1} d^{-1}.

Green pasture on offer is not of static composition. A significant modifying factor is the rate of pasture growth and its relationship to consumption. Young green material of high nutritive value is constantly before the animal if growth exceeds consumption; if consumption exceeds growth the nutritive value of the ingested material must decrease because of the incorporation of more senescing material. Thus at Coolum, Queensland (lat. 27° S, 1500 mm rainfall), cattle LWG was poorly related to both DM and GDM forage allowance (Ebersohn & Moir, 1984) and in a later study (Ebersohn & Moir & Duncalfe, 1985) the positive relationship between LWG and rate of pasture growth was emphasised, since a high pasture growth rate was linked to a lower % non-green grass. Rate of LWG was also negatively correlated with the cell wall content of green grass leaf, indicating the variation in nutritive value of green leaf.

Some scientists when using rotational grazing have attempted to equalise the grazing intensity on different pasture treatments by varying stock density to achieve comparable yields of pasture residue when the animals were removed from each paddock. It should be

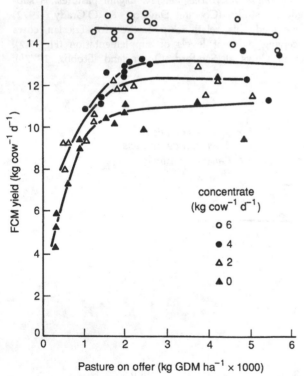

Fig. 7.7. Relationship between milk yield and GDM on offer of *Panicum maximum* var. *trichoglume*/*Neonotonia wightii* for different levels of concentrate feeding. (From Cowan *et al.*, 1977.)

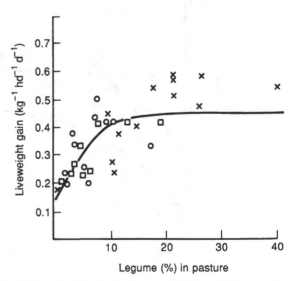

Fig. 7.8. Relationship between LWG and % legume in pasture (see Fig. 7.6). (From Watson & Whiteman, 1981*a*.)

recognised that the amount of pasture present at the beginning of grazing and the rate of pasture growth during the pasture grazing period also control the rate of leaf removal by the animal. At Lismore, New South Wales, Murtagh, Kaiser & Huett (1980) grazed *P. clandestinum* pastures for one week on and three weeks off, and varied cow density to achieve a target residue after grazing of 380 kg dry green leaf (DGL) ha^{-1}. At any particular cow density (Fig. 7.9) the rate of leaf removal was strongly influenced by the level of leaf present before grazing and grown during the grazing period.

Sward height is an excellent indicator of forage allowance for *L. perenne* in Britain, and is widely used to adjust SR by varying the area grazed and the area reserved for conservation (Leaver, 1982). As yet this technique has not been found to be applicable in the tropics.

The study of pasture–animal production relationships gained greatly from the fixed SR grazing experiments which reflected actual farm conditions, where seasonal underutilisation and overutilisation commonly occur in the same year. In order to apply the results of these findings more widely in farm practice it is necessary to understand better the effects of variation in the instantaneous availability of forage on animal performance, and the environmental factors which are simultaneously controlling pasture growth and senescence, and whose influence is modified by the level of utilisation. Predictive relationships which can then be used in management practice have to take account of a multiplicity of factors which influence the level of pasture intake and the nutritive value of the diet ingested from a particular pasture; these include pasture genotype, leafiness, sward structure, rates of growth and senescence, N content and mineral availability.

The bottom line is the crude rule of thumb:

$$Desirable\ forage\ allowance =$$
$$\frac{Intake\ as\ a\ fraction\ of\ body\ weight}{Desired\ ratio\ of\ utilisation}$$

Thus for a notional 2.5% daily intake as a fraction of body weight, and pasture utilisation of 40%, forage allowance = 6.25 kg 100 kg^{-1} LW d^{-1}.

7.2
The interaction of fertiliser input and stocking rate

The application of fertiliser which rectifies soil mineral deficiencies or specifically increases legume N fixation is a powerful tool for increasing pasture growth. The investment in fertiliser input is wasted unless the synchrony between SR and the new levels of pasture availability is maintained. The interaction between defoliation and plant mineral requirement for growth was discussed in Section 4.5.3, and this section considers the levels of animal response to different levels of forage availability and quality, as induced by variation in fertiliser practice.

(i) Pasture growth is universally responsive to N supply and high levels of animal production are feasible if SR is increased to match the increased pasture growth. At Lismore, New South Wales, the SR giving maximum LWG ha^{-1} on *P. clandestinum* pastures increased in a particular year from 3.4 weaners ha^{-1} to 9.8 weaners ha^{-1} as N level increased from 0 to 672 kg N ha^{-1}; this increased LWG ha^{-1} from 380 to 1065 kg (Mears & Humphreys, 1974b). At Palmira, Colombia (4° N, 1000 m a.s.l.), irrigated *D. decumbens* pasture was grazed at varying SR from 3.3 to 9.2 steers ha^{-1} over the same range of N levels (Fig. 7.10); LWG increased from 1.9 to 4.9 kg d^{-1} as SR was increased in synchrony with N application. At Nova Odessa, Brazil, the desirable SR or *P. purpureum* pastures was 2.5, 3.3 and 4.1 beasts ha^{-1} at N application levels of 50, 100 and 150 kg N (Lourenco *et al.*, 1978). Meléndez, Pérez & Alvarez (1977) found a similar positive relationship between SR and N inputs for animal production, and responses to N were due to SR and not to changes in LWG d^{-1}. Again, in Perico, Cuba, stocking rates of 1.5,

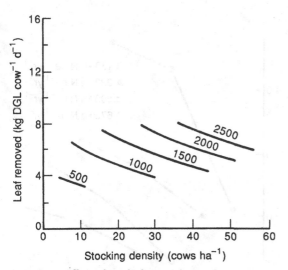

Fig. 7.9. Effect of stock density during the week of grazing on rate of leaf removal of *Pennisetum clandestinum* pastures, as influenced by the sum of leaf yield present before grazing and leaf grown during grazing (kg GL ha^{-1}), as labelled on the curves. (From Murtagh, Kaiser & Huett, 1980.)

3.0 and 4.0 calves ha^{-1} were recommended at N application levels of 0, 80 and 160 kg ha^{-1} respectively (Alfonso, Hernández & Batista, 1986).

For milk production at Atherton, Queensland, response to N fertiliser on mixed grass–legume pasture was greater at 2.5 cows ha^{-1} than at 1.3 cows ha^{-1} (Cowan & Stobbs, 1976). On *P. maximum* pastures (Fig. 7.3) the response in milk yield to 400 N relative to 200 N appeared to be greater at intermediate SR (Davison *et al.*, 1985).

Skill and experience of the local environment are necessary if SR is to be matched to N response. At Brisbane, Australia, *P. dilatatum* pastures grew better when fertilised with N, but no response in LWG occurred, because at SR of 2.5 or 5.0 heifers ha^{-1} there were insufficient animals present to consume the additional feed, and the loss of *T. repens* as N was applied decreased nutritive value (Moir *et al.*, 1969). At Coolum, Queensland (lat. 27° S, 1500 mm rainfall), steers grazing *D. decumbens* pasture at SR 4.9–9.9 beasts ha^{-1} gave no additional LWG response to 480 kg N ha^{-1} relative to 288 kg N ha^{-1}, which was adequate for pasture growth in that particular situation (Tierney & Goward, 1983*a*). The positive association between SR and N level breaks down at very high SR which degrades the pasture to the point where the pasture growth capacity is reduced.

An additional problem is the year-to-year variation in response which affects the determination of the optimum combination of SR and N level. Part of this variation is predictable if there are time trends indicating residual effects of previous applications on productivity, but often variations are due to climatic change or to unforeseen animal health problems. In Zimbabwe Barnes (1982) noted for *C. aethiopicus* pastures that N × SR combinations which resulted in high gross margins in one year gave only small margins or losses in subsequent years. The stocking rates giving maximum LWG ha^{-1} for different N inputs quoted previously for the Mears & Humphreys (1974*b*) study were 8–79% greater for particular N levels in the subsequent year.

The CIAT experiment at Palmira (Fig. 7.10) was used to develop a response surface assuming a base level of production on unirrigated, unfertilised *D. decumbens*. This shows elegantly the diminishing responses of kg LWG kg^{-1} N applied to both increasing SR and increasing N application level (Fig. 7.11). It is from the construction of the biological model of production that economic models may be built to accommodate the varying costs and returns involved in such intensive production systems.

(ii)   A balanced fertiliser policy which avoids extreme soil acidification and induced deficiencies of other elements such as P and K is necessary to maximise response to higher levels of N. The requirement for P may increase at the greater levels of production established with high N supply. This changed requirement may not be obvious immediately but may nevertheless be necessary for the sustainability of the system. This is illustrated by a long-term study (Davison *et al.*, 1989) at Atherton,

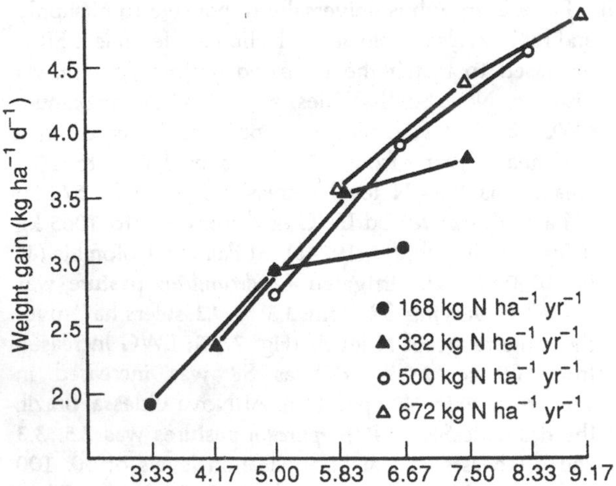

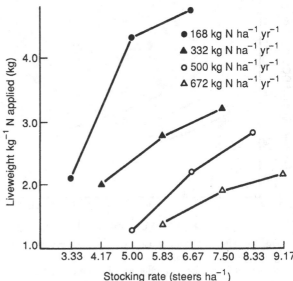

Fig. 7.10. Relationships of LWG ha^{-1} to SR, as influenced by level of N application on irrigated pastures. (From CIAT, 1976.)

Fig. 7.11. Relationship of LWG to applied N at varying SR. (From CIAT, 1976.)

Queensland, in which *P. maximum* pastures were not undergrazed at a stocking rate of 2.6 Friesian cows ha^{-1} (see Fig. 7.3). N fertiliser rate was 100 or 300 kg N ha^{-1} yr^{-1}, and maintenance P fertiliser as a single superphosphate was applied at 0, 22.5, or 45 kg P ha^{-1} yr^{-1}; Ca and S levels of application were balanced for all treatments. The response to the higher level of N averaged 7 kg milk kg^{-1} N. No response in milk yield occurred to P application in the 100 N treatments. At 300 N a significant response in milk yield to maintenance P fertiliser (Fig. 7.12) first appeared in year 6 of the experiment (1987). Leaf P % was influenced by P application in year 2, four years earlier; in a separate study control animals did not respond to P supplementation. The major effect of P fertiliser appeared to be mediated through additional production of green leaf, which was reflected in a higher content of leaf and of N in the diet of animals grazing pastures receiving 22.5 or 45 kg P.

There is a considerable literature on the interaction of SR and P level in determining animal production from tropical pastures. It is usual to obtain a response in animal LWG to P application at higher SR. For example, pastures of *B. decumbens* or *P. maximum* with *S. guianensis* and *M. atropurpureum* were grazed with varying applications of P fertiliser (Winter, Edye & Williams, 1977) at Heathlands, Queensland. At 0.7 beasts ha^{-1} maintenance P of 10 kg ha^{-1} yr^{-1} was adequate for maximum LWG, but LWG hd^{-1} showed increasing responses to higher levels of P as SR increased to 1.2, 1.7 and 2.2 beasts ha^{-1} (Table 7.2). In Fiji LWG on pastures of *P. polystachyon* with *M. atropurpureum* and *S. guianensis* receiving 440 kg single superphosphate ha^{-1} only exceeded that from 220 kg ha^{-1} if SR was increased above 2 beasts ha^{-1} (Partridge, 1986).

The interaction of P level and SR is sometimes mediated through effects on the grass–legume balance. This was evident in natural pastures dominated by *H. contortus* and oversown with *S. humilis* at Rodds Bay, Queensland (lat. 24° S, 810 mm annual rainfall). A long-term study was undertaken in which the range of SR was

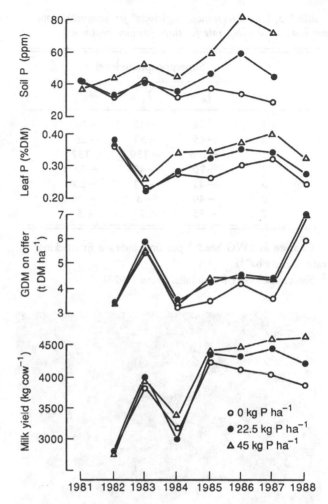

Fig. 7.12. Levels of soil P, autumn leaf P, green dry matter on offer and milk yield as varied by application rates of 0, 22.5 and 45 kg P ha^{-1}. (From Davison *et al.* 1989.)

Table 7.2. *Effect of stocking rate and annual fertiliser phosphorus rate on liveweight gain (kg head^{-1} day^{-1})*

P rate (kg ha^{-1} yr^{-1})	Stocking rate (beasts ha^{-1})			
	0.7	1.2	1.7	2.2
10	0.812	0.702	0.428	0.398
20	0.778	0.714	0.738	0.306
40	0.814	0.731	0.750	0.703

Source: Winter, Edye & Williams (1977).

Table 7.3. *Linear regression coefficients[a] for liveweight gain per head on stocking rate for three fertiliser treatments*

Year	Superphosphate level[b]		
	F_0	F_1	F_2
1	−94	−42	−76
2	−45	−53	−26
3	−128	−159	−132
4	−70	−42	−36
5	−41	−21	−26
6	−49	−3	−1
7	−55	+77	+5

[a] Change in LWG head^{-1} per unit increase in stocking rate (beasts ha^{-1}).

[b] *Source*: See text for details. Shaw (1978).

Fig. 7.13. Heavily grazed *Stylosanthes humilis* seeding well.

gradually increased to 0.55–1.65 beasts ha^{-1} (Shaw, 1978). The fertiliser treatments were: F_0, no fertiliser, F_1 125 kg and F_2 250 kg single superphosphate ha^{-1} yr^{-1} plus an additional 250 kg ha^{-1} initially. Rate of LWG in different years varied from 0.27 to 0.47 kg ha^{-1} d^{-1} in the F_0 treatment and 0.37 to 0.71 kg ha^{-1} d^{-1} in the F_2 treatment. There was a linear relationship between LWG hd^{-1} d^{-1} and SR, and the regression coefficient, b, varied considerably between years (Table 7.3). After year 3 it decreased in the fertilised treatments, and in year 7 actually became positive: LWG per head increased with increasing SR. This was associated with the response of the legume *S. humilis*, which is a low-growing shade-intolerant plant (Fig. 7.13) which increased its contribution to the sward under high SR, and was also in this instance favoured by P application. There was a highly significant linear relation between LWG per head and yield ($\log_{10}$ kg ha^{-1}) of *S. humilis*, irrespective of treatment. At Beerwah, Queensland (lat. 27° S, 1630 mm rainfall), there was a similar positive correlation between LWG and legume content of pasture, which was favoured by high P supply, but the converse SR effect occurred on the trailing legumes *D. intortum* and *D. uncinatum* (Evans & Bryan, 1973).

These examples indicate that fertiliser practice may not only control the amount of the feed supply, necessitating SR adjustment on this account, but may also influence animal production through changes in botanical composition (Chapter 5) and nutritive value.

7.3
Conclusions

(i)   The choice of the optimum SR is highly specific to farm and paddock location, the type of enterprise, and the skills, goals and resources of the farm manager. Some of the principal factors which favour a high SR are:

(1) Benign environment for pasture growth or the application of farm inputs which increase pasture growth. This is also related to the reliability of rainfall, and little seasonality of its occurrence.

(2) Pasture accessibility over the whole property, through the dispersed availability of stock water and shade, and the control of predators. Good nutritive value, so that the need for selectivity to maintain individual animal performance is less acute, will also promote high levels of utilisation.

(3) Sward characteristics which confer resistance to weed invasion and soil erosion at high SR.

(4) Mixed herd composition, which facilitates year to year adjustment of SR. This also bears on whether the

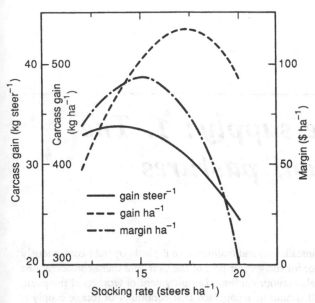

Fig. 7.14. Relationships between SR, carcass gain per steer and per ha, and gross margin for irrigated pastures in Zimbabwe. (From Carew, 1976.)

product output is relatively insensitive to nutritional stress. Sheep are compulsive wool growers and LWG is less sensitive to nutritional shortfall than reproduction or milk secretion. Access to markets which facilitate stock transfer permits more ready adjustment of SR.

(5) An enterprise in which resources are available which can be used to overcome seasonal occurrence of animal stress. Acceptance of risk by the manager is a related factor.

(ii) This chapter has dealt with the way in which SR influences the biological model of production. A manager can use a hard data base in this area to model economic practice in terms of his own long-term goals as economic conditions fluctuate. Managers have a varying acceptance of risk and a varying capacity to deal successfully with changing seasonal conditions. Many farmers stock conservatively because of their wish to minimise risk, their pastoral concern for their individual animals, their knowledge of the fluctuation in pasture availability from year to year, their desire to preserve the grazing resource upon which future production depends, and their interest in the conservation of environmental quality. This will usually lead to the adoption of a farm SR much less than the SR giving maximum output per hectare.

The simplest case is a fattening operation where young animals are purchased and sold at the end of the growing season. For example a quadratic relationship between SR and LWG hd^{-1} d^{-1} occurred on irrigated pastures in Zimbabwe (Fig. 7.14; Carew, 1976). Maximum LWG ha^{-1} occurred at 17.4 steers ha^{-1}. However, at costs and prices then operating, gross margin was maximal at the lesser SR of 14.8 steers ha^{-1}. The curve of response in gross margin to SR is often flatter near the optimum than in Fig. 7.14, and the manager might accept a SR of 12 ha^{-1} without great loss of profit and with considerable gain in peace of mind. A similar study for the Ord River irrigation project, Western Australia (Izac, Anaman & Jones, 1990) suggested that LWG ha^{-1} was optimal at 7 beasts ha^{-1} with 312 kg N fertiliser ha^{-1} input, but economic profits were optimal at 4.9 beasts ha^{-1} with 312 kg N fertiliser ha^{-1} application. Much depends on whether there is a premium for quality, which favours high individual animal performance and low SR. Many pasture managers will choose a farm SR perhaps 40% of the SR giving maximum output per unit area. In predominantly subsistence economies human population pressure on the land inevitably leads to high SR, and the adjustment of human population pressure may be the only biological solution to the environmental problems created by excessive SR.

8

Continuity of forage supply: 1. The integration of different pastures

Discontinuities in forage supply are avoided or mitigated by pasture managers in order to (1) minimise animal stress by maintaining body condition and animal survival, (2) promote successful reproductive activity, and (3) sustain growth, milk secretion or the production of fibre. This policy may be directed to meeting a particular market, such as urban demand for fresh milk produced in the dry or cool season, or cattle in high condition at the beginning of the wet season. There may be a special farm demand, such as the need for draught animals to be in working order at the beginning of the ploughing season.

Social pressure may operate on the farmer to avoid seasonal shortfalls in production. The owners of milk or meat processing facilities minimise the costs of machinery installation and maintenance if plants operate continuously or for the greater part of the year, and these considerations also favour continued employment of workers at the plant. It should be recognised that continuity of forage supply is only achieved through added farm costs, and the farm incentive to maintain continuous output depends upon price adjustment for season of milk or beef production, or a premium payable for meat class and quality.

There are five approaches to the maintenance of continuity of forage supply:
(1) The seasonal pattern of animal demand for nutrients may be synchronised with the seasonal availability of forage, by adjusting purchase and sale of animals and time of mating.
(2) Plant species may be added whose seasonal growth rhythm or whose maintenance of nutritive value complements the seasonal shortfall in the feed availability of the existing species.
(3) The environment in which the plant grows may be changed by the application of fertiliser or by the manipulation of moisture supply to grow additional feed in periods when the base pastures are poorly available.
(4) Forage which is surplus to animal requirements may be conserved as hay, haylage or silage.
(5) Energy or protein supplements may be produced or purchased from off-farm sources.

The latter two options are considered in Chapter 9.

A simplified diagram (Fig. 8.1) of a common field situation in a wet/dry climate shows that pasture growth rate, expressed as the net balance between growth and (senescence plus consumption), is negative when wet season rains commence at point A, since pasture deterioration and consumption exceed new growth. At this time there may be an apparent sharp loss in body weight, which is mainly due to a reduction in gut contents and not body tissues (McLean *et al.*, 1983). Growth rate continues to increase as *L* builds up and environmental conditions are

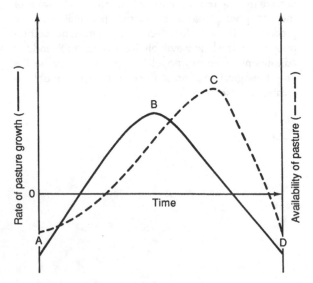

Fig. 8.1. Schematic representation of seasonal variation in pasture growth and availability under a fixed stocking regime.

favourable, reaching a peak at point B, and declining thereafter. Under fixed stocking conditions pasture availability continues to increase beyond B to point C, since pasture growth rate still exceeds consumption and senescence. Pasture growth rate decreases to point D, becoming negative, and pasture availability (and quality) also decrease. The objective of the pasture manager is to minimise the peaks and troughs of pasture availability and quality in order to synchronise the latter with animal requirements.

8.1
Seasonal variation in animal requirements

(i) Pasture managers endeavour to present the amount and quality of pasture to the various categories of animals on the farm which will meet their differing physiological requirements for particular targets of production. These targets may be expressed in terms of particular levels of survival, rate of gain, conception and birth, or milk production. It is assumed that measures are taken to

Table 8.1. *Daily metabolisable energy (ME) allowances for growing and fattening cattle (MJ head^{-1})*

Liveweight (kg)	ME in ration (MJ kg^{-1})	Rate of gain (kg d^{-1})				
		0	0.25	0.50	0.75	1.00
100	8	17	24			
	10	17	22	29		
	12	17	21	27	33	
150	8	22	29			
	10	22	28	35		
	12	22	27	33	40	48
200	8	27	35			
	10	27	34	41	51	
	12	27	33	39	47	56
250	8	31	40	51		
	10	31	38	47	57	
	12	31	37	44	52	63
300	8	36	46	57		
	10	36	44	53	64	
	12	36	43	50	59	70
350	8	40	51	63		
	10	40	48	58	70	84
	12	40	47	55	65	77
400	8	45	56	70		
	10	45	54	65	77	93
	12	45	53	61	72	85
450	8	49	61	75		
	10	49	59	70	83	
	12	49	57	67	78	91
500	8	54	67	82		
	10	54	64	76	91	
	12	54	63	73	85	99
550	8	59	73	89		
	10	59	70	83	98	
	12	59	68	79	91	107
600	8	63	77	94		
	10	63	75	88	104	
	12	63	73	84	97	114

Source: MAFF (1975).

promote animal health, so that the animal product represents more truly the genetic potential of the animal for pasture conversion.

The development of seasonal budgeting of pasture needs is based primarily on estimates of energy supply, but must also take into account requirements of digestible protein and minerals. The most simplistic approach is to estimate daily DM requirements for maximum intake as the product of a fraction of body weight (often 2.5%) and the desired percentage utilisation of pasture on offer (for example, 70%).

The ME of the diet offered is conventionally determined by the product of its total energy value (c. 18 MJ kg^{-1} OM), and its apparent digestibility (%) corrected for urinary and methane components (DE × 0.81). A guide for animal ME requirements for maintenance and rate of gain according to LW (Table 8.1) is based on UK standards (MAFF, 1975). Feeds of high energy value require less MJ per head than lower quality feeds to attain the same rate of gain. For example, a 300 kg beef animal gaining 0.5 kg d^{-1} requires 57 MJ hd^{-1} d^{-1}, when grazing *P. clandestinum* providing ME of 8 MJ kg^{-1} pasture; this would represent 7.1 kg intake, or 10.7 kg d^{-1} available pasture at 67% utilisation. The figure for ME of pasture at 8 MJ kg^{-1} assumes a total energy value of 18 MJ kg^{-1}, which at 55% digestibility delivers 9.9 MJ kg^{-1} of DE; this when corrected by a factor of 0.81 suggests 8 MJ kg^{-1} of ME.

The maintenance requirements of dry stock are less than those of pregnant animals; an empty cow of 500 kg LW may require ME of 54 MJ d^{-1} compared with 61 MJ d^{-1} when six months pregnant, and 74 MJ d^{-1} at nine months (MAFF, 1975). This would increase to 132 MJ d^{-1} if producing 15 kg milk d^{-1} (Table 8.2) and suggests in this instance a variation by a factor of 2.4 in seasonal energy requirements according to physiological stage and production target. Particular attention may be paid to the nutrition of first-calf heifers; Kerr, Bird & Buchanan (1985) found that 1 kg additional LW at calving of Friesian heifers led to 23 kg total additional milk over the first three lactations.

Dairy cows in milk require a high plane of nutrition if the maintenance component of the forage is not to dominate budgeted requirements. For a 500 kg cow, the maintenance component of the total energy required represents 68% of feed intake at 5 kg milk d^{-1}, and 51% of feed intake at 10 kg milk d^{-1} (Table 8.2). Estimated energy needs also take into account whether lactating cows are losing, maintaining or increasing body weight. Cow appetite tends to increase in early lactation; for mid and late lactation a conventional rule of thumb is to estimate maximum DM intake (kg d^{-1}) as 0.025 LW (kg) plus 0.1 milk yield (kg d^{-1}).

The seasonal energy requirements on the farm may therefore be estimated by simple arithmetic with a hand-held calculator by the manager who knows the liveweight, physiological stage and production of his animals. The object is to bring this into synchrony with what is known of the nutritive value, availability and carrying capacity of the forages which can be grown. The apparent shortfalls are validated by experience in

Table 8.2. *Daily metabolisable energy allowances for three breeds of dairy cattle (MJ head^{-1})*

Breed	Liveweight change	Maintenance	Milk yield (kg d^{-1}) 5	10	15
Jersey					
(363 kg,	Losing 0.5 kg d^{-1}	–	58	88	118
49 g kg^{-1} BF,	No weight change	41	74	102	132
95 g kg^{-1} SNF)	Gaining 0.5 kg d^{-1}	–	89	119	149
Ayrshire					
(500 kg,	Losing 0.5 kg d^{-1}	–	66	92	118
38 g kg^{-1} BF,	No weight change	54	80	106	132
89 kg^{-1} SNF)	Gaining 0.5 kg d^{-1}	–	97	123	149
Friesian					
(590 kg,	Losing 0.5 kg d^{-1}	–	73	97	122
36 g kg^{-1} BF,	No weight change	62	87	111	136
86 g kg^{-1} SNF)	Gaining 0.5 kg d^{-1}	–	104	128	153

Source: MAFF (1975). BF, butter fat; SNF, solids-not-fat.

successive years, and this knowledge provides the basis for choosing the approaches which follow.

(ii) The simplest adjustments may be made in a pasture system based wholly on purchase and sale of growing or fattening stock. Young animals are mainly purchased at the beginning of the growing season and progressively sold as the feed supply deteriorates at the end of the growing season (Fig. 8.1). The severity of conditions during the dry or cool season determines what animals are carried through on the pasture until the subsequent growing season.

The manager has flexibility in determining the class of animals purchased, especially if the fattening period may extend beyond one season. At Utchee Creek, Queensland (lat. 18 °S, 3650 mm rainfall), successive drafts of cattle were purchased and fattened on *P. maximum/C. pubescens* pastures to a target carcass weight which varied from 220 to 260 kg (Round, Mellor & Hibberd, 1982). The final finishing period, when fat deposition increases relative to muscle growth, is regarded as energetically more costly than earlier growth stages. However, in this study (Table 8.3) the overall rate of gain to target weight was positively related to age of introduction to improved pastures. The steers had previously been grown on inferior native pasture, and it is probable that the extent of compensatory gain (see Section 9.2.2) was greater in larger animals.

The stocking policy of enterprises directed to growth and fattening of store cattle has to be sensitive to price fluctuations. The worst case is where cattle prices are high at the beginning of the growing season, because of high demand, and cattle prices are low at the end of the season, because of surplus cattle on the market and a low demand.

Models are available which use linear programming to offer the manager the option for monthly purchase which maximises gross margin at given prices of inputs and outputs for particular farms whose balance of land classes, pasture types, expected animals responses and management policy are specified (Teitzel, Monypenny

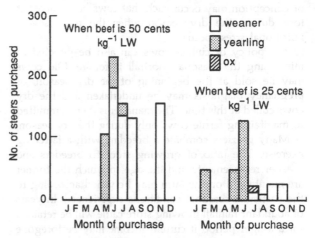

Fig. 8.2. Monthly programme of purchases of steers of differing age for two levels of pricing. (From Teitzel *et al.*, 1986.)

& Rogers, 1986). An illustration is given for a rather poorly drained property of 300 ha on the humid tropical coast of Queensland (Fig. 8.2). The assumptions in the model include a premium price of 5 cents kg^{-1} LW for heavier animals sold at 'export' weight, and a further premium of 5 cents kg^{-1} LW for export animals sold during February, March, September and October. When beef is 50 cents kg^{-1} LW, optimum purchases sum to 791 head for the year, whilst at the lower price of 25 cents kg^{-1} LW a lower SR with total purchases of 323 cattle is appropriate.

(iii) Time of mating is a powerful tool in influencing the degree of synchrony between physiological demand and pasture availability. Cattle which calve near to the beginning of the main growing season lactate during a period of pasture abundance, and if the calves are weaned at the beginning of the dry or cool season the lower energy requirement of the cow during early pregnancy coincides with poorer pasture conditions. In extensive cattle raising enterprises the loss of body condition may limit cow conception to alternate years,

Table 8.3. *Performance of steers introduced to improved pastures at varying ages*

Class of stock	Initial weight (kg)	Final weight (kg)	Days on pasture	Rate of gain (kg head^{-1} d^{-1})
Weaners	179	434	491	0.52
Yearlings	275	426	233	0.65
Two-year-olds	349	470	163	0.74

Source: Round *et al.* (1982).

or conception may occur such that cows have calves at foot during the dry season when the most severe nutritional stress occur.

The policy of culling cows can also be directed to minimising the seasonal shortfall in forage. Old cows may be sold at the beginning of the dry season, or pregnancy diagnosis may be undertaken and the dry cows culled at this time. The manager is then committed to maintaining fertile cows only during the dry season.

Many graziers combine a breeding with a fattening exercise. The ratio of growing stock to breeders and breeder replacements and the age to which the former are maintained on the farm may be varied according to the particular seasonal conditions experienced. In years of abundant rainfall growing store cattle will be retained for a longer period if current income may be foregone to achieve the future benefit which accrues from more

effective pasture utilisation. In years of subnormal rainfall the retention of breeder replacements will have priority and more young males may be sold.

Naturally, better synchrony of animal requirements and feed availability is attained if flexible stocking policies may be adopted as discussed in Chapter 7.

8.2
Systems of integrated pasture use

8.2.1
Pastures of differing seasonal utility
The growth of pasture plants under varying conditions of temperature and moisture supply, and the rate at which nutritive value deteriorates as the growing season advances, are varied according to genetic makeup. The

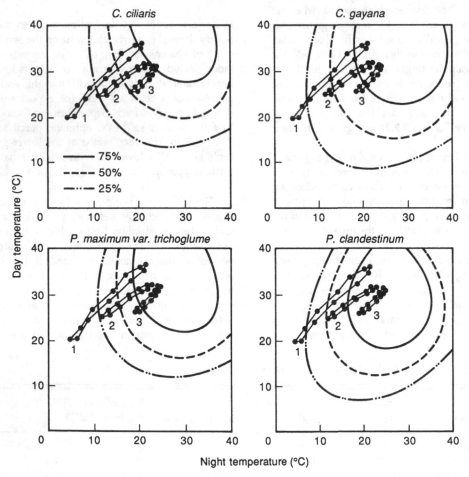

Fig. 8.3. Lines of equal growth rate for the levels of 25%, 50% and 75% of maximum growth rate for four tropical grasses as related to day and night temperatures and to mean monthly maximum/minimum temperatures for (1) Charleville, lat. 27° S, (2) Ayr, lat. 20° S, and (3) Cooktown, lat. 16° S. (From Ivory & Whiteman, 1978.)

pasture manager seeks plants with a long growing season and superior nutritive value as standing hay. It should be recognised that these attributes do not necessarily confer any ecological advantage and programmes of plant improvement probably select against the trends of plant evolution. Ecological success is associated with the capacity of plants to grow quickly at the outset of the growing season and to grow tall, thus interfering with the availability of environmental growth factors – light, water and nutrients – to their neighbours. Plant replacement, at least in open bunch grass savanna systems, is favoured by earliness of flowering and seed formation. These attributes of rapid growth, height and early flowering require investment in plant structural support and may be inimical to pasture quality (for example, Stobbs, 1975). Continuity of forage supply in the paddock is best offered by plants able to maintain growth as climatic conditions become cooler or drier.

(i) The C_4 grasses have highest growth rates under very warm conditions which affect adversely OM digestibility, as described in Section 6.1.1. Our interest at this point is in their capacity to grow at cool temperatures. Ivory and Whiteman (1978) plotted lines of equal growth rate of grasses as a proportion of their maximum growth rate for different combinations of day and night temperature (Fig. 8.3). This diagram also shows the monthly temperatures of three sites in Queensland which are plotted for comparison. The temperature of the coldest month varies from 20–4 °C at Charleville to 26–19 °C at Cooktown. At the coldest site 25 % of maximum growth potential is theoretically possible in at least nine months of the year for *P. clandestinum* and *C. gayana*, but only for seven months of the year for *C. ciliaris* and *P. maximum*. Growth ceased at *c.* 8 °C in the former two species, and at *c.* 12 °C in the latter two species, which have a shorter season of growth under conditions not limited by drought.

A second example illustrates the seasonal change in carrying capacity of *P. clandestinum*, as related to growth rate (Murtagh, Kaiser, Huett & Hughes, 1980). This study was conducted near Lismore, Australia (lat. 29 °S, 1660 mm rainfall), and operated for 10 months of the year when growth was reliably active (Fig. 8.4). Pastures were grazed on a one week on/three week rest system, and sufficient cows were grazed to reduce residual pasture at the end of grazing to 380 kg dry green leaf ha^{-1}, or to 290 kg ha^{-1} in the more heavily stocked *NH* treatment. Carrying capacity for each month, which reflected leaf growth, was maintained at a low level throughout the 10-month period if N fertiliser was not applied (treatment *K*). The carrying capacity of the N fertilised pastures (treatment *N*) increased from *c.* 2 cows ha^{-1} in September to *c.* 9 cows ha^{-1} in Feb-

ruary–March, and decreased to *c.* 4 cows ha^{-1} in May–June. The application of heavier SR in the *NH* treatment increased carrying capacity in spring and summer, but there was no advantage in the autumn, suggesting that the potential SR of the pasture had been attained. Monthly temperatures increased from *c.* 21–12 °C (max.–min.) in September to *c.* 26–19 °C in January–February, and decreased to *c.* 19–12 °C in June. In the same experiment the autumn growth rates of *P. clandestinum* were superior to those of other naturalised grasses, such as *A. compressus* and *P. dilatatum*. The growth rate in June of the latter, when fertilised with N, was 45 % of that of *P. clandestinum*.

The climatic adaptation of the principal grasses and legumes planted in the tropics is summarised in Humphreys (1981*b*, Table 10.1). The C_4 grasses with the best tolerance of cool conditions include *C. gayana*, *P. dilatatum*, *P. clandestinum* and *S. sphacelata* var. *sericea*, whilst amongst the tropical legumes *D. intortum*, *D. uncinatum*, *L. bainesii*, *N. wightii* and *S. guianensis* var. *intermedia* are pre-eminent. The cool season C_3 grasses, such as *Avena sativa*, *L. multiflorum* and *L. perenne*, and temperate legumes are grown under irrigation or after fallowing (Section 8.2.4); some of these, such as *Medicago sativa*, *M. scutellata*, *M. truncatula*, *Lotus pedunculatus*, and *T. repens* are adapted to subtropical and highland tropical regions.

(ii) The period of cattle LWG may be related to the number of 'green days', or days in which green material is present. The capacity of plants to continue growing and retaining green material under dry conditions is often a function of a deep rooting habit, which provides access to subsoil moisture which is unavailable to more shallow rooted species. Available water storage in the soil is a primary determinant of the length of the growing season and the susceptibility of pasture to

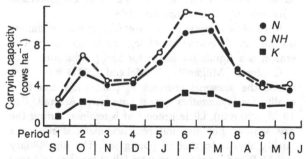

Fig. 8.4. Seasonal carrying capacity (cows ha^{-1}) of *Pennisetum clandestinum* pastures. *N*, high N fertiliser; *NH*, high N fertiliser and low residual pasture after grazing; *K*, no N fertiliser. (From Murtagh, Kaiser, Huett & Hughes, 1980).

interruption of growth due to water stress (McCown, 1973). Within this constraint herbage plants operate with varying success to maintain growth and to persist. Planted grasses in the tropics with superior characteristics in this respect include *A. gayanus, C. ciliaris, C. dactylon, Panicum antidotale, P. coloratum* var. *makarikariense, P. maximum* var. *trichoglume* and *Sorghum almum*. Planted herbage legumes grow in more humid areas, but the most drought resistant include *Centrosema pascuorum, M. atropurpureum, N. wightii, S. guianensis* var. *intermedia, S. hamata,* and *S. scabra.*

S. scabra may be regarded as a low shrub, and it is to the deep-rooted shrubs and trees to which the pasture manager looks for dry season feed, as elaborated in Section 8.2.2.

(iii)   All tropical pasture plants decrease in digestibility as they age. In Section 6.1.1 attention was drawn to the slower rate of decline in nutritive value of the legumes, and of C_4 grasses of the genera *Brachiaria, Setaria* and *Digitaria,* in contrast to many species of *Panicum, Chloris* and *Hyparrhenia* (Norton, 1982).

In subtropical regions the deterioration of standing pasture during winter is exacerbated by both frost and rain. Mixtures of various pasture legumes with *P. plicatulum* were reserved from grazing and accumulated as standing forage from March to June at Samford, Australia (lat. 27 °S, 1100 mm rainfall). In successive years 10 and 11 frosts occurred in the period June–August (Jones, 1967). The average loss of DM yield over the period was 0.18–0.21 d^{-1}. However, losses of N were greater, and averaged 0.53 % d^{-1} in a wet winter with 19 wet days, and 0.31 % d^{-1} in a dry winter with 3 wet days. The rate of loss in pastures containing *M. atropurpureum* was substantial, but *L. bainesii* continued green through the winter. In Gainesville, Florida (lat. 30 °N), *Hemarthria altissima* cv. Bigalata maintained better IVOMD as accumulated autumn–winter feed than cv. Redalta or cv. Greenalta (Quesenberry & Ocumpaugh, 1980).

In tropical regions with a wet/dry monsoon climate the rate of deterioration of legume forage in the dry season is a significant factor for beef production.

Gardener, Megarrity & McLeod (1982) have described the seasonal changes in plant parts and their quality for *Stylosanthes* spp. at Lansdown, Australia (lat. 19 °S, 870 mm). Little green leaf is retained during the dry season, but the inflorescence material retains a quality similar to that of young leaf. The digestibility and N, P and S content of stem fall to low levels. Rapid senescence and detachment of leaf tissue occurs in *S. hamata* until the main rains resume where the dry season is reliably dry, and deferment of grazing until the dry season is a pragmatic option because moulding of the

leaf material delayed (McCown, Wall & Harrison, 1981). Leaf senescence is slow in locations with intermittent rain and high humidity, but moulding of the dead material and its consequent rejection by cattle occur. A leaf 'Discoloration Index' (DI) from yellow to dark greyish brown (0 to 7) was developed to indicate the degree of mouldiness of litter and standing hay; cattle refused to eat litter of DI 6, and the value of leaf as a fodder approaches zero at DI 5.

Whilst fungal growth was initiated by small falls of rain, effects of these were modified by site location. These differences were mainly associated with the relative humidity persisting after rain, which influenced the rate of drying of rain-wetted leaf litter. The amount of rainfall was also poorly related to mouldiness; changes in DI were poorly related to dewfall (McCown & Wall, 1981). Falls of more than 2 mm had a priming effect on spore germination, but the start of deterioration could be related to the duration of the period following rain when humidity exceeded 95 % (Fig. 8.5).

Cattle will eat green grass in preference to legume, as discussed in Section 5.3.2, and rain during the dry season which is sufficient to generate green grass compensates for the effect of rain in promoting mouldiness and deterioration of standing legume. The worst situation arises from small falls of rain which promote deterioration but not grass growth.

(iv)   Many examples may be found in the literature of the seasonal variation in SR for different pasture species and locations. In Sao Paulo, Brazil, Bianchine *et al.* (1987) found that *P. maximum* var. *trichoglume/N. wightii* pastures gave maximum LWG ha^{-1} at SR of 1.33 AU ha^{-1} in the wet season, and 0.55 AU ha^{-1} in the dry season; *P. maximum/D. uncinatum* gave superior gains in the dry season. At Utchee Creek, Queensland,

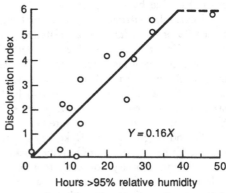

Fig. 8.5. The relationship between duration of the period with relative humidity exceeding 95 % following rainfall exceeding 2 mm and the Discoloration Index of *Stylosanthes hamata*. (From McCown & Wall, 1981.)

the addition of *N. wightii* cv. Tinaroo to *P. maximum/C. pubescens* mixtures improved LWG in autumn–winter, but not in the spring–summer period (Mellor, Hibberd & Grof, 1973). At Walkamin, Queensland, *N. wightii* cv. Tinaroo also gave superior winter growth when irrigated in mixture with *B. mutica*, yielding 860 kg LWG ha^{-1} yr^{-1} but *D. intortum* provided higher yields in spring and summer (Miller & Van Der List, 1977). At Carimagua, Colombia, SR of *A. gayanus/P. phaseoloides* pasture was 2 beasts ha^{-1} in the wet season, and 1 beast ha^{-1} in the dry season (Böhnert *et al.*, 1985).

In some cases the integration of different pastures has successfully raised production, or has maximised the output from one resource (for example, N fertilised pasture) that is more expensive than another forage resource. J. K. Teitzel (pers. comm.) compared the performance of cattle on different grass–legume mixtures at Utchee Creek, Australia (lat. 18 °S, 3650 mm), with the output from the mixtures integrated with 25% of the area planted to *B. decumbens* and fertilised with N. The latter pasture has a higher carrying capacity in the cool season than the grass–legume mixtures, and relieving the pressure on these enables a high SR to be sustained throughout the year. *P. maximum* gave 550 kg LW ha^{-1} yr^{-1} at 3.0 beasts ha^{-1} (Table 8.4); if the grazing of this same mixture were integrated with N fertilised *B. decumbens*, output was raised to 650 kg LW ha^{-1} yr^{-1} at 3.5 beasts ha^{-1}.

A more complex example of the integration of different feed sources for the fattening of weaner cattle derives from a long-term study of a system tested at Gayndah, Australia (lat. 26 °S, 730 mm). In this undulating landscape native pastures on the rocky upper slopes are based on *H. contortus*, *Bothriochloa bladhii* and *Dichanthium sericeum*, and the lower slopes and flats

might be planted to pasture or rotated with crops. Cows and calves are grazed on native pastures, and weaner cattle born around December are available at the end of May at *c.* 181 kg LW. About 31% of the annual rainfall falls in the six coolest months; animals grazing native pastures or sown pastures in the summer gain at similar rates per head but the latter show to advantage in the cool season. In this system *P. maximum* var. *trichoglume* pastures were reserved for grazing from June to November (Table 8.5; Robbins & Bushell, 1983). Yearlings were then grazed on native pasture until mid-April; native pasture grazed in the summer only could sustain a higher SR than if they were required for year-round grazing. One-sixth of the native pasture area was planted to *L. leucocephala* and this was used in the early autumn. Over the summer months an area was cropped to *S. bicolor* for grain and *L. purpureus* for hay, and cattle grazed the stubbles of these crops until the end of May. They were then finished in a feed lot on *L. purpureus* chaff and milled sorghum grain. The figures in Table 8.5 are derived from six fattening cycles and indicate that an animal averaging 426 kg LW might be produced at age 21 months; LWG output from weaning varied from 213 to 278 kg hd^{-1} in individual years. A unit of 25 ha crop, 25 ha sown pasture and 60 ha native pasture would fatten 60 weaners yr^{-1} with an output of 134 kg LWG ha^{-1} yr^{-1}. This represents an increase in production by a factor of *c.* 4 relative to that attainable from native pastures alone. An additional area of *c.* 220 ha of native pasture might be devoted to calf production, assuming a 90% calf crop and SR of 0.3 cows ha^{-1}. This example shows how land classes of differing potential can be developed and integrated to give a relatively intensive fattening programme incorporating more extensive components.

Table 8.4. *Productivity of pasture systems based on grass/legume pastures or grass/legume integrated with N-fertilised grass pastures*

Pasture system	Safe SR (beasts ha^{-1})	Liveweight gain (kg ha^{-1} yr^{-1})
1 *Panicum maximum/Centrosema pubescens*	2.7	490
2 *P. maximum* cv. Makueni/*Centrosema schiedianum* cv. Belalto	3.0	550
3 75% *P. maximum/C. pubescens* plus 25% *Brachiaria decumbens*+N	3.3	600
4 75% *P. maximum* cv. Makueni/*C. schiedianum* cv. Belalto plus 25% *B. decumbens*+N	3.5	650
5 75% *P. maximum/C. schiedianum* cv. Belalto plus 25% *B. decumbens*+N	3.4	630

Source: J. K. Teitzel (pers. comm.).

(v)   The above physical model was established after many years of experiments which suggested production parameters in a rather variable climatic environment. The planned integration of pastures of varying utility can proceed if realistic estimates of production components are available.

The simplest case is where a base pasture B, deficient in one period (1),is integrated with a special purpose pasture A, which is deficient during period (1) but which is inferior at other times of the year (period 2). It is desired to maintain a constant SR on the farm year round, and realistic stocking rates are formulated. Let a_1 and a_2 be the SRs on pasture A during periods 1 and 2 respectively, and let b_1 and b_2 be the SRs on pasture B during periods 1 and 2. The farm overall SR is designated f beasts ha^{-1}, and $A + B = 1$. Myers (1967) provides the following solutions:

$$A = \frac{b_2 - b_1}{(a_1 - a_2) + (b_2 - b_1)}$$

$$f = \frac{a_1 b_2 - a_2 b_1}{(a_1 - a_2) + (b_2 - b_1)}$$

This calculation is illustrated for a subtropical dairy farm in the southern hemisphere in which raingrown *C. gayana* pasture (B) is fertilised with 300 N and carries 2.5 cows ha^{-1} from mid-December to the end of May (period 2) and 1 cow ha^{-1} from June to mid-December (period 1). Irrigated *Lolium* pastures carry 5 cows ha^{-1} from June to mid-December, and 0.5 cows ha^{-1} from mid-December to the end of May. Then:

$$A = \frac{2.5 - 1}{(5 - 0.5) + (2.5 - 1)} = 0.25 \text{ and } B = 0.75$$

$$f = \frac{(5 \times 2.5) - (0.4 \times 1)}{(5 - 0.5) + (2.5 - 1)} = 2 \text{ beasts ha}^{-1}$$

Thus on a 16 ha farm, 32 cows would be carried on 12 ha *C. gayana* and 4 ha irrigated *Lolium*. This example assumes dry stock are carried on a separate area of the farm.

More sophisticated modelling is needed to take account of climatic variation from year to year, choice of SR, and price fluctuations. McKeon, Rickert & Scattini (1986) have modelled plant production, which determines the level of utilisation at particular SR, and potential LWG as modified by forage quality. Production was simulated over 30 years of climatic data; it is important to recognise that the 'average season' often overestimates average production, which is usually lower if the projection data are based on the performance in a series of individual years.

The sample chosen is from a subtropical situation where forage oats might be grown on a fallow for winter grazing (June to August) and used in conjunction with native pastures in an attempt to produce cattle suitable for an export market requirement of 580 kg LW at 42 months. Steers were bought at 6 months, and the oldest animals were used to graze oats; younger animals were also given access to oats in good years with surplus oats. In poor years animals received purchased feed. It was desired to maintain native pasture utilisation below 40%, which might be considered critical for the long-term productivity of these pastures. The simulation is designed to show the effects on cattle production of varying SR (expressed as weaner equivalents ha^{-1}) and the proportion of the farm cultivated to oats.

Fig. 8.6a shows that it was more difficult to attain 580 kg export weight as SR increased and the proportion of oats decreased. When these production figures were translated to the economic indicator of gross margin (Fig. 8.6c) combinations of 10–15% oats with overall property SR of 0.75–0.8 weaner equivalents ha^{-1} maximised gross margin; this SR was just below

Table 8.5. *Cattle liveweight gain (LWG) and stocking rate (SR) on a sequence of forages*

Forage	Period	SR (beasts ha^{-1})	LWG (kg head^{-1})
Sown pasture	June–Aug	2.5	17
Sown pasture	Sept–Nov	2.5	36
Native pasture	Dec–Feb	1.1	56
Native pasture + *Leucaena leucocephala*	Mar–mid Apr	1.0	29
Crop residues	Mid Apr–May	2.5	21
Crop products	June–Aug	2.5	86
Total	15 months		245

Source: Robbins & Bushell (1983).

40% native pasture utilisation, and the coefficient of variation (Fig. 8.6*d*) was low relative to other combinations of choices.

These diagrams illustrate the results of modelling exercises which may be used to help the manager make decisions in his own farm situation about the best options for integrating forages of differing seasonal utility to achieve particular production targets.

8.2.2
Special purpose legume stands

The viewpoint of this book is that grazing of legume stands in the dry season (Fig. 8.7) is the most widely applicable and pragmatic approach to the problem of maintaining continuity of forage supply in wet/dry climates. This approach should be fully explored before attempting pasture conservation, which has inherent practical difficulties in the tropics (Chapter 9), or before purchasing or growing grain for this purpose, which raises questions of social equity and possibly of environmental hazard.

(i) Many deep-rooted shrubs maintain growth and retain green foliage during the dry season. In Bali, Indonesia, Nitis (1985) advocates a 'three-strata' system of animal feeding. This comprises (1) a herbage layer, including *P. maximum* var. *trichoglume*, *C. ciliaris*, *S. hamata*, *S. guianensis* and *C. puhescens*; (2) a shrub layer of *L. leucocephala* and *G. sepium* on the border of the herbaceous layer; and (3) a tree layer of *Hibiscus tillaceous*, *Lannea corromandilica* and *Ficus peacelli* interplanted with the shrubs. The herbage layer provides the main feed base during the wet and early dry season, the shrubs are mainly used during the dry season, and foliage and young shoots of the tree species are cut towards the end of the dry season, especially if it is abnormally lengthy.

L. leucocephala has been the most widely planted shrub species. This arises from high acceptability of the leaf material, high level of N fixation, and superior drought resistance. There are many instances of successful animal production systems integrating the use of *L. leucocephala*. In Fiji Partridge & Ranacou (1974) devoted 0%, 10% or 20% of a *Dichanthium caricosum* pasture to the establishment of fertilised *L. leucocephala*; annual steer LWG averaged 110, 170 and 270 kg ha^{-1} when the areas were grazed in conjunction. This result illustrates how an investment in a relatively small area of property improvement might increase the efficiency of pasture use from a wider area.

Similarly, at Gayndah, Queensland, Cooksley (1983) established *L. leucocephala* in rows 3 m apart adjacent to native pasture areas of *H. contortus* and *B. bladhii*. Cattle weaned in May were offered 0.2 ha legume and 1.0 ha native pasture per head (0.83 beasts ha^{-1} overall), and 12 months later the SR was halved. *Leucaena* was grazed rotationally during the winter and spring. The average LWG over 2 years of four successive drafts was 231 kg ha^{-1} for heifers, and 267 kg hd^{-1} for steers, which gave a mean weight at 30 months of 430 kg. This weight was attained at least 12 months earlier than occurred with animals grazing native pastures alone. In this subhumid, subtropical climate the yields of edible *L. leucocephala* were *c.* 800 kg DM ha^{-1}, compared with 3300 kg ha^{-1} under irrigated conditions in northwestern Australia (Cooksley, Prinsen & Paton, 1988); yields were greater on deep, fertile soils, but as these were located in topographically low situations, frost damage at these was greater. Better performance at this site occurred if cattle were provided with rumen bacteria capable of degrading DHP (Section 6.1.4), and these cattle grazing *L. leucocephala* gained 1.1 kg hd^{-1} d^{-1} for 19 weeks of test (Quirk *et al.*, 1988).

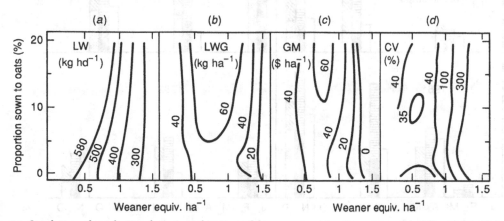

Fig. 8.6. Simulation of production from a combination of forage oats and native pasture for different SR, showing (*a*) final liveweight per head, (*b*) LWG ha^{-1}, (*c*) gross margin (Aus$ ha^{-1}) and (*d*) coefficient of variation. (From McKeon *et al.*, 1986.)

Positive effects of *L. leucocephala* on LWG are also reported from Bolivia. At Guabirpa (lat. 17 °S), *L. leucocephala* was established as 0%, 9%, 16% and 22% of an area of degraded *H. rufa* pastures (Paterson, Samur & Sauma, 1982). Access to *L. leucocephala* did not influence LWG during the wet season, but increased it substantially during the dry season, when 9% *L. leucocephala* area raised animal production *c.* 80% and *c.* 50% in successive years. SR could be increased through the reservation of the legume area for dry

season grazing. The expected responses to legume depend partly upon the character and availability of the base pasture, and differences will be less on good pasture. At Estación Experimental Agrícola de Saavedra (lat. 17 °S) better quality *H. rufa* and *B. decumbens* pastures were grazed by zebu–criollo steers. These pastures were still protein deficient, and during the dry season N content of the pasture on offer decreased from 1.0% to 0.7% for *H. rufa* and 1.2% to 0.8% for *B. decumbens*. Access to 20% *L. leucocephala* area, mixed with *N. wightii*, increased LWG 27% (Paterson *et al.*, 1983). Legume protein banks have also had success in Cuba: at Perico, Matanzas, a mixture of *L. leucocephala* with herbaceous legumes increased LWG 51% relative to natural pasture alone where N content in the dry season decreased to 0.8% (Hernández, Alfonso & Duquesne, 1986).

L. leucocephala has a limited edaphic range, is intolerant of waterlogging, and has fallen victim to the predations of the psyllid *Heteropsylla cubana* in the Pacific, Asia and Australia (Bray & Sands, 1987). Many other shrubs or sub-shrubs are available for dry season grazing. At a site in Nova Odessa, Brazil (lat. 21 °S, 1450 mm rainfall), Lourenco *et al.* (1984) offered steers of 350–400 kg LW 0%, 18%, 33% or 51% of the area planted to *Cajanus cajan*, grazed in conjunction with *H. rufa* pasture; SR was 1.5 and 2.4 beasts ha⁻¹. Diet selection

Fig. 8.7. A legume fodder bank, *Leucaena*

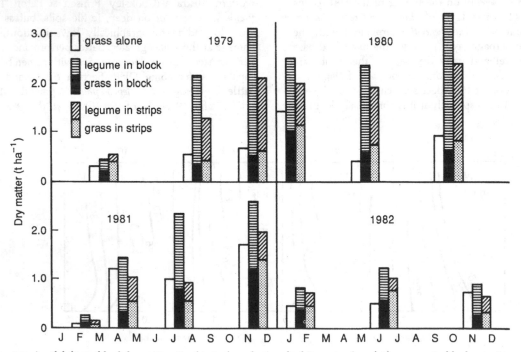

Fig. 8.8. Availability of leaf dry matter *Brachiaria decumbens* and of *Pueraria phaseoloides* grown in blocks or strips at Carimagua, Colombia. (From Tergas *et al.*, 1984.)

for *C. cajan* was high (up to 80%) in the first two months of the dry season (June–July) but decreased subsequently as leaf availability decreased. LWG in this first period varied from 0.2 kg hd^{-1} d^{-1} for steers without access to legume to 0.75 kg hd^{-1} d^{-1} for steers with 33% *C. cajan* area, and thereafter steers in all treatments performed similarly. *Gliricidia sepium* (syn. *G. maculata*) has been widely planted as a living fence in the Caribbean, in Sri Lanka and other Asian regions, and in African countries. It has been mainly used as a cut fodder, and in mixture with other feeds has promoted percentage lambing, ewe and lamb survival, and lamb growth (Chadhokar & Kantharaju, 1980) and also shown utility for dairying (Chadhokar & Lecamwasam, 1982). *Sesbania sesban* is well adapted to wet situations; some lines have low tannin content and high digestibility. Other planted shrub legumes include *Acacia villosa, A. leucophlea, Albizia lebbek, A. falcataria, Calliandra calothyrsus, Enterolobium cyclocarpum, Erythrina* spp., *Samanea saman, Sesbania grandiflora, Tamarindus indica* and *Tephrosia* spp. (Skerman, 1977; R. C. Gutteridge, pers. comm.). Otsyina & McKell (1985) have reviewed the role of browse in the nutrition of livestock in Africa.

(ii) The concept of the protein bank has also been based on herbage legumes. At Carimagua, Colombia (lat. 5 °N, 2160 mm rainfall), Tergas *et al.* (1984) compared cattle production from *B. decumbens* pastures alone, or from *B. decumbens* with *P. phaseoloides* covering *c.* 30% area as either blocks or strips. SR was *c.* 1.85 beasts ha^{-1} in the rainy season and 1.25 beasts ha^{-1} in the dry season. Feed availability over four years (Fig. 8.8) was consistently greater in the systems containing legumes. However, during the rainy season leaf of *B. decumbens* averaged 1.4% N and was more abundant in the grass alone treatment when cattle preferred grass; LWG was similar in all treatments. In the dry season rate of gain averaged 0.23, 0.36 and 0.50 kg hd^{-1} d^{-1} in the grass alone, legume blocks, and legume strip treatments respectively. This led to annual gains of 145, 158 and 183 kg hd^{-1} in these same treatments, indicating an advantage of 26% for strip sowing.

Many similar examples might be cited. At Estación Experimental Agrícola de Saavedra in Bolivia, Paterson, Sauma & Samur (1979) found that the addition of *N. wightii* to *P. maximum* var. *trichoglume* pastures increased LWG of young bulls in the dry season by *c.* 40 kg hd^{-1}. *M. axillare* on 25% area also increased milk production and LWG of cows (Paterson & Samur, 1982). In Cuba at Perico, Matanzas (lat. 23 °N, 1390 mm rainfall), supplementing a natural pasture of *Dichanthium* spp. and *P. notatum* with 33% area of *N. wightii* and *M. atropurpureum* (51–67% botanical composition) increased LWG from 0.24 to 0.33 kg hd^{-1} d^{-1} (Hernández,

Alfonso & Duquesne, 1988). In the subhumid zone of West Africa grassland workers have promoted the establishment of areas of *S. hamata* as fodder banks to supplement ruminant production from crop residues and natural pasture. There have been long established fodder banks at Kurmin Biri, and the demographic changes in 16 herds having access or not to fodder banks was monitored (ILCA, 1987). The data have been summarized in terms of the percentage of total entries to the herd or exits from it contributed by events (Table 8.6). Herders with access to fodder banks maintained their stocks numbers by births alone, and transferred more stock out to relatives. Herders without access to fodder banks had to rely on purchases and transfers in from relatives and others to maintain herd number, and made more 'salvage' sales, which were often associated with nutritional stress.

(iii) The last example indicates less satisfactory performance, especially with respect to mortality, than that exemplified by the animal production experiments quoted in this section. It is quite customary to devalue production responses by *c.* 40% when moving from the research station to the farm. In the case of the present topic, the performance of the legume fodder bank may be constrained by avoidable management factors or by the extrapolation of a demonstrated technique to an inappropriate farm situation. The farmer may plant *L. leucocephala* in a poorly drained area, or on an acid soil with high aluminium content. Planting rate may be suboptimal, and the specific *Bradyrhizobium* may be unavailable in the soil of a new area, leading to delayed

Table 8.6. *Performance of herds with and without fodder banks of* Stylosanthes hamata *in Kurmon Biri, Niger*

Component	With fodder bank	Without fodder bank
Entries (% of total)		
Births	83	52
Purchases	5	16
Transfers in	11	31
Exits (% of total)		
Deaths	24	23
Transfers out	19	9
Sales	58	65
Births (% of exits excluding transfers out)	114	63

Source: ILCA (1987).

nodulation. Grass weeds may not be adequately controlled in the seedling phase, unrecognised mineral deficiencies may limit the growth and N fixation of the seedling, and the legumes may be grazed too early in their establishment phase, due to feed shortfall on the farm or to the depredations of feral herbivores. In some societies the planting of forage legumes may not be seen to be pertinent in the terms the farmer understands; pasture may be viewed as a waste product or as the outcome of natural phenomena unrelated to farming intervention. This historical basis of grassland use is such that changed cultural perceptions of grassland are basic to its improvement. Legume adoption in farm practice is slow in all tropical countries. The social factors responsible are outside the scope of this book; the concomitant need is to develop technologies which are sufficiently robust to attract validation in farm practice.

(iv)   The provision of dry or cool season legume forage may be organised in different systems (Paterson, Proverbs & Keoghan, 1987). Examples of the establishment of legumes in blocks, in strips, or in full combination with grasses have been given, and these may be cut or grazed. The concept of the living fence is especially useful in smallholder practice (Sumberg, 1983). Chadhokar (1982) observed that *G. sepium* in Sri Lanka planted as 400 m fence around 1 ha of land yielded *c.* 1.1 t DM yr^{-1} of green leaf if individual plants were harvested at intervals of three months. Since these contain *c.* 4% N, a fence of this size would meet the supplementary protein requirements of at least two milking cows in dry season. A block planting gave 9.2 t ha^{-1} yr^{-1} of leaf DM.

Alley farming, in which a leguminous shrub is grown in wide rows and crops are cultivated in the inter-row space, has gained acceptance in many tropical countries. The shrubs, if planted on the contour, diminish runoff and erosion, and the green branches laid on or incorporated in the soil maintain an enhanced soil fertility, eventually promoting superior crop yields. In mixed farming systems the manager has the option of using the leguminous shoots to enhance crop yields, or of feeding these to ruminant stock. One economic analysis in humid West Africa (Sumberg *et al.*, 1987) suggests that supplementation of goats or sheep with *L. leucocephala* or *G. sepium* requires a response of *c.* 30–40% in net production per dam before it becomes competitive with the practice of using the legume shoots to assist maize production.

The inter-row spacing of block plantings will influence production, and although longer cutting intervals increase the proportion of wood to leaf, these enhance the total biomass grown. This is illustrated for a study

at Ibadan in the humid zone of Nigeria, where a high yield of 38 t DM ha^{-1} yr^{-1} (Fig. 8.9) was attained with 0.5 m inter-row spacing of *L. leucocephala* harvested at 12 weeks intervals. The decrease in yield associated with wide row spacing tended to lessen beyond 1.5 m, due to greater compensatory individual plant growth and survival at the 2 m spacing. Mortality was greater in frequently cut trees.

The cutting management of shrubs needs to take account of the genetic capacity for basal tillering, which is high in bred lines of *L. leucocephala*. In other species, such as some lines of *G. sepium*, early controlled defoliation is necessary to stimulate branching, since buds may be elevated above cutting height and a late first cut after one year's growth may cause mortality.

The concluding note it is desired to emphasise in this section is the benefit to continuity of forage supply and to N fixation which accrues from wet season deferment of use. Field examples have been given of the common lack of response to legume access during the main growing season, when ruminants prefer to eat green grass, and when green grass of good nutritive value is available for grazing. Wet season utilisation depresses the accumulation of yield in the fodder bank. This is illustrated (Fig. 8.10) for *L. leucocephala* at Samford, Queensland, which was cut at 30, 60 and 120 cm height in February (Isaraseenee *et al.*, 1984). Peak accumulation of edible DM was 2.4 t ha^{-1} if cut at 30 cm, and

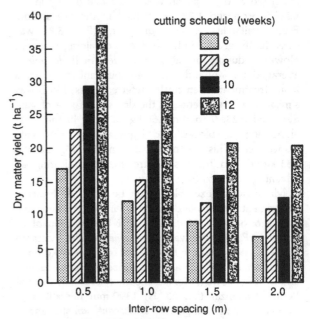

Fig. 8.9. Effects of cutting frequency and inter-row spacing on fodder DM yield of *Leucaena leucocephala* at Ibadan, Nigeria. (From ILCA, 1987.)

4.2 t ha^{-1} if cut at 120 cm. Severe frost in July caused leaf fall and a decrease in presentation yield to similar levels for all treatments, suggesting that some before July utilisation was desirable in this subtropical environment. It was also emphasised in Section 4.6.2 that N fixation is positively dependent upon rate of growth; lenient use during the main growing season maximises leaf canopy and the opportunity for greater N fixation.

8.2.3
Fodder crops and crop residues

(i) Farms in many tropical regions have developed a seasonal sequence of feeds to minimise animal stress. The central question is how additional forages may be introduced into the system in ways which are compatible with the availability of land and farm labour, or which replace existing forage sources with those which will give higher performance or provide better utilisation of existing low quality feeds. These solutions avoid the import of concentrates or grain.

Case studies of mixed annual cropping systems containing ruminants were given in Section 2.1, and Fig. 2.3 illustrates a sequence of feeds used in raising cattle

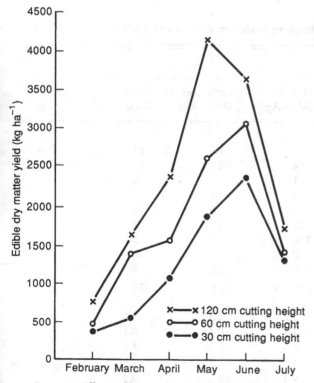

Fig. 8.10. Effects of cutting height on edible DM of *Leucaena leucocephala* at Samford, Queensland. (From Isaraseenee *et al.*, 1984.)

in Batangas, Philippines. Table 8.5 showed a feeding sequence for weaner cattle which included crop grazing and crop products, whilst Fig. 8.6 indicates a combination of forage oats with native pasture. A diagram of the seasonal change in dry season feeding sources in Niger State, Nigeria, shows a varying dependence upon sorghum stalks, cowpea stalks, natural grasses, rice straw, groundnut hay and other feeds (Mohamed-Saleem & Von Kaufmann, 1989).

A further example is taken from a new dairying program (Gibson, 1987) in north-east Thailand (lat. 17 °N, *c.* 1100–1200 mm rainfall, 200 m a.s.l.). This is primarily a rain-fed rice-growing area, with the poor upland soils devoted to cassava (*Manihot esculenta*), sugarcane (*Saccharum officinarum*) and roselle (kenaf, *Hibiscus sabdariffa* var. *altissima*). *M. atropurpureum* and *S. hamata* are well adapted to the region, and contribute sufficient OM and N substantially to increase subsequent crop yields, provided the legume ley is fertilised with P, S and perhaps K. It is significant that the fertiliser needs to be applied to the pasture ley and not to the following crop. Eleven farmers undertook to plant 0.64 ha of fertilised *M. atropurpureum/S. hamata* pasture, and acquired two adult milking cows, mostly with a minimum of 50% Friesian heritage.

The composition of the feed supply in the different seasons (Table 8.7) fluctuated in its dependence upon hand feeding; ley pastures were heavily used during the main wet season and the early cool, dry season, whilst the rice paddies were mainly grazed in the hot dry season when other grazing was in short supply. Recently harvested sugarcane fields contributed appreciably in the dry season. The volunteer weed component of the cropped areas was a significant feed source.

The main hand-fed roughage source in the dry season was untreated rice straw and sugarcane tops. Rice bran was fed year-round and was produced on the farm or purchased from local mills. Dried cassava chips were grown and processed on the farm as a high energy source, whose use diminished in the early wet season when other feed sources were available. *L. leucocephala* was used in the dry season and leguminous *Crotolaria juncea* hay was produced from residual soil moisture in the paddy areas (Kessler & Shelton, 1980). Minerals were also fed.

In the early phase of this project, milk (4.4% butter fat) averaged 5.4 litres cow^{-1} d^{-1} with a lactation length of 295 days and a calving interval averaging 384 days. Total milk production per farm (2 cows) was 3165 litres yr^{-1}, of which 2595 litres were sold for US$833, and the balance was mainly fed to calves, who were reared without purchase of concentrates. Dairying is a burgeoning industry in Thailand; this example displays a high degree of farm self-sufficiency, whilst the incorporation of the legume ley in the cropping system contributes to the maintenance of the latter and to the control of erosion.

(ii) Feed sources include grazed or cut forage crops, crop residues and crop by-products. There are many general reviews available which deal with the production and utilisation of such feed sources in the tropics (Coombe, 1980, for crop residues; Devendra, 1981, for South-east Asia; Ranjhan & Chadhokar, 1984, for Sri Lanka; Wanapat & Devendra, 1985, for the South-east Asia; Hermans, 1986, for Bangledesh; Humphreys, 1986a, for rice systems in South-east Asia; Dixon, 1987, for fibrous residues; Little & Said, 1987, for fibrous by-products in Africa; Cobbina, 1988, for Ghana; Nnadi & Haque, 1988, for legume sources in Africa; Reed *et al.*, 1988, for crop residues).

C_4 grasses have been developed as forage crops, and exhibit high rates of growth during the main growing season. Under irrigation in north-western Australia, LWG on *Sorghum* spp. varies from 0.3 to 0.7 kg hd^{-1} d^{-1} at SR of 4–6.4 beasts ha^{-1} (Blunt & Fisher, 1973). Under rain-grown conditions in subhumid central Queensland *Sorghum* spp. crops produced LWG of 0.6–0.8 kg hd^{-1} at SR of 3 beasts ha^{-1}, giving total LWG of 143–162 kg ha^{-1} (French, O'Rourke & Cameron, 1988a). It is estimated that in these circumstances yield increases for each mm of available moisture by 30 kg DM ha^{-1}. *Sorghum* spp. and *Pennisetum americanum* often give similar levels of LWG and milk production (Clark, Hemken & Vandersall, 1965) but *P. americanum* is a tall crop, and difficult to manage since rapid adjustments of SR are needed to maintain a relatively even forage allowance (McCartor & Rouquette, 1977). These grass crops have limited utility for cool or dry season use (Norman & Phillips, 1968). *Saccharum officinarum* has a longer season of availability than annual grass crops. The tops represent a satisfactory energy source (ME 8.45 MJ kg^{-1}) but a poor source of N (digestible crude protein 0.7%, Devendra, 1983). A good deal of research has been devoted to the use of cane as a ruminant feed, especially in conjunction with other protein sources (Ferreiro, Preston & Sutherland, 1977; Preston & Leng, 1978). Derinding of cane for cattle feeding has not been justified under farm conditions, where whole chopped cane appears to be equally satisfactory.

Tropical forage legumes retain their nutritive value with age better than the C_4 grasses, and reference has been made to the use of *C. juncea* grown for hay on

Table 8.7. *Seasonal feeding regimes of crop products and pastures for dairy production in north-east Thailand*

Feed	Season			
	Cool dry (Nov–Feb)	Hot dry (Mar–Apr)	Early wet (May–July)	Main wet (Aug–Oct)
Hand feed (kg DM $month^{-1}$ $farm^{-1}$)				
Rice bran	36	33	45	51
Cassava tubers	64	84	27	61
Cassava tops	3	3	3	24
Sugarcane tops	58	62	0	7
Crotolaria juncea	28	23	7	2
Leucaena leucocephala	24	25	17	3
Rice straw	144	126	0	0
Maize stover	0	0	0	20
Weeds	4	10	22	9
Total	361	366	121	177
Grazing (hours d^{-1})				
Legume	6.6	1.2	4.3	7.3
Sugarcane fields	1.2	1.5	0.2	0.1
Rice paddies	1.2	5.4	3.2	1.4
Bush and village	0.2	1.7	1.5	0.4
Total	9.3	9.8	9.2	9.2

Source: Gibson (1987).

residual moisture in the rice paddy (Kessler & Shelton, 1980). Relatively late flowering legume crops such as *L. purpureus* cv. Rongai extend the grazing season into cool or dry conditions, and changes in growth, sward structure and nutritive value were discussed previously in Section 6.2 (Hendricksen & Minson, 1980, 1985). An alternative approach is to sow forage legumes such as *Stylosanthes* spp. in an established cash or food crop; the delay in forage legume sowing time minimises competition with the established crop and ensures higher quality dry season feed (Shelton & Humphreys, 1975, for rice; Wilaipon, Gutteridge & Chutikul, 1981, for cassava; Mohammed-Saleem, 1982, for sorghum).

A further option is to manipulate by fallowing the seasonal availability of moisture and to sow a C_3 forage crop in the cool season; this impinges on Section 8.2.4. In subtropical southern Queensland fallowing on clay soils gives about 30% efficiency of moisture conservation; continual removal of weeds by cultivation and the consequent reduction in evapotranspiration leads to 30% of the rain falling during the fallowing period being subsequently used by a winter crop of *Avena sativa* or *A. strigosa*. The use of a C_3 grass leads to higher rates of LWG than usually attained with C_4 crops, and a common expectation for weaners is 0.9–1.0 kg hd^{-1} d^{-1} for 90–100 d on *A. sativa*, leading to a total LWG of 220–330 kg ha^{-1}, depending upon moisture availability. In drier and hotter areas the planting of rain-grown winter crops becomes too unreliable in farm practice, as occurs at Theodore, Queensland (lat. 25 °S, 700 mm rainfall, of which 230 mm falls in the cool season, French *et al.*, 1988*b*). Under irrigation in north-western Australia, LWG from oats sown in May was *c.* 330–370 kg ha^{-1} at 4.3 beasts ha^{-1} or 330–480 kg ha^{-1} at 5.7 beasts ha^{-1} (Blunt & Fisher, 1976).

(iii)   Residues of crops grown for cash or food form the basic feed supply for many animal production systems in the tropics (Reed *et al.*, 1988). The residues of grass crops are usually low in digestible nutrients and high in fibre. This problem has not been assisted by the concentration of plant breeders on improving grain yields, which have often been associated with the development of short varieties with high silica content in the straw and low total yield of digestible nutrients from the residues. The mean N content of the rice straw grown at Los Banos, Philippines, varied from 0.77% to 1.16% according to cultivar (Table 8.8, Roxas *et al.*, 1983). Neutral detergent fibre (NDF) ranges were 54–71%, lignin 5–12%, silica 16–22% and IVDMD 31–42%, being negatively correlated with lignin and NDF. Quality also varied with N inputs to the crop. Farmers who value straw yield have rejected short straw

varieties. For example, in the wet rice lands of Burma, farmers prefer cultivars of straw length *c.* 1.5 m in areas with deeper water, *c.* 1.2 m in areas of medium water depth to 0.6 m, and *c.* 1 m in the dry zone irrigated rice areas where water depth is controlled (P. B. Escuro, pers. comm.). The residues of other grass crops, such as *Sorghum* spp. (Arias, López & Aurrecoechea, 1980), barely provide a maintenance diet and require supplementation with higher value feeds if production is to be attained.

Pulse crops provide higher value residues, and the incorporation of pulses as relay crops or intercrops raises potential animal production. For example, in Venezuela groundnut (*Arachis hypogaea*) straw contained 2.3–2.4% N, and gave 52–56% OM digestibility and 56–81 g kg$^{-0.75}$ intake (Velásquez & Gonzalez, 1972); Combellas, Centeno & Mazzani (1971) report similar values.

Many non-leguminous crops give residues of superior nutritive value to that of the grasses, and in some instances (as with cassava) green tops may be harvested during the life of the crop with little detriment to the eventual economic yield. Considerable research in many tropical countries has assessed the value of cassava tops and roots (Meyreles, MacLeod & Preston, 1977; Devendra, 1979*b*; Ffoulkes & Preston, 1979; Araújo & Languidey, 1982; Geoffroy & Barreto-Velez, 1983*a, b*). The banana (*Musa* spp.) leaves have high digestibility (Ffoulkes & Preston, 1978), but the pseudostem requires to be fed with other protein and mineral sources (Ffoulkes *et al.*, 1978; Geoffroy *et al.*, 1978, Geoffroy & Despois, 1978; Ruiz & Rowe, 1980). Sweet potato (*Ipomea batata*) tops gave 79% digestibility and 83 g kg$^{-0.75}$ intake in Venezuela, and appeared to provide by-pass nutrients when fed with sugarcane (Ffoulkes, Hovell & Preston, 1978). Crops of kenaf *Hibiscus cannabinus* (Puentes, 1974) and sunflower *Helianthus annus* (Santana, Caceres & Rivero, 1986) have also received attention as feed sources.

Table 8.8. *Nitrogen (%) of rice straw grown in wet season, Los Banos*

Cultivar	Fertiliser nitrogen level (kg ha^{-1})				
	0	30	60	120	Mean
H4	0.63	0.80	0.79	0.92	0.79
IR8	1.01	1.17	1.13	1.26	1.14
IR36	1.11	1.09	1.19	1.23	1.16
IR42	0.89	0.84	0.95	0.95	0.91
Mean	0.91	0.98	1.02	1.06	

Source: Roxas *et al.* (1983).

(iv) The range of crop by-products in use or having potential use is large. The adoption of these in farm practice is contingent only in part on their nutritive value, since questions of season and continuity of supply, and the cost of transport and wastage in transport and feeding are often overriding. Closeness of location of the farm to the manufacturing site is desirable. By-products may be grouped conveniently as (1) high energy, (2) high protein, and (3) high roughage feeds.

The use of banana fruit as a high energy supplement is illustrated from work at Turrialba, Costa Rica (Cubillos, Vohnout & Jiménez, 1975). Small green bananas are a waste product where banana fruit is

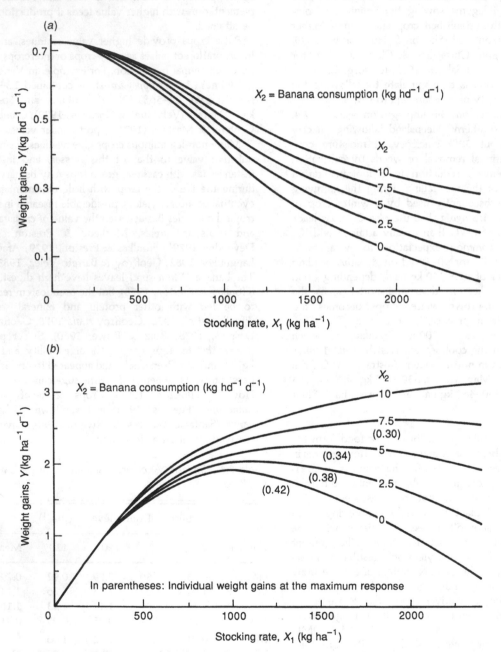

Fig. 8.11. Effects of stocking rate on *Panicum maximum* and level of banana consumption on (*a*) LWG hd^{-1} d^{-1} and (*b*) LWG ha^{-1}. (From Cubillos *et al.*, 1975.)

marketed; the fruit is low in protein and minerals but high in energy, and may contain 15% starch on a fresh basis. Steers were grazed on *P. maximum* pastures at varying SR, expressed as kg body mass ha^{-1} (X_1). At low SR without supplementation maximum gains (Y) were 0.72 kg hd^{-1} d^{-1} (Fig. 8.11*a*), and feeding bananas at 0–10 kg hd^{-1} d^{-1} (X_2) did not increase animal performance. Response to banana supplementation increased as SR increased and the forage allowance of *P. maximum* decreased; the substitutionary effect, which is discussed subsequently in Sections 9.2.1 and 9.3.1, lessened and enabled SR to be increased without detriment to individual animal performance. The potential maximum LWG per unit area (Fig. 8.11*b*) occurred at increasing SR as banana supplementation increased; at zero supplementation maximum LWG ha^{-1} occurred at SR of *c*. 1000 kg body mass ha^{-1}, and with limited supplementation at the level of 2.5 kg bananas hd^{-1} d^{-1} a SR of 1300 kg ha^{-1} gave increased LWG ha^{-1}. This illustrates how sensitive the response to feeding crop by-products is to the availability of other feeds.

Other high energy by-products include molasses (Section 9.3.1), bagasse, citrus pulp (Kirk *et al.*, 1974) and pineapple waste (Chen, Chen & Chung, 1972, Kellems *et al.*, 1979).

Copra meal (Natural & Perez, 1977) and meal from cotton seed, rapeseed, sesame and soybean are commonly used high protein by-products. Spent tea leaf (Jayasuriya & Panditharatne, 1978) may contain 5% N; it has a high polyphenol content but may be fed to ruminants at levels of up to 18% of the ration.

High fibre feeds are more commonly supplied from the on-farm sources of straw and stubble than from manufacturing plants. Sisal by-products obtained after removal of the long fibres (Ferreiro, Preston & Herrera, 1979) may be used by cattle in combination with higher energy and protein feeds.

Some scientists are seized by the potential of discarded agricultural wastes which may not currently enter the ruminant feeding system. In Java Rahardjo *et al.* (1981), having referred to many of the by-products listed above, add sources such as potato skin, soya sauce waste, soy crud waste, 'Kempong tahu' waste, kapok seed press cake, 'Emping' waste, sago waste, 'Nyamplung' oil waste, corn press cake, canned fruit waste, 'Mete' wine waste, 'Mete' skins, 'Perdag' fish waste, canned fish waste, skin and head of shrimps, and corn and rice bran. Seasonal shortages of conventional feed sources may be overcome by access to these materials.

8.2.4
Strategies of irrigation and fertiliser use

(i) Considerations of economics and of social equity dictate that irrigation water will be used primarily in the tropics for the production of cash and food crops; its use on pastures is generally restricted to intensive dairy systems. It has a significant role in these, and high levels of milk production are feasible in the lowland tropics if irrigation is used to maintain continuity of forage supply and ensure the continual availability of freshly grown leaf material. At Ayr, Queensland (lat. 20 °S, 1090 mm rainfall), *D. decumbens* pastures were fertilised every six weeks to a total of 672 kg N ha^{-1} yr^{-1} and irrigated to avoid moisture stress. This led to a mean milk output over three years from Friesian cows at SR of 7.9 cows ha^{-1} of 25630 kg milk ha^{-1} yr^{-1}; these cows were supplemented with molasses/urea at 3.6 kg hd^{-1} d^{-1}. Similar production per unit area was achieved at a lighter SR of 5.9 cows ha^{-1} (Chopping *et al.*, 1976).

Production in the cool season at this lowland site was still limited by the genetic potential for growth of *D. decumbens*, and the better cool season growth and the higher nutritive value of rye grass and clover prompted their evaluation (Chopping *et al.*, 1982). All pastures were well irrigated and received 25–50 mm at 7–14 day intervals according to need. A comparison was made of (1) *C. dactylon* stocked at 7 cows ha^{-1}; (2) *C. dactylon* lightly grazed in April–May, in the same fashion as the areas established to temperate pastures; (3) *C. dactylon* oversown with *L. multiflorum* and *L. multiflorum* × *perenne*; (4) *C. dactylon* oversown with *Trifolium subterraneum* and *T. repens*; and (5) *C. dactylon* oversown with *Lolium* and *Trifolium* spp. Milk production (Fig. 8.12) was substantially increased by oversowing, and level of individual cow secretion reached 15 litres d^{-1} on clover pastures. Milk production during the winter–spring period increased from 12000 kg ha^{-1} to 15000 kg ha^{-1} in the oversown clover treatment. This demonstrated that the successful adaptation of annual temperate grasses and clovers under irrigation might extend at a lowland site to lat. 20°.

A common option for dairymen in the subtropics is the choice of irrigated clover based on rye grass + N pastures for winter and spring grazing as discussed in Section 2.4.2; a seasonal growth curve is illustrated in Fig. 2.7. High seeding rates are desirable, and in central and southern Queensland a conventional *Lolium* mixture might be seeded at 40 kg ha^{-1} in April (Murphy & Whiteman, 1981) or a mixture of 15 kg *T. subterraneum* cv. Clare, 5 kg *T. repens* cv. Haifa or cv. Ladino, 5 kg *T. pratense*, plus 5 kg *L. multiflorum* might be seeded in March or April. The contrasting performance in central Queensland of these alternatives is shown in Table 8.9

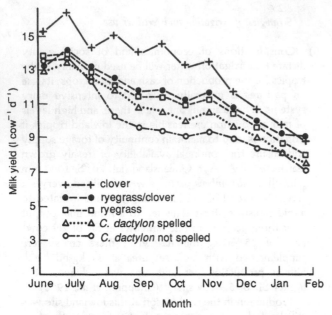

Fig.8.12. Milk production of Friesian cows grazing irrigated *Cynodon dactylon* pastures oversown or not with ryegrass and/or clover. (From Chopping *et al.*, 1982.)

(Chopping, Lowe & Clark, 1983). The winter growth of *L. multiflorum* is superior to that of clover, but milk production from clover is superior to that of *L. multiflorum* in spring. It is suggested that the better nutritive value of clover leads to similar levels of milk production if forage allowance of clover is only 70% of that of rye grass. Rye grass pastures require higher inputs of N, which might amount to 8×50 kg N ha^{-1} per season, and this reduces gross margin relative to that of clover; the latter requires special inputs for control of

weeds and prevention of bloat. Some irrigated temperate pastures give yields exceeding 16 t DM ha^{-1} (Robbins & Faulkner, 1983). In south-east Queensland at Mutdapilly (lat. 28 °S) pastures during the winter and spring months might receive irrigations of 25–40 mm every 14 days according to need, with a total irrigation input of 6 million litres ha^{-1}.

This technology of irrigated temperate pastures has proved to be robust and profitable and has been widely adopted in subtropical intensive dairy systems.

(ii)   Swamp grasslands provide green pasture in the dry season, and pasture managers may fence out their lowlands to reserve these for dry season grazing when green leaf is otherwise unavailable. A further development of this option is to create swamp lands through the construction of wide banks on the contour which will trap runoff and extend the growing season on the ponded areas (Wildin & Chapman, 1987). This technology requires clay soils which retain moisture and is best developed on gently sloping land, since this maximises the area of ponded pasture relative to unit length of bank; slopes up to 0.5% are considered desirable. Banks might be constructed of *c.* 1.3 m height, which would provide 0.6 m water depth with an allowance of 0.4 m freeboard and 0.3 m for settling; banks are best spaced at vertical intervals equal to the depth of ponding. Notional catchment areas in central Queensland for 1 ha of ponded pasture would range from 2.0 ha at St Lawrence (980 mm rainfall) to 7.4 ha at Clermont (640 mm rainfall).

Traditionally *B. mutica* is planted in poorly drained areas, but this species is only productive in less than 0.6 m water. *Echinochloa polystachya* is adapted to deeper zones of water whilst *Hymenachne amplexicaulis* may be planted in areas subject to deep flooding of 1 m or more. The pasture legumes *Aeschynomene americana*, *M.*

Table 8.9. *Production from irrigated* Lolium multiflorum *and* Trifolium *spp. in central Queensland*

Month	Milk production (litres cow^{-1} d^{-1})		Pasture production (kg DM ha^{-1})	
	L. multiflorum	*Trifolium* spp.	*L. multiflorum*	*Trifolium* spp.
July	15.4	13.9	1870	630
August	15.9	16.1	2390	1640
September	14.7	16.0	1740	1800
October	12.9	14.6	1710	1650
November	10.7	12.5	1650	1700
Mean	13.9	14.6	1870	1480

Source: Chopping *et al.* (1983).

lathyroides and *Sesbania sesban* may be established on the fringe areas of the ponds. The presence of free water reduces the incidence of frosting in subtropical regions. Grazing of ponded pastures has led to LWG of up to 100 kg hd^{-1} during the dry season and earlier marketing of fattening cattle. This promising development has been validated in central Queensland and requires wider testing.

(iii)   A seasonal shortfall in forage availability may arise from mineral deficiencies limiting the current rate of growth or limiting the previous accumulation of forage for use in the season of shortfall. Fig. 8.4 in Section 8.2.1 shows how the variation in seasonal carrying capacity of *P. clandestinum* is accentuated under conditions of high N supply. The higher carrying capacity of the N fertilised pastures in the autumn and spring relative to the unfertilised pasture should also be recognised. Strategic applications of N fertiliser in late summer and late winter can therefore be used to increase the forage supply. The response per unit of N applied may average 30–50 kg DM per kg N in the main growing season when feed may be abundant; the pasture manager may accept a lower response level of 10–25 kg DM per kg N in autumn and spring if better forage utilisation is expected and nutritive value better maintained. The use of a higher level of fertilisation on *P. maximum* pastures (300 N vs. 100 N) at Atherton, Queensland, was mainly justified by the higher levels of milk production attained in the autumn and winter, associated with differences in the level of green leaf on offer (Davison *et al.*, 1988).

In Chapingo, Mexico, LWG during the dry season varied from 36 to 218 kg ha^{-1} as N level was increased from 0 to 300 kg ha^{-1}, whilst the grazing season was extended from 56 to 113 days respectively (Roman, 1979). At Palmira, Colombia, dry season SR and the rate of LWG was similarly increased by raising N supply (Vélez & Escobar, 1970). In the Aracatuba district of Sao Paulo, Brazil, Quinn *et al.* (1970) obtained similar total LWG from summer or winter applications of fertiliser N, but winter LWG, both in terms of SR and production perhead, was increased by winter N application and this may have special market or physiological advantages. Table 8.4 indicates how the integration of N fertilised grass pastures, predominantly for winter use, enhanced cattle production.

Other nutrient deficiencies may also shorten the growing season. Blunt & Humphreys (1970) found at Mt Cotton, Queensland (lat. 28 °S, 1430 mm rainfall) that the growth of *D. intortum*/*S. sphacelata* var. *sericea* swards during April–May was enhanced by previous application of 45 or 89 kg P ha^{-1}, relative to growth at lower P application rates.

This chapter has illustrated the multiplicity of solutions available to the manager when determining how to avoid seasonal animal stress. The cheapest solution is to manipulate animal requirements through purchase, sales and mating policies so that needs are in synchrony with the natural supply of forage. The first level of input to the pasture system, and the one most readily adopted by farmers, is to modify the feed supply by changing some of the plant species grown. The introduction of legumes will also change the soil N status. The second level of input is more costly, and may require the purchase of fertiliser or the use of irrigation water to change further the environment in which the pasture grows. These solutions become pragmatic when the manager has at his disposal a hard physical data base which can predict reliably the performance of animals in the various forage situations developed. This decision process will fail unless the interrelationships that exist among the farm system, and between the constituent elements and the farm environment, are recognised.

9

Continuity of forage supply: 2. Pasture conservation and supplementary feeding

A seasonal shortfall in forage availability is best overcome by providing green forage for grazing, using the techniques described in Chapter 8. The secondary approach, which managers need only consider if these techniques are not feasible, is fodder conservation and the feeding of supplements. Pasture conservation is rarely practised in the tropics, although conservation of forage crops is employed in intensive dairy systems and conserved crop residues play a significant role in rice-based agricultural systems. Agricultural advisers who advocate fodder conservation may feel aggrieved by the rejection of their counsel by farmers in the tropics but usually this rejection is soundly based. Fodder conservation adds to the efficiency of the farm system if the following conditions are met:

(1) The negative effects on animal production of the seasonal shortfall in forage availability is real and cannot be overcome by other less costly or more efficient approaches to pasture management. Tropical scientists educated in Europe and North America, where fodder conservation is essential to animal survival through the winter, may fall victim to a cultural imperialism which implies that fodder conservation is an intrinsic component of any sophisticated system of animal production.

(2) The energetics of fodder conservation, taking into account losses during the processes of conservation, transport, storage and feeding of tropical forages, are superior to the energetics of the animal drawing upon surplus body reserves during the period of forage shortfall.

(3) The animal response to supplementary feeding, despite any substitution of 'supplement' for pasture, is greater than any penalty incurred by withdrawing pasture from grazing for the purposes of conservation, and greater than the compensatory gain which occurs in non-supplemented animals during the subsequent post-feeding period.

9.1
Pasture conservation techniques

The efficiency of the techniques available for the conservation of tropical forages is a central constraint to the incorporation of fodder conservation in the farm system. Climatic risks limit hay making until the end of the main growing season, when forage is of low quality. The ensilage of higher quality material at an earlier stage of development also has intrinsic problems in the tropics. The nutritive value of the forage or the price received for animal products may not merit the energetic or financial expense involved in pelleting or drying materials, or in chemical treatment, which may also involve hazards at the farm level. Technology is being developed which mitigates these problems.

9.1.1
Hay

Forage which is cut, dried and stored retains its nutritive value better than standing forage; the essential feature of hay is its low moisture content which inhibits the growth of micro-organisms.

(i) The stage of cutting is crucial in determining the nutritive value of the hay and the risk of spoilage during field curing; the management objective is to achieve a compromise between earliness of cut and likely weather damage. Hay which is spoilt by rain also retards the growth of the sward over which it lies. The decrease of nutritive value with age is a central management consideration which was discussed in Section 6.1.1. Table 9.1 illustrates the decrease in N concentration and the increase in acid detergent fibre (ADF), cellulose and lignin as a hay cut of *P. notatum* cv. Pensacola was delayed from 6 weeks to 10 or 14 weeks after a previous cut at Gainesville, Florida (lat. 30° N) (Moore *et al.*, 1971). These hays were artificially dried at 50 °C and may be expected to have had a higher nutritive value than field-cured hays, as represented by the means of values for two commercial hays of unknown age at cutting (Table 9.1).

These hays were fed to sheep; the 6-week hay exhibited a high value of 61.3% of digestibility and 61.5 units of OM intake, giving a digestible OM intake of 37.7 g kg $W^{-0.75}$ d^{-1} or *c.* 0.68 MJ kg $W^{-0.75}$ d^{-1}.

Unfortunately the latter values are not attained in field practice with C_4 grass hays. In regions with bimodal rainfall it may be feasible to make hay at an early growth stage after the 'small rains' when sunny weather appropriate for field curing is available, but in most tropical situations weather which is reliably dry occurs at the end of the main rains when the grass hay made will have a nutritive value similar to or poorer than the commercial hays shown in Table 9.1. In this study there was a high positive and linear association between %N and both N and OM digestibility, and a strong negative link between % acid detergent fibre and OM digestibility. In these circumstances the increased DM yield of late cut grass hays does not result in increased yield of digestible OM, unless the hays are fed with supplements, as discussed in Section 9.3.1. Nutritive value is better maintained with age in leguminous hays, as has been shown in Campo Grande, Brazil, for *M. atropurpureum* (Lima *et al.*, 1972) and for *N. wightii* (Lima & Souto, 1972).

(ii) Height of hay cut should be judged in the recognition that the lowest strata in the sward have the lowest nutritive value, and that contamination with soil both decreases animal acceptance and increases handling cost. The higher DM yield attained by close cutting may not result in increased DOM intake. Additionally, drying in the field is facilitated by the aeration the forage receives when perched above plant crowns. Uneven soil surfaces and the occurrence of stones increase machinery wear and breakdown. Sod-forming grasses are usually not cut below 3 cm height whereas bunch grasses are cut at 7 cm or higher. Mechanised cutting is commonly carried out with disc, drum, flail or horizontal rotary mowers. The use of reciprocating blade mowers has decreased because of blockage and maintenance difficulties. Hand cutting is the norm in labour-intensive tropical systems.

(iii) Rapidity of curing in the field is sought in the interests of retention of nutritive value, completeness of conserved forage recovery, and minimisation of weather damage. There are various mechanical devices available to facilitate drying. Simple hand turning of the swathe with a fork reduces drying time, since the moister forage is exposed and the swathe is aerated. Mechanised tedding may be employed; this is most effective if applied when the swathe is at a high moisture content soon after cutting. Catchpoole (1969) was able to reduce the field drying time of *Setaria sphacelata* var. *sericea* hay crops yielding over 4.5 t ha^{-1} to 50–55 h by tedding, and shorter periods are feasible in hotter, drier conditions.

Leaves dry more rapidly than stems, and may abscind

Table 9.1. *Effects of provenance on attributes of* Paspalum notatum *hays in Florida*

Attribute	Age at cutting (weeks)			Commercial hays
	6	10	14	
Chemical composition (%)				
Nitrogen	1.22	0.93	1.01	0.98
Acid detergent fibre	40.9	42.7	45.3	46.6
Cellulose	34.9	36.3	37.0	37.2
Lignin	3.7	4.5	5.4	5.6
Digestibility (%)				
In vitro organic matter	59.9	52.7	43.2	42.9
Organic matter	61.3	53.3	48.7	46.4
Nitrogen	48.2	35.2	31.9	25.0
Cellulose	72.4	64.3	63.8	61.9
Intake (g kg $W^{-0.75}$ d^{-1})				
Organic matter	61.5	45.8	41.2	44.3
Digestible organic matter	37.7	24.4	20.0	20.6

Source: Moore *et al.* (1971).

and be lost on the ground before stem material is sufficiently dry for the hay to be taken in from the field. Conditioning, in which the swathe is passed between two rollers (steel, rubber, plastic or a combination of these) is designed to crush the stems and promote more uniform drying (Siewerdt, 1980). It is most effective if applied immediately following cutting, and is sometimes combined with the mower as one piece of machinery. Moisture is lost through the plant cuticle, since stomates close soon after cutting, and crushing or tearing facilitates water loss. This advantage is also claimed for the bruising action of flail cutting. However, these losses are partially offset by the greater absorption of moisture as humidity increases at night, or of rain (Catchpoole, 1969).

Hand placement of drying forage on racks or fences, or in small field ricks assists the drying process.

Spraying with chemicals to increase rate of drying represents a more sophisticated development which has been little adopted in tropical practice as yet. Potassium carbonate is sprayed on some temperate pasture hays, and Norton & Gondipon (1984) have shown that sodium carbonate, sodium hydroxide and potassium hydroxide also exert similar effects on tropical plants. The time to reach 23% moisture content decreased with increasing concentration from 0.1 M to 0.2 M, and the *in vitro* digestibility of material dried in less than 19.4 h was higher than that of untreated forage.

Plant species differ in this suitability for hay making, as related to variation in their rate of moisture loss. *C. gayana.*, *P. coloratum* and *P. maximum* var. *trichoglume*

dried more rapidly in Aichi, Japan, than the thicker stemmed *P. dichotomiflorum* and *Echinochloa frumentacea* (Kanbe *et al.*, 1984); a negative correlation was established between rate of drying and stem diameter. In Brazil water concentration of hays drying in an artificial chamber followed an exponential function (Fig. 9.1; Costa & Gomide, 1989) which was linearly related to leafiness; *H. rufa* and *P. maximum* were superior in this respect.

Hays may be stored at 20–25% moisture content, although drying to *c.* 10% is preferable if long-term deterioration is to be avoided. Ammonia may be injected to enhance N content, and this also acts as a preservative for high-moisture hays, as discussed in Section 9.1.4 (Grotheer *et al.*, 1985). The storage of untreated hay with a high moisture content may lead to the proliferation of thermophilic bacteria and a resultant decrease in nutritive value and stock acceptance. Temperature during storage increases with moisture content (Costa & Gomide, 1989) and in the worst case the hay stack ignites.

(iv) Loose hay may be hand carried and stacked. Alternatively mechanised baling produces a compressed product of longer storage life and more convenient characteristics for handling in transport to the hay stack, stack building and feeding out. The conventional rectangular bale was *c.* 26 × 46 × 90 cm and weighed 15–20 kg (Murdoch, 1980). Currently large round bales weighing 400–500 kg are more commonly used; they may be moved by a single spike mounted on a tractor which is thrust into the centre of the bale, and these shed water if standing in the field. Such bales may be stored in the paddock for seasonal feeding out in regions in which the dry season is reliably dry. Aggregating bales into a stack reduces losses, and covering the stack or providing indoor storage further reduces losses during storage.

(v) The utility of hay making is evaluated in terms of the nutritive value of the hay and of the losses involved in conservation. Losses during the process of hay making may be as little as 10% DM (Catchpoole, 1969), but greater losses are more common. Gomide & da Cruz (1986) reported DM losses of 14–62% in making hay from *P. maximum* and *H. rufa* in Vicosa, Brazil, and these values excluded the occasions when heavy rain occurred during curing. Another Brazilian study (Ribeiro *et al.*, 1980) with *N. wightii* noted field losses during hay making of 5–39% in DM and 10–45% in N. Storage for 180 days (1) in the field uncovered, (2) under plastic sheeting or (3) in a barn led to further losses of respectively 25%, 31% and 15% for DM, and 28%, 26% and 13% for N; losses of digestible DM were

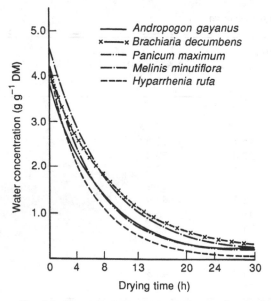

Fig. 9.1. Change in water concentration with time for five tropical grasses (From Costa & Gomide, 1989.)

considerably higher and represented 28 % even when stored in the barn. The levels of these losses, coupled with the low quality of the C_4 grass hays, explain the reluctance of many farmers to adopt hay making.

9.1.2
Silage and haylage
Silage made from tropical pastures is inferior in quality to that made from temperate pastures (Davies, 1963; Jarrige, Demarquilly & Dulphy, 1982), and the advisory officer who wrote 'silage stands supreme ... because of its succulence and appetite stimulating properties' (Anon. 1957) was whistling against the wind. Even when special care (wilting, formic acid addition) is taken with *P. purpureum* silage, the *ad lib* intake is only of the order of *c.* 32–35 g kg $W^{-0.75}$ (Lavezzo *et al.*, 1989). More success is achieved with the special cultivation of fodder crops such as *Zea mays*, but this practice competes with the use of land for food or cash crops, and the silage produced is inferior to that manufactured in temperate areas, due to the higher temperature of fermentation in the tropics and the lower inputs of fertiliser used. What are the circumstances which favour the production of silage having an acceptable standard?

(i)   Silage is the product of a series of processes by which cut forage of high moisture content is fermented to produce a stable feed which resists further breakdown in anaerobic storage. The objective is to retain or augment the nutrients present in the original forage and deliver a silage acceptable to livestock; this is usually achieved through an anaerobic fermentation dominated by lactic acid bacteria.

Water-soluble carbohydrates (WSC) are the primary substrate for the multiplication of lactic acid bacteria, which are initially present at low density in the forage, but which multiply rapidly during the initial ensilage process (Murdoch, 1980). Organic acids are generated which lower the pH of the material, and which inhibit the development of undesirable micro-organisms such as clostridia. These may be proteolytic, or alternatively saccharolytic clostridia and coliform bacteria may compete with lactic acid bacteria for WSC and increase the butyric acid content of the silage. The fast development of anaerobic conditions and a rapid fall in pH constrains the activity of the organisms which lead to spoilage.

Standards for silage quality reflect different aspects of these processes. After reviewing the literature Tjandraatmadja (1989) suggests that stable, acceptable silages usually have pH 4.2 or less, lactic acid as 50 % or more of the total organic acids, butyric acid not greater than *c.* 0.5 % DM, and NH_3-N less than 10 % of total N.

(ii)   Quality is first determined by the chemical composition of the forage ensiled. The choice of time of cut is crucial. Very young forage will have high moisture content and low yield; *Sorghum almum* ensiled at four to seven weeks after planting gave unstable silages with high content of volatile bases (Catchpoole, 1972*a*). Older material has higher DM yield but may exhibit higher fibre and lower WSC (on a DM basis) and N concentrations. In Cuba Esperance *et al.* (1980) compared the quality of regrowths of *D. decumbens* aged 42 or 64 days. The former had crude fibre 30 % (cf. 35 %), 1.3 % N (cf. 1.0 %), and 21 % DM (cf. 27 %), and ensiled at a lower temperature with less loss of N in the ensilage process. Digestibility and intake were greater in the 42 day silage, which produced 7.5 kg milk cow^{-1} d^{-1} relative to 6.1 kg from the 64 day silage.

The disadvantage of the low N content of C_4 grasses may be overcome by the addition of legumes (Xande, 1978) to yield a silage which will lead to production above body maintenance. As discussed earlier in Chapter 6, a minimum N content of 1.1–1.3 % is suggested if intake is not to be limited by this factor.

A mixed pasture with a good legume content will give a silage with satisfactory N content. Barker & Levitt (1969) report a series of silages made from *P. maximum* var. *trichoglume/N. wightii* pasture in which *N. wightii* comprised 11–60 % of the sward; these silages had N content of 1.8–4.2 %.

Alternatively, pure grass swards may be mixed with legume material grown separately. Adding 33 % (fresh matter basis) of *L. leucocephala*, *G. sepium* or *Vigna unguiculata* to a 12-week regrowth of *D. decumbens* cv. Pangola produced silages of acceptable N content (Table 9.2; Tjandraatmadja, 1989). The fodders were all fine chopped and ensiled with 4 % molasses in large drums and these artificial silages were of high quality. Cowpea material had lower DM and N content than the other legumes, and this is reflected in the silage which resulted (Table 9.2). Effluent losses were negligible. Addition of legume reduced butyric acid content and slightly increased pH, but values were still below 4.2. The levels of NH_3-N were satisfactorily low, and lactic acid dominated the fermentation process.

The above results might seem optimistic in the light of other findings. *M. atropurpureum* has produced poor quality silage unless 8 % molasses were added (Catchpoole, 1970) whilst at Okayama in Japan silage quality made from *C. gayana* with *M. atropurpureum* or *S. guianensis* was negatively correlated with the ratio of legumes in the mixture; this appeared to be linked to legume buffering capacity (Uchida & Kitamura, 1987*b*).

A further option is to grow a grass fodder crop either in direct association with a fodder legume or in separate blocks; in either situation the culture of a legume reduces overall DM yield in the short term but increases N yield and content. For example, in Vicosa, Brazil, maize alone gave a fodder with 0.85 % N whilst maize grown in association with *Gylcine max, L. purpureus, C. cajan, Stylozobium aterrimum* or *C. juncea* produced fodder with 1.25–1.73 % N; all these mixtures produced silages with pH 3.8–4.2 (Obeid & Cruz, 1989). *V. unguiculata* with maize produced good silage in Uttar Pradesh, India (Verma & Mojumdar, 1985) but *V. unguiculata* with *P. americanum* in Porto Alegre, Brazil, evidenced problems of low silage quality (Filho & Lopez, 1979).

(iii) Various management approaches are directed to controlling the ensilage process. Wilting the forage before ensiling is advocated as a means of increasing the water-soluble carbohydrate on a fresh weight basis and reducing losses from effluent during storage; rapid wilting is expected in the tropics, indicating the desirability of this practice (Wilkinson, 1983b). There is wide experimental support for this custom. Catchpoole (1972b) compared *S. sphacelata* var. *sericea* wilted for 0, 6 or 26 h; 5-week regrowths dried quickly to 17, 31, or 42 % DM respectively, whilst 8-week regrowths dried more slowly to 19, 24, and 34 % DM respectively over the same periods. The unwilted fodders produced unstable silages and secondary fermentation which developed butyric acid. This might be linked to the water activity, which is a measure of the osmotic pressure within the plant cell. Butyric acid is not usually produced when at a particular pH water activity is sufficiently low at ensiling or shortly after to prevent the conversion of lactic acid into butyric acid. The curved dashed line in Fig. 9.2 indicates the division where butyric acid growth is expected; wilting for 26 h of both regrowths led to stable silages with sufficiently low water activity to control butyric acid development. From these data and from work with *C. gayana* Catchpoole (1972b) recommended pre-wilting to DM values exceeding 35 %. Similarly, Davies (1963) in Zambia found 3 h wilting of *C. gayana* to 30 % DM reduced nutrient losses, whilst *C. dactylon* wilted to 35 % DM decreased losses of carotene and xanthophyll (McHan *et al.,* 1979). Wilting promoted better fermentation and preservation of *P. purpureum* in Brazil (Farias & Gomide, 1973; Lavezzo *et al.,* 1989), and of *P. maximum* in Cuba (Ojeda & Cáceres, 1981). However, wilting is not a cure-all for every aspect of ensilage practice, and negative effects have been reported (Filho & Mühlbach, 1986; Uchida & Kitamura, 1987a).

(iv) Mechanical treatment of the crop which controls the degree of maceration influences fermentation; small particle size also facilitates compression and the exclusion of air from the silo. In Sri Lanka Panditharatne *et al.* (1988) found that chopping *P. maximum* to *c.* 1.5 cm length before ensiling increased the digestibility and intake of DM, N, and cell wall fractions relative to that of unchopped material, and this was attributed to the better release of fermentable substrates. Many

Table 9.2. *Chemical composition of silages*

Component (g kg^{-1} DM)	Pangola	Pangola + *Leucaena*	Pangola + *Gliricidia*	Pangola + cowpea
Dry matter (g kg^{-1})	297	321	319	289
Neutral detergent fibre	624	517	522	583
Acid detergent fibre	388	340	338	377
Hemicellulose	236	177	184	209
Cellulose	342	283	282	330
Lignin	45	56	56	45
Ash	80	74	75	78
Water-soluble carbohydrates	45.7	32.4	44.2	41.2
Total nitrogen (TN)	9.5	17.4	16.1	12.0
NH$_3$-N (g kg^{-1} TN)	90	85	78	92
pH	3.99	4.14	4.09	4.04
Lactic acid	41.4	47.6	48.9	51
Acetic acid	9.3	8.5	8.5	12.8
Butyric acid	5.5	3.5	0.3	1.7
Ethanol	13.6	13.7	11.2	19.2

Source: Tjandraatmadja (1989).

tropical forages are bulky, fibrous and springy; Levitt, Hegarty & Radel (1964) found difficulty in producing silage of reasonable quality from *P. dilatatum* and this was partly attributed to compaction difficulties; chopping to 2.5–5 cm produced a better silage than flail-harvested material of 15–30 cm length. Rapidity of filling also favours the exclusion of air and reduces the amount of surface spoilage. Compaction of trench silos is accomplished by repeated tractor movement, or the surface may be weighted with soil or with discarded tractor and vehicle tyres, or discarded masonry. Bulk density of tropical silages is consistently lower than that of temperate silages (Catchpoole & Henzell, 1971).

(v)   The exclusion of oxygen is a central problem for the ensilage of tropical forages, and has not been overcome in field conditions. The significance of air exclusion is illustrated by a study in which silages made within airtight containers (films with oxygen transmission of 1 ml d^{-1} m^{-2}) were contrasted with those manufactured under polyethylene films of 50 μm thickness, which were weakly permeable to oxygen (Tjandraatmadja, 1989). At different temperatures of incubation stable silages of low pH were produced from *S. bicolor*

fermented in anaerobic conditions (Fig. 9.3). By contrast silages produced in polyethylene bags were unstable and showed increasing pH (5.5–7.7) with period of incubation and significant DM loss. Lactic acid production was negatively correlated with the pH of the material. The exposure of plant material to air delays the release of plant cell sap and the onset of reduction in pH, which may be insufficient to preclude clostridial fermentation.

A further feature of these data was the prevalence of acetic acid fermentation in the permeable polyethylene bags. Many of the experiments with tropical silages have been conducted in these conditions, and the conclusion of Catchpoole & Henzell (1971) that tropical silages are predominantly of the acetic (rather than the lactic) acid type may be associated with the great difficulty of achieving anaerobic conditions when ensiling bulky tropical feeds. Hamilton *et al.* (1978) could not exclude air from flail-harvested *S. sphacelata* var. *sericea* even in a polythene-covered stack with a vacuum pump applied, and considerable surface wastage occurred.

(vi)   Compression of the silage stack is directed to promoting a cold fermentation, since temperatures rise under aerobic conditions. In good farm practice a

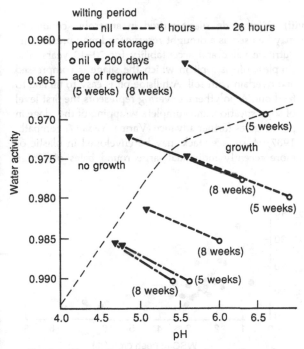

Fig. 9.2. Change in water activity during ensilage of *Setaria sphacelata* var. *sericea* at two maturities and three wilting periods, together with areas of suggested growth or non-growth for butyric acid bacteria. (From Catchpoole, 1972*b*.)

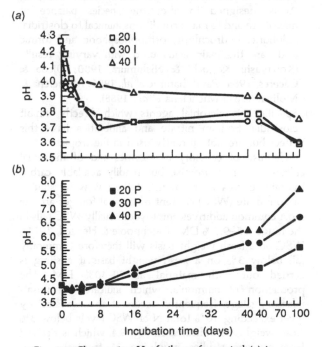

Fig. 9.3. Changes in pH of silages fermented (*a*) in oxygen impermeable bags (I) and (*b*) polyethylene bags (P) at 20, 30 and 40 °C. (From Tjandraatmadja, 1989.)

thermometer is plunged below the surface of the plant material being ensiled in a trench silo, and additional rolling is applied if the temperature exceeds *c.* 35–40 °C. The ambient temperature conditions also influence the quality of the silage produced, as is cogently illustrated in Fig. 9.3. The silages developed under both aerobic and anaerobic conditions had much lower pH values if incubated at 20 °C, and this indicates a further problem in achieving high quality silage in the tropics, especially with respect to the occurrence of heat-damaged protein (Wilkinson, 1983*a*).

(vii)   Silage additives have been developed to direct the processes of fermentation and preservation in particular directions. Woolford (1985) classifies these as:

(1) Fermentation stimulants, designed to stimulate rapid increase or dominance of lactic acid bacteria. Molasses, (Catchpoole, 1966; Holm, 1974; Alli *et al.*, 1984) cassava meal, (Ferreira, Silva & Gomide, 1974; Panditharatne *et al.*, 1986) coconut milk and also cellulolytic and amylolytic enzymes and microbial cultures (McHan, Burdick & Wilson, 1984; McHan, 1986; Lusi, Ojeda & Ramirez, 1986) fall into this category.

(2) Fermentation inhibitors, added to inhibit microflora overall. Formaldehyde and sodium metabisulphite (Levitt, Taylor & Hegarty, 1962; Ojeda & Varsolomiev, 1982; Lavezzo, Lavezzo & Silveira, 1984) are examples.

(3) Acids, designed to alter the species balance of microflora and to make conditions inimical to clostridia. Sulphuric, hydrochloric, orthophosphoric and formic acids are the main acids used with varying results (Kobayashi, Koguchi & Nishimura, 1980; Ojeda & Cáceres, 1984; Panditharatne *et al.*, 1986; Michelena & Molina, 1987; Michelena *et al.*, 1988).

(4) Specific anti-microbial agents which directly inhibit clostridia. Sodium nitrate and antibiotics have this effect but are not currently used in the tropics.

The use of cellulolytic enzymes and of microbial cultures exhibits promise but readily available carbohydrate sources which augment the low water-soluble carbohydrate (WSC) content of tropical forages are the most common additives employed. Usually WSC falls in the range 2.5–9.9% DM (Catchpoole & Henzell, 1971); WSC on a fresh weight basis will therefore commonly fall below 3% on a fresh weight basis if ensiling is carried out with material at 20–40% DM. The production of ammonia, which indicates an unsatisfactory silage, frequently occurs at an unsatisfactory level exceeding 10% total N at WSC levels below 3% fresh weight, as indicated in Fig. 9.4, which is derived from various studies involving 31 tropical grasses (Wilkinson, 1983*a*). Tjandraatmadja (1989) achieved good fermentation at lower WSC than the suggested

3% WSC fresh weight critical value, but this occurred under anaerobic laboratory conditions and other carbohydrate sources may also be involved.

The effects of molasses addition are illustrated for whole chopped *L. leucocephala* grown at Trinidad, West Indies (Alli *et al.*, 1984). *Leucaena* was ensiled at a high DM content of 40.5%, but WSC was only 6.3% DM; the addition of molasses at 2.3 or 4.5% raised WSC to 9.7 or 13.0% DM respectively. This led to reduced pH in the silage produced after 28 days (Table 9.3); volatile N was reduced by molasses addition but levels were satisfactory in the three silages. Non-protein nitrogen was also reduced, indicating that addition of molasses increased proteolysis. Lactic acid and acetic acid production was higher in the treated silages, which was consistent with the reduced population of yeasts and moulds found in the silages receiving molasses.

The need for molasses addition obviously depends upon the substrate quality and availability in the forage being ensiled. The rate of addition of 4% molasses on a fresh weight basis is commonly recommended, but successful levels have varied from 1.8% (Levitt *et al.*, 1962) to 8% for *M. atropurpureum* and *D. intortum* (Catchpoole, 1970); in the latter study *L. bainesii* required no molasses, indicating the need to develop specific recommendations.

(viii)   Well-made silage stored in anaerobic conditions may be used as a drought reserve, and may preserve its nutritive value and acceptability for 20–40 years. The simplest silo is a pit in which the silage is compressed and overlain with soil. A hillside pit is easy to fill and to feed out. A polythene covering represents the first level of sophistication, and complete wrapping of the silage in plastic gives further advance (Varma, Yadav & Sampath, 1987). A surface stack may be enveloped in plastic or more recently individual large round bales of forage

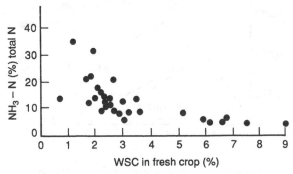

Fig. 9.4. Relationship between the content of water-soluble carbohydrates (WSC) in fresh tropical grasses and the proportion of total N as NH_3-N. (From Wilkinson, 1983*a*.)

may be 'silo-wrapped'. Silage may be compressed in the 'Silopress' system into long plastic 'sausages'. Tower silos traditionally minimise losses. These may be further modified to provide a sliding internal sealed top; with the provision of vacuum storage a 'Harvestore' (Fig. 9.5) enables nutritive value to be well-preserved. This capital-intensive installation is especially suited to the conservation of 'haylage', a fodder intermediate in

Table 9.3. *Effects of molasses addition on composition of* Leucaena leucocephala *silage after 28 days*

Indicator	Level of molasses addition (fresh weight)		
	0%	2.3%	4.5%
pH	4.7	4.3	4.1
Dry matter (%)	38	38	38
Total N (%)	2.8	2.9	2.9
Volatile N (% of total N)	5.6	4.9	4.9
Non-protein (% of total N)	15.5	13.3	11.9
Lactic acid (%)	2.0	4.1	5.0
Acetic acid (%)	0.4	0.6	0.7
Propionic acid (%)	0.2	0.5	0.6

Source: Alli *et al.* (1984).

Fig. 9.5. Dairy cow grazing *Pennisetum americanum* with Harvestore in background.

moisture content between hay and silage. Chopped material is unloaded from a trailer mechanically and is blown to the top of the tower; feeding out from a mobile delivery tube is also mechanised. This complex is suited to large-scale intensive dairy systems.

The losses incurred through the use of silage occur sequentially in the field (respiration, mechanical losses), in the silo (respiration, fermentation, effluent and aerobic deterioration), and in the feeding out-process (mechanical losses, further spoilage, rejection). Pizarro & Vera (1980) have developed a model for predicting yield losses, and animal response to maize silage. They estimate typical DM losses from maize ensiled in bunker silos as 6% in the field, 14% in the silo, and 5% during removal, leading to a total of 25%. Farm experience with perennial pasture silages suggests rather higher losses as the norm, and these need to be considered with the animal responses discussed subsequently in Section 9.3.

9.1.3
Drying and pelleting

(i) Artificial dehydration of pastures can be carried out at a young growth stage which leads to a product of high value. Heating the chopped material, either in a slow drying process at 150–250 °C or in a rapid drying rotary drum at 600–1000 °C (Murdoch, 1980), gives low DM losses of c. 3–10% in temperate grasses and good conservation of nutrients. The energy cost is related to the initial moisture content; oil consumption in one study was $168\,l\,t^{-1}$ at 70% moisture, and increased exponentially to $367\,l\,t^{-1}$ at 85% moisture, which points to the advantage of pre-wilting (Manby & Shepperson, 1975).

The thrust of environmental policies and the need to conserve fossil fuels in tropical countries run counter to the adoption of pasture drying, which is also more suited to capital-intensive, large-scale enterprises than the smallholder animal production systems which predominate. Some *L. leucocephala* leaf is artificially dried and the main use of artificial drying is in the stock feed industry, whose outputs are focused on non-ruminant production or on intensive dairying in a restricted number of subtropical regions.

(ii) Grinding and pelleting a forage increases intake, especially for feeds of low nutritive value and high fibre content (Minson, 1982). This is associated with a faster rate of passage of forage particles from the rumen, and this may reduce digestibility.

This is illustrated from a study at Belle Glade, Florida (Coleman *et al.*, 1978), in which *B. mutica* at four (2.3% N) or eight (1.8% N) weeks of age was fed to steers, either as green-chop or following drying at 170 °C, grinding in a mill with a 3.2-mm screen and pelleting

Table 9.4. *Influence of forage species, physical form and maturity on voluntary intake, nutrient digestibility and rate of passage of* Brachiaria mutica

Indicator	Green chopped		Pelleted	
	4 weeks	8 weeks	4 weeks	8 weeks
Intake (g kg $W^{-0.75}$ d^{-1})				
Organic matter (OM)	80.1	57.2	100.9	75.0
Digestible OM	53.4	35.0	61.0	36.1
Digestibility (%)				
Dry matter	66.4	60.7	59.6	47.4
Organic matter	66.9	61.0	60.3	47.7
Crude protein	71.8	69.8	55.2	53.5
Neutral detergent fibre	70.1	63.5	63.5	47.5
Acid detergent fibre	69.6	63.0	59.7	41.6
Cellulose	74.3	68.5	66.0	49.8
Lignin	47.1	40.2	31.8	27.7
Rate of passage				
Mean time (h)	45.5	51.9	41.2	42.6
Rate (% h^{-1})	4.3	3.3	4.6	4.0

Source: Coleman *et al.* (1978).

with water and steam using a 9.5 mm die. Intake (Table 9.4) of *B. mutica* at both stages of development was increased substantially by pelleting. However, the intake of digestible OM was similar in the 8-week pelleted or green-chopped feeds, since digestibility was due to changes in cell wall digestibility and not to changes in digestibility of the N and non-structural carbohydrate in cell contents. The faster rate of passage (Table 9.4) of ground fibrous constituents might be linked to the decrease in digestibility.

In other studies (Chapman *et al.*, 1972; Hennessy & Williamson, 1976) pelleting of C_4 grasses led to better rates of liveweight gain, associated with increased intake. As with artificial drying, pelleting is a capital-intensive process which is expensive for support energy. Wilkins (1982) has shown for temperate grass hay that the ratio of liveweight gain to support energy in the system is much reduced by pelleting.

9.1.4
Alkali and other chemical treatment of residues

(i) The great quantity of low-quality standing grass hay and of the crop residues upon which many animal production systems depend (Chapter 2) constitute an important biological resource in the tropics. Con-siderable research has been undertaken on every continent to develop technologies which would lead to more effective use of these low N, high fibre fodders. The principal technologies are directed to the improvement of nutritive value through (1) alkali treatment designed to degrade cell wall content, and (2) treatment with ammonia or urea to provide fibre breakdown, N augmentation and better preservation of high moisture hay.

The techniques developed have met these objectives, but have been little adopted in farm practice. Usually this has resulted from the high cost or inaccessibility of materials in relation to the low value of the conserved product and to the prices for animal products (Holm, 1972; Tessema & Emojong 1984); questions of safety are involved with sodium hydroxide and other substances, or sophisticated equipment and supplies may be needed, as is the case with the injection of anhydrous ammonia which is, in any event, difficult to distribute evenly in a compressed hay. The level of support energy is also involved (Wilkins, 1982). A perspective on this topic is given by the recommendation of an international workshop (Reed *et al.* 1988) that 'chemical treatment of crop residues has limited applicability in tropical and subtropical countries. Emphasis should be transferred to exploiting genetic variation in crop residue quality'.

Table 9.5. *Voluntary intake and digestibility and N retention of cattle given untreated and urea-treated hay with or without supplementary urea*

Item	Treated + urea	Treated alone	Untreated + urea	Untreated alone
Intake of DM (g kg $W^{-0.75}$ d^{-1})	80.2	75.2	77.8	61.1
Dry matter digestibility (%)	68.0	64.0	66.8	53.6
Intake of organic matter (g kg $W^{-0.75}$ d^{-1})	74.5	70.1	73.8	57.5
Organic matter digestibility (%)	69.7	63.9	67.3	56.4
Neutral detergent fibre digestibility (%)	77.0	73.6	75.0	63.6
Acid detergent fibre digestibility (%)	71.4	67.5	67.5	56.8
N apparently retained (g d^{-1})	37.3	26.3	18.7	−9.3
Apparently retained N (% apparently absorbed)	63.7	63.3	82.0	−

Source: Lufadeju, Olayiwole & Umunna (1987).

(ii)   Treatment of hays and straws with alkalis such as sodium, potassium or calcium hydroxide is directed to solubilising lignin and/or hemicellulose and their linkages (Doyle, Devendra & Pearce, 1986). The increased swelling capacity of the cell walls facilitates access of microbial enzymes and greater digestion of structural elements. Spencer, Akin & Rigsby (1984) observed tissue disruption in *C. dactylon*, swelling of the vascular, sclerenchyma and epidermal cells, and break-down of the bonding in the intercellular layers. These types of effects lead to greater nutrient digestibility and intake of grasses such as *Hyparrhenia* spp. (Gihad, 1979) or of crops such as rice (Devendra, 1979a) or maize (Benitez, Bravo & Gómez, 1984). Either a wet or a dry process may be used. Negative reports also occur (Thiago, Leibholz & Kellaway, 1981), and it should be recognised that usually alkali-treatment converts a submaintenance fodder to a maintenance fodder only.

(iii)   Treatment with ammonia may have similar effects to those exerted by other alkalis, and other positive effects. Ben-Ghedalia *et al.* (1988) noted swelling of the cell wall fibrils in *C. dactylon*, saponification of ester bonds and loosening of the cell wall, which were regarded as more significant than the solubilisation of cell wall material. Digestibility was also increased in *M. minutiflora* and *B. decumbens* (Reis & Garcia, 1989). The fungistatic effect of injecting anhydrous ammonia has been demonstrated for high-moisture hay whose keeping-quality would otherwise be poor; Grotheer *et al.* (1985) reported ammonia treatment lowered total aerobic bacterial and fungal counts in large round bales of *C. dactylon* hay stored at 68% DM and increased digestibility. Similar results occurred with *P. maximum* hay treated with urea (Silanikove *et al.*, 1988); naturally occurring urease leads to the production of ammonia. However, some reports indicate no benefit to animal performance from ammonia treatment (Lara, Parra & Neher, 1985; Llanos, Parra & Neher, 1985).

Treatment of hay with urea is simpler for smallholders than the use of anhydrous ammonia, and urea both influences cell wall content and digestibility and corrects a common N-deficiency in these fodders (Ulloa, Watkins & Craig, 1985). Addition of urine is a less sophisticated approach (Saadulah, Haque & Dolberg, 1980). Effects of urea treatment were contrasted with urea supplementation in Zaria, Nigeria (Lufadeju, Olayiwole & Umunna, 1987). *A. gayanus* hay was treated with 50 g urea kg^{-1} hay as a 6.7% solution, covered with a polythene sheet in a pit, and left for 20 days. Aeration for 2 days was provided after removal of the sheet to minimise the content of volatile ammonia. Treated or untreated hay was fed to heifers, who either received a urea supplement of 12.5 g urea kg^{-1} hay intake (to a maximum to 50 g urea hd^{-1} d^{-1}) or not.

Treatment of hay with urea increased N content from 0.55 to 1.78% but did not significantly influence the content of other main constituents. Digestibility of DM, OM, neutral detergent fibre and acid detergent fibre was substantially increased by both urea treatment and urea supplementation (Table 9.5), but the latter effects did not differ significantly. Apparent retention of N was also increased by both hay treatment and urea supplementation, and the combined effects were greater than for urea supplementation alone.

Lufadeju *et al.* (1987) conclude that most of the positive animal response could be achieved more simply than hay treatment by providing a urea supplement. This might be achieved by spraying the hay at the time of feeding with urea at the rate of 1% DM intake.

Other potential developments include the use of oxidative reagents, acids, gamma irradiation, biological treatment and fractionation (Wilkins, 1982; Doyle *et al.*, 1986).

9.2
Behaviour of supplemented animals

The utilisation of fibrous materials may be improved by using materials which 'provide additional nutrients to ensure an appropriate substrate environment in which fibre fermentation rate is increased, (or ...) which provide an alternative source of nutrients for the animal which does not adversely interact with the processes of fermentation of the roughage' (Egan, Frederick & Dixon, 1987). Two factors which need to be considered when evaluating the fodder conservation systems are (1) the possibility that the additional feed offered is substitutionary rather than supplementary, and (2) the persistence of supplementation effects on animal performance after supplementary feeding has ceased.

9.2.1
Substitutionary feeding

Animals whose consumption of pasture is limited by the low N concentration in the pasture may increase their pasture intake if an N supplement is fed, as discussed in Section 6.1. However, if a supplement of good energy and protein content is fed, animals reduce their pasture intake. This phenomenon also occurs if the feed supplement is preferred to standing pasture for other reasons which confer acceptability. Thus in Malaysia weaner cattle receiving fresh *P. maximum* reduced their forage intake according to the level of concentrate fed, and the ratio of feed conversion to LWG decreased with increasing amount of concentrate provided (Chik, Hassan & Idris, 1984). Similarly in Venezuela heifers grazing *C. ciliaris* pastures at a forage allowance of 5 kg DM per 100 kg LW reduced their intake of pasture if a concentrate mixture was fed (Combellas, Baker & Hodgson, 1979). Concentrate levels were 0, 3 or 6 kg heifer^{-1} d^{-1}; herbage OM intake was reduced by 0.64 kg in the rainy season and 0.42 kg in the dry season for each kg concentrate OM eaten. The response to supplementation was only 0.27 kg extra milk kg^{-1} concentrate.

There are few tropical experiments in which the substitutionary effect of hay or silage has been estimated. Harvey *et al.* (1963) noted that feeding *P. dilatatum* silage reduced the duration of grazing. A similar effect was observed in Friesian cows grazing *P. maximum* var. *trichoglume/N. wightii* pastures at 2.5 beasts ha^{-1} at Atherton, Queensland (Lat. 17° S, 1300 mm rainfall, 700 m a.s.l.). During an 8-week winter period cows were fed maize silage at two levels, with or without a meat-and-bone meal supplement (Davison, Marschke & Brown, 1982). Some of the highest milk yields per cow recorded from tropical pastures with silage were evident, and responses both to a high level of maize silage and to meat-and-bonemeal supplement occurred (Table 9.6); the response to this protein supplement was greater at the higher level of silage feeding. Content of solids-not-fat in milk and LW change were increased by the higher level of silage. These responses need to be evaluated in terms which take account of the additional land area allocated to maize silage production.

Pasture yields were similar for all treatments at the beginning of the experiment, averaging 3480 kg DM ha^{-1} with 47% legume content. However, on this relatively high quality pasture, presentation yield after eight weeks was 670 kg ha^{-1} greater on the paddocks where a low silage level was fed, indicating reduced pasture intake on the high silage treatment. Grazing time (Table 9.6) was decreased *c.* 40 min kg^{-1} supplement fed.

The degree of substitutionary feeding will be influenced not only by the relative acceptability of the feeds offered, but by the level of pasture availability. Less substitution will occur as forage allowance decreases, and the effect disappears when no pasture is available.

9.2.2
Persistent effects of supplementation

The interpretation of the results of much of the research undertaken with pasture conservation is hampered by the failure of scientists to view it in the context of the whole farm situation. Animal response during a feeding experiment needs to be qualified by the effects of withdrawing land for the production of the supplement on animal performance during that earlier period, and by the effects of feeding the supplement on the subsequent performance of the supplemented animals. This is usually positive for milk production and for reproduction, but negative for liveweight gain.

Table 9.6. *Milk yield and composition, milk fatty acids, liveweight change and grazing time of cows during and after a period of supplementary feeding*

Attribute	Low silage[a]		High silage[a]	
	U	S	U	S
Total intake of supplement (kg DM cow^{-1} d^{-1})	3.0	3.0	7.0	7.9
Milk yield (kg cow^{-1} d^{-1})				
Weeks 1–8	14.3	15.0	15.3	16.6
Weeks 9–16	13.6	14.0	15.0	15.0
Fat (%)	3.66	3.88	3.56	3.23
Solids-not-fat (%)	8.38	8.42	8.54	8.4
Milk fatty acids (molar % C4–C16)				
Week 1	53.8	57.3	57.1	57.1
Week 8	54.6	51.5	61.5	56.6
Liveweight change (kg)				
Week 1–8	−15.0	−15.4	2.0	11.4
Grazing time (min d^{-1})	497	505	343	293

[a] U, nil meat-and-bonemeal; S, meat-and-bonemeal provided in the ratio 5:1 (silage to meat-and-bonemeal) on a dry matter basis.
Source: Davison *et al.* (1982).

The persistent effect of earlier supplementation on milk yield was positive for the eight weeks after cessation of feeding in the experiment of Davison *et al.* (1982); meat-and-bonemeal supplementation had no persistent effect but the higher level of maize silage feeding increased milk yield by 1–2 kg milk cow^{-1} d^{-1} (Table 9.6). During this later period the level of pasture on offer decreased due to moisture stress, and the higher liveweight and body reserves in the well-fed cows helped to maintain milk secretion. Similar effects of high body reserves might be expected to promote conception and reduce abortion in both beef and dairy cows.

On the other hand, a prevalent phenomenon in growing cattle is that of compensatory weight gain. Fig. 9.6 illustrates diagrammatically a situation where cattle grazing grass pastures are gaining 0.5 kg hd^{-1} d^{-1} to point A. As feed quality deteriorates the rate of gain decreases to point B, and hay is fed to maintain LWG at 0.25 kg hd^{-1} d^{-1} to point C, when pasture availability and quality has recovered sufficiently to give continued LWG at 0.5 kg hd^{-1} d^{-1}. If animals are not fed from point B they might lose 20 kg to point C, when a difference between the two groups in LWG of 40 kg hd^{-1} would exist. However, this difference is reduced to 20 kg at point D, since the unsupplemented animals gain more quickly than the previously supplemented group. If the animals are slaughtered at point C no opportunity for compensatory weight gain is present.

The reasons for compensatory gain are not fully understood. The major factor appears to be increased appetite when feed is abundant for animals previously underfed; a temporary increase in gut contents together with increased efficiency of growth resulting from lower maintenance requirements and a lower energy requirement for muscle rather than fat deposition may also be involved. The relationship between the difference in LWG between supplemented and unsupplemented groups and the subsequent compensatory gain appears to be linear. The coefficient of reduction in wet season gain (kg) per kg of previous dry season advantage was 0.19 in north-west Australia (Robinson, 1967), 0.22 in south-east Queensland (Addison, 1970) and 0.52 in Tanzania (Walker, 1969). The extent of the phenomenon is greater in old than in young stock; in Zambia Smith & Hodnett (1962) noted regression coefficients of 0.37, 0.70 and 0.82 in cattle aged 2, 3 and 4 years respectively. In Kenya young steers gained 0.1 kg hd^{-1} d^{-1} on *C. gayana* silage; the advantage of higher growth rates achieved by concentrate supplementation were subsequently lost during wet season grazing (Thomas, 1977).

9.3
Production systems involving supplementation

The focus of this chapter is on the production potential of systems involving pasture conservation. However, as described previously, some intensive systems use silage based on fodder crops such as maize. There is also widespread interest in rectifying the seasonal nutrient deficiencies of pasture by providing grazing animals with high energy and/or protein supplements. Antoni Padilla *et al.* (1983) in Puerto Rico suggest that high levels of concentrate feeding on well-fertilised pastures can enable SR to be raised to 5 cows ha^{-1}. This conflicts with the considerable movement in many tropical countries away from high level concentrate feeding and the adoption of alternative approaches, as described in Chapter 8, since purchase or production of concentrates (1) competes with the production of food and cash crops, (2) requires a high level of support energy and/or foreign exchange, (3) involves nutrient transfer and creates environmental pollution, and (4) is more efficiently applied to monogastric than ruminant production. The emphasis given to this question is obviously influenced by the cost and availability of local concentrate sources and the returns for animal products. Systems involving lot-feeding are outside the scope of this book.

9.3.1
Responses to energy and protein supplementation

Much early attention of scientists was given to rectifying N deficiency in animals grazing tropical pastures (Elliott & Fokkema, 1960; Milford & Haydock, 1965; Haggar, 1971; Minson, 1971*c*) because the widespread N deficiency could be overcome relatively simply in the field and was seen as the key to increasing energy intake. More recently the intransigent problem of the low energy

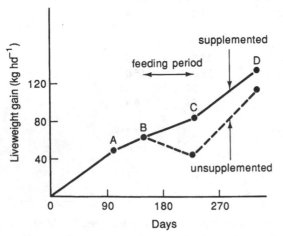

Fig. 9.6. Diagrammatic change with time in liveweight gain of cattle supplemented or unsupplemented with hay.

availability of C_4 grasses, as discussed in Section 6.1, has gained greater recognition as the major constraint to the performance of grazing animals in many tropical situations (Romero & Siebert, 1980).

(i) The pasture manager seeks to identify the types of grazing situation in which a response to supplements is expected, and the nature of the supplements required. These are principally (1) high energy supplements, such as molasses, bran or grain, (2) protein supplements, such as cotton seed meal or waste meat or fish meal, (3) non-protein nitrogen (NPN) sources, such as urea or biuret

and (4) sulphur, sometimes fed as sodium sulphate, or as a constituent of other supplements; molasses has a significant S content of *c.* 0.7%.

Siebert & Hunter (182) have developed a pragmatic summary of the expected responses in LWG and DE intake to supplementation of sheep and cattle according to the availability and quality of the pasture on offer (Table 9.7). Energy supplementation gives a positive response if pasture availability is low, especially with high fibre pasture; at high levels of pasture availability some response to energy supplementation may occur if the pasture is fibrous.

Table 9.7. *Expected response[a] in energy intake to various supplements according to pasture characteristics*

Pasture attribute	Type of supplement			
	Energy	Protein	NPN[b]+S	S
Availability low				
Fibre low				
Protein low				
N/S low	+	+	+	0
N/S high	+	+	+	+
Protein high				
N/S low	+	0	0	0
N/S high	+	0	0	0
Fibre high				
Protein low				
N/S low	+	0	0	0
N/S high	+	0	0	0
Protein high				
N/S low	+ +	+	0	0
N/S high	+ +	+	0	0
Availability high				
Fibre low				
Protein low				
N/S low	0	+ + +	+ +	0
N/S high	0	+ + +	+ +	+
Protein high				
N/S low	0	0	0	0
N/S high	0	0	0	0
Fibre high				
Protein low				
N/S low	+	+ +	+	0
N/S high	+	+ +	+	+
Protein high				
N/S low	+	+	0	0
N/S high	+	+	0	0

[a]Nil, 0; small, + ; medium, + + ; large, + + +.
[b]NPN, non-protein nitrogen.
Source: Siebert & Hunter (1982).

Protein supplementation, naturally enough, will be most effective if pasture protein is low, but responses can occur in the unusual combination of high fibre, high protein diets. The primary effect is to increase intake. Supplementation with NPN is rarely effective under conditions of low forage availability, and responses are most prevalent with high forage availability and low N content, provided a ready source of available carbohydrate (such as molasses) is present. A wide N/S ratio in the feed may restrict the capacity of an NPN supplement to stimulate the synthesis of protein by rumen organisms; the limit may be 10:1 in sheep, or 15:1 in cattle (Winks, O'Rourke & McLennan, 1982). On sulphur deficient soils which produce a wide N/S ratio in the feed grown it may therefore be necessary to augment S supply.

(ii) Energy supplementation is used in intensive dairy systems for cows whose genetic potential for milk production exceeds that of the capacity of the pasture on offer. It is common to find that quite leafy C_4 grasses do not satisfy the full nutritional requirements of lactating cows. Provision of energy supplements stabilises production during adverse weather conditions, and raises the overall level of milk secretion.

This is illustrated for Friesian cows and first-calf heifers grazing *P. maximum/N. wightii* pastures at Atherton, Queensland (lat. 17° S, 1300 mm rainfall, 700 m a.s.l.) at 4 cows ha^{-1} (Fig. 9.7; Cowan, Davison & O'Grady, 1977). The level of milk production was linearly related to the level of a maize–soybean meal concentrate supplied over the range 0–6 kg hd^{-1} d^{-1}, and the response was 1.03 kg milk kg^{-1} concentrate. The persistency of lactation was clearly influenced by concentrate feeding and fell away quite steeply with time after calving in the unsupplemented animals, which also had a shorter lactation; length of lactation varied from 222 to 277 days according to level of concentrate fed. Fat content was negatively related to concentrate fed (4.0–3.4%) but solids-not-fat was increased by concentrate from 8.1 to 8.6%.

Liveweight of cows (Fig. 9.7) increased with level of concentrate fed and this assists both the persistency of lactation as the quality of pasture on offer decreases in the winter and facilitates subsequent conception (Yazman *et al.*, 1982). Pasture availability was increased by concentrate feeding, and substitutionary feeding was strongly evident; it was estimated that pasture intake decreased by 0.9 kg DM for each kg of concentrate fed.

The level of response to concentrate in this study was relatively high; short-term studies which do not take into account the effect of supplementation on the length and persistency of lactation record lower values of *c.* 0.3–0.6 kg milk kg^{-1} concentrate. In a subsequent study

(Cowan & Davison, 1978) the response was 0.63 kg milk kg^{-1} maize grain, and 0.5 kg milk kg^{-1} molasses. It is also inevitable that the linear response only operates within the genetic potential of the cow for milk production, and a quadratic or asymptotic response occurs over a greater feeding range or under conditions where other factors limit the response, as occurred on *M. minutiflora* pastures in Vicosa, Brazil (Vilela *et al.*, 1980). Some added advantage in efficiency of supplementation may accrue from raising the level of concentrate feeding early in lactation and reducing it after *c.* 100 days when a high level of secretion has been established (Davison, Jarrett & Martin, 1985).

The choice of supplementation level is decided first by the shape of the biological response. At Garubo in

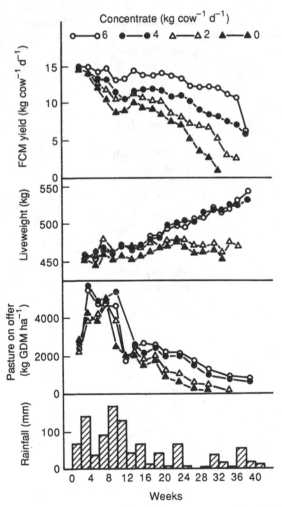

Fig. 9.7. Yield of milk, liveweight of cows, and available pasture in relation to level of concentrate fed. (From Cowan, Davison & O'Grady, 1977.)

Puerto Rico (lat. 18° N, 1930 mm rainfall) supplementation at moderate levels, such as 1 kg concentrate per 2 kg milk in excess of 10 kg produced gave more efficient responses than supplementing at higher levels, such as 1 kg concentrate per 2 kg milk irrespective of daily milk yield, although the latter treatment increased milk yield and other attributes above that attained with moderate supplementation (Yazman *et al.*, 1982). Similarly, at Muak Lek, Thailand (15° N, 1090 mm rainfall, 220 mm a.s.l.) efficiency of conversion of concentrate for cows grazing leafy *P. maximum/S. hamata* pasture decreased at high levels of feeding, and supplementation of 1 kg concentrate per 3 kg milk produced gave better economic returns than higher levels of supplementation (Lekchom *et al.*, 1989).

Energy supplementation of growing animals is usually uneconomic except in protected economies where the product does not have to compete at world prices. As indicated in Table 9.7, the best responses to energy supplementation occur when animals are offered high fibre diets. At Gainesville, Florida, *C. dactylon* hays of varying quality were fed to sheep with and without supplemental corn and soybean meal (Golding *et al.*, 1976). The depression in voluntary DE intake from hay was 75%, 42%, 45% and 2% of grain DE consumed for hays of 4, 6, 8 or 10 week maturity. Consequently the greatest increase in total DE intake occurred when supplementing the lowest quality hay. In Nueva Ecija, central Luzon, Phillipines, calves were grazing *Themeda triandra* pasture during the dry season, when N content of the pasture decreased from 0.5% to 0.4% (Cruz & Calub, 1981). LWG on this low quality pasture was increased from 0.02 to 0.43 kg hd^{-1} d^{-1} by feeding concentrate, or to 0.60 kg hd^{-1} d^{-1} by feeding concentrate plus silage.

On a mixed grass pasture near Nairobi, Kenya, Kayongo-Male, Karue & Mutinga (1977) found that supplementing heifer calves with concentrate increased LWG from 0.30 to 0.46 kg hd^{-1} d^{-1}. The level of concentrate was fed according to LW and averaged *c.* 1.7 kg hd^{-1} d^{-1}, or a response of *c.* 1 kg LWG per 10 kg concentrate. On irrigated *D. decumbens* pasture, supplementation of dairy heifers with 1.4 kg maize gain hd^{-1} d^{-1} increased LWG from 0.28 to 0.53 kg hd^{-1} d^{-1}, or a response of *c.* 1 kg LWG per 5.6 kg grain (Deans *et al.*, 1976). Such a feeding policy leads to earlier maturity and reproductive activity in young females.

(iii) Feed intake of ruminants is reduced by protein deficiency which limits microbial growth and the digestion of OM in the rumen, and hence the rate of removal of digested material from the rumen (Hunter & Siebert, 1987). Lesser amounts of amino acids are available in the intestine to satisfy tissue requirement for

protein if protein synthesis by organisms is limited by low protein intake from the pasture. This deficiency may be overcome by supplying protein supplements which are protected from rumen degradation, and in recent decades the value of 'by-pass' protein' (Ferguson, Hemsley & Reis, 1967) and the solubility of protein have received attention, as mentioned in Section 6.1.2. Protein supplements are therefore considered in two categories: (1) those which are readily degradable in the rumen, and (2) those which have attributes giving partial protection or which have been treated to enhance their protection.

Hunter & Siebert (1987) treated casein with formaldehyde to reduce rumen degradation and compared its effects with rumen degradable N and S fed to steers receiving a high roughage diet of *H. contortus* hay. The DM intake of young Aberdeen Angus steers was enhanced 22% by treated casein relative to a rumen-degradable supplement, but older animals were unresponsive. Young animals were also unresponsive if *H. contortus* were replaced by better quality *D. decumbens* hay, and Brahman animals were less responsive than Aberdeen Angus.

Processes of feed treatment to reduce rumen degradation have not been widely adopted for supplementing animals grazing tropical pasture, but an awareness of protein quality has been generated in the assessment of pasture plants (Section 6.1.2) and of supplementary protein sources. For example, fish meal and soybean meal supplements are less degraded in the rumen than peanut and lupin meals (Doyle, 1987). Cottonseed meal gives LWG in steers superior to that of a biuret supplement balanced for energy content (Addison, Cameron & Blight, 1984).

(iv) The use of urea as non-protein nitrogen supplement when fed with molasses enhances the intake of high fibre, low N pasture, and these are common dry season supplements in savanna regions. They are effective in improving the utilisation of otherwise wasted forage, in minimising seasonal weight loss, and enhancing survival. Rarely does this practice provide good LWG, and it is best applied in extensive beef production systems based on native pasture. Legume supplementation, as discussed in Chapter 8, usually provides additional digestible energy and better LWG. Responses to urea are also conditional upon the S status of the diet, as mentioned at the beginning of this section.

Effects on intake are illustrated by a study (Table 9.8) with native pasture hay (*Heteropogon contortus, Bothriochloa bladhii* and *Dichanthium* spp., 0.5% N) fed to weaner steers and supplemented with molasses (227 g hd^{-1} d^{-1}), urea (57 g hd^{-1} d^{-1}) or urea and molasses (Ernst, Limpus & O'Rourke, 1975). The combination of urea and molasses increased OM intake by 66%, did not affect OM digestibility, but improved the N status of the diet. Winks *et al.* (1982) suggest molasses supplementation may be reduced to 115 g hd^{-1} d^{-1} without detriment to the urea response.

Urea supplementation also has positive effects on the onset of subsequent oestrus activity, which is well related to small changes in body fat content (Siebert & Field, 1975). Earlier time of calving and superior cow survival follows from the better body condition of supplemented cows, and at Millaroo, Queensland (lat. 20° S, 920 mm rainfall), mean calving interval was decreased by supplementation from 441 to 403 days and from 404 to 355 days in successive years (Holroyd, Allan & O'Rourke, 1977). However, in later years supplementation responses were less apparent (Holroyd *et al.*, 1983). The utility of urea/molasses supplementation varies greatly with pasture and seasonal conditions, and subsequent compensatory weight gain may cause earlier responses to disappear completely (Entwistle & Knights, 1974; Winks *et al.*, 1982; Graham *et al.*, 1983). The effect of urea/molasses supplementation in increasing the capacity of animals to utilise dry season roughage and to survive better may lead to increased SR, which can cause degradation of botanical composition and soil erosion unless conservative management policies are adopted.

9.3.2
Adjustment of stocking rate
Animal responses to pasture conservation only occur within a range of SR specific to the farm feed supply and its seasonal variation in relation to the animals grazed. At low SR substitution eliminates or reduces the response; at high SR there is less surplus feed to conserve, animals suffer a penalty to performance when pasture or crop areas are reserved for conservation, and the practice of hay or silage making may aggravate the shortfall in feed supply (Hutchinson, 1966).

(i) Reference was made in Sections 9.2.1 and 9.3.1 to substitutionary effects when supplements are fed. Additional stock may be carried when high levels of substitution occur, or the level of pasture wastage will increase. Care needs to be exercised in raising SR, especially if energy supplementation of lactating cows is practised. For example, Friesian cows grazing *P. maximum* var. *trichoglume*/*N. wightii* pastures grazing at various SR from 1.3 to 2.5 cows ha^{-1} were supplemented with grain at 3.6 kg cow^{-1} d^{-1} for the first 50 days of lactation (Cowan, Byford & Stobbs, 1975). A high level response of 564 kg milk cow^{-1} occurred at 1.3 cows ha^{-1} but the response was only 70 kg milk cow^{-1} at 2.5 cows ha^{-1}, since reduced pasture availability at the highest SR caused premature drying-off of cows.

Care needs to be exercised in evaluating the scientific literature of this topic. Hay and silage making had no significant effect on the milk production of Friesian cows grazed at the relatively low rate of 1.3 cows^{-1} ha^{-1} on these pastures over a three-year period (Cowan *et al.*, 1974); however, Tucker *et al.* (1972) report from the same experiment a small but non-significant 15% positive response in milk production to hay and silage feeding in one drought year.

(ii) Animal stress is more acute at high SR, and it seems ironic that fodder conservation should be least successful in these circumstances, unless fodder is imported from off-farm sources.

The negative association between production per

Table 9.8. *Effects of supplementation on hay intake and digestibility*

	Supplement			
Attribute	Nil	Molasses	Urea	Molasses plus urea
Intake (g hd^{-1} d^{-1})				
Hay organic matter	1360	1745	1560	2260
Total organic matter	1360	1880	1560	2390
Total N	7	10	30	35
Digestibility (%)				
Organic matter	51.4	49.3	51.6	51.0
N	−40.2	−35.0	+58.1	+51.5

Source: Ernst *et al.* (1975).

animal and SR was discussed in Chapter 7, and reservations were expressed about the generality of this observation at high levels of forage allowance. Where the reservation of feed for conservation increases SR and reduces forage allowance, the penalty to animal performance must be more than compensated by the subsequent response to feeding conserved forage. For example, LWG of cattle grazing *S. almum./M. sativa* pasture at Gatton, Queensland (lat. 28° S) at 2.0 steers ha⁻¹ was unaffected by hay conservation and feeding; at a SR of 3.3 steers ha⁻¹ hay conservation and feeding reduced annual LWG by 27 kg hd⁻¹, due to decreased steer growth in the period from April–June, when hay conservation had reduced pasture availability (Smith, 1967).

Some estimate of the carrying capacity of the pasture during the main growing season is needed to decide how much pasture is truly surplus to animal requirements. Fig. 9.8 illustrates the situation for a pasture which will carry 3 beasts ha⁻¹ with maximum individual animal gains during the growing season; line A indicates the area which may be reserved from stock and conserved without penalty to animal performance. At an overall SR of 2 beasts ha⁻¹ 33 % of the pasture may be reserved without penalty, whilst at 1 beast ha⁻¹ 67 % may be reserved without reducing SR below 3 beasts ha⁻¹ during the main growing season. Above line A penalties to animal performance would need to be taken into account. Line B indicates the area of pasture which needs to be conserved to maintain animal body weight in all seasons. It is suggested that maximum opportunity for increasing animal performance occurs in the shaded area of the segment of the diagram where

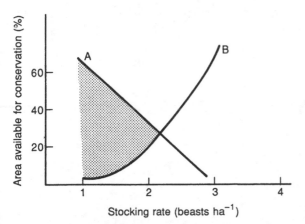

Fig. 9.8. Relation between stocking rate and (line A) area available for conservation and (line B) area required to be conserved to maintain animal liveweight. See text for explanation of shaded area. (Suggested by Bishop & Birrell, 1973.)

penalty to animal performance is avoided but sufficient conserved feed is available to meet animal needs.

Comparative studies are needed in the tropics to evaluate the efficiency of energy catabolism of body reserves during periods of stress in relation to the energetic efficiency of the process of pasture conservation, storage and feeding out. This type of study has been useful in evaluating the combination of SR and pasture conservation practice in a temperate environment (Hutchinson, 1971).

9.3.3
Integrated production responses to conservation

There are many tropical studies of animal response to silages made from fodder crops (for example, Roverso *et al.*, 1975, Pizarro, Escuder & Andrade, 1981; Torres *et al.*, 1982; Adebowale, 1983; Rivero, *et al.*, 1984) and to various imported supplements, as discussed in Section 9.3.1. There are few comprehensive studies examining the value of pasture hay and silage conservation on self-contained tropical farms.

The response to fodder conservation in terms of animal survival is consistently positive, and managers seeking to preserve the nucleus of their breeding stock in erratic climates often rely on long-term fodder conservation.

The responses of milk production to maize or other good quality silages are good, but less evident from pasture; for example, in Cuba benefits to milk production from conservation of *D. decumbens* were 9–13 % in commercial practice (Esperance, 1984).

Responses in beef production systems have usually not been sufficiently great to encourage the adoption of pasture conservation. At Samford, Queensland (lat. 27° S, 1100 mm rainfall), cattle grazed N fertilised *C. gayana* or *S. sphacelata* var. *sericea* pastures at 3.8 or 5.0 beasts ha⁻¹ (Jones, 1976). One half of the paddocks were cut for hay without interrupting continuous grazing of the whole area, and hay (1.5 % N) was fed in the late winter and spring. The results are considered in terms of the location of the paddocks (Fig. 9.9). There was no response to conservation on the hillslope pastures, which were relatively free of frost. On the alluvial terrace pastures, whose quality deteriorated badly in the winter, LWG was increased by 30 kg hd⁻¹ or 22 % by conservation.

P. maximum var. *trichoglume/M. sativa* pastures reserved for winter and spring grazing at Gayndah, Queensland (lat. 26° S, 730 mm) produced LWG in weaner steers of 73 and 97 kg hd⁻¹ respectively from non-conserved and conserved treatments at 1.5 beasts ha⁻¹; at 2.5 beasts ha⁻¹ only 48 and 55 kg hd⁻¹ respectively were recorded (Scateni, 1966). Conservation of pastures at Lansdown, Queensland (lat. 19° S, 870 mm), *H. contortus/S. humilis* increased calf birth weight from 26 to 29 kg hd⁻¹ and weaning weight from 185 to 194 kg in a

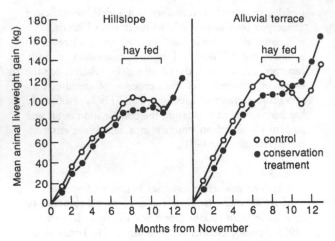

Fig. 9.9. Liveweight gain of cattle unsupplemented or fed with hay on hillslope or alluvial terrace pastures. (From Jones, 1976.)

drought year but did not influence performance in two other years (Edye, Ritson & Haydock, 1972). Final cow LW was 497 and 516 kg in non-conserved and conserved treatments (Edye *et al.*, 1971). In Adelaide River, Australia (lat. 13° S, 1270 mm rainfall), the response of weaner and yearling steers to fodder rolls of *S. humilis*/native pasture averaged 13 kg hd^{-1} (Sturtz & Parker, 1974). In Cuba reserving 25% of the area of *C. dactylon* pasture for haymaking did not significantly affect LWG of yearling steers during the wet season, and increased LWG during the dry season by 31% (Valdes, Molina & Garcia, 1988).

There have been attempts to simulate the effects of pasture conservation on animal output. Such studies will inevitably be received with scepticism until they are based on and validated in farm practice. Sullivan *et al.* (1980) estimate the probable increase in milk output during the dry season from the conservation of native grass hay in the Shinyango Region of Tanzania (840 mm rainfall); these estimates seem optimistic when the feeding of native grass hay in Zambia did not decrease liveweight loss in cattle (Smith, 1961).

Some special social circumstances may favour the adoption of hay conservation. The reservation of sloping, eroding lands as stock exclusion areas is being promoted in Shewa and Harerge Provinces, Ethiopia; these oversown pastures have recovered considerable vigour in the absence of grazing and exhibit improved ground cover, whilst their use for hay cut at the end of the main rains in November has facilitated the adoption of this system by associations of peasants.

The data presented in this chapter indicate clearly that responses in animal production to pasture conservation in the tropics are small relative to the considerable advances which result from the introduction of improved pasture species and from investment in fertiliser. Irrigation, where available, also represents a preferred option. Wider adoption of pasture conservation may follow advances in the nutritive value of the material available for conservation, especially through the incorporation of legumes, and is contingent upon managers exercising control of SR and recognising the reduction in risk associated with conservation policies.

10

Systems of rotational, deferred and mixed grazing

10.1
Rotational and deferred grazing

(i) The number of animals grazed on the farm and the ways in which the farmer arranges a sequence of feeds to meet their physiological needs are the primary determinants of animal output and of the sustainability of the ecosystem. The question of the stocking method – how the animals are moved about the paddocks and the number of paddocks established on the farm – is quite secondary and usually unimportant for production. It does not receive much space in this book, but it is necessary to present some evidence which indicates the paucity of gains from investment in rotational grazing systems and the positive harm to the farm enterprise which they often occasion.

Some aspects of stocking method have already been discussed. The decrease in nutritive value due to ageing of pasture was described (Section 6.1.1) and this bears on rotational grazing, since more aged feed is presented to the animal than occurs in continuous grazing practice. The favourable effects of slashing and mowing on the leafiness of the pasture and structure are more than counterbalanced by the deleterious effects of reduced total leaf availability on animal performance (Section 6.2). The negative effects of night corralling and the loss of advantages due to curtailing animal selectivity in zero grazing or cut and remove systems were mentioned in Section 6.3, whilst problems of nutrient wastage and pollution associated with cut-and-remove systems received attention in Section 3.1.3. The benefits of added selective grazing available to the leader animals in a leader–follower system were mentioned in Section 6.3. The desirability of seasonal variability in animal density in continuously grazed paddocks was referred to in Section 7.1, when discussing the relationship between forage availability and animal response, and in Section 5.2.7 with respect to the control of botanical composition.

Continuous grazing in the tropics implies that animals are present in the paddock throughout the year, or at least throughout a grazing season if the period of pasture utilisation is limited. Rotational grazing implies the movement of animals through a sequence of paddocks in which the period of grazing in each paddock and the next interval before the next grazing of that paddock may be defined. Deferred grazing is a form of rotational grazing involving a long rest for particular paddocks. In strip grazing animals are offered a defined and narrow width of pasture designed to meet daily requirements. This is usually achieved with an electric fence and a back wire to prevent fouling or grazing the pasture area previously used; tethering, especially if it involves a fixed wire, has a similar effect. Creep grazing (Section 6.3) provides access to additional feed for young or smaller animals.

Constraints to animal production are defined variously and the principal limitations in many tropical situations are overcome by replacing trees with herbage, by adding improved pasture species such as adapted legumes, and by fertilising pastures to remove nutrient deficiencies. Animal production may be increased by a factor of as much as ten if animals are added to consume the augmented feed supply (Tothill, 1974). By contrast, the gain to animal production from manipulating stocking method is more likely to be in the range 0–15%, and may be negative.

The results from a sample of 60 grazing experiments carried out in the tropics and subtropics (Table 10.1) indicates the clear superiority of continuous grazing relative to other grazing systems scientists have devised. This table attempted to exclude the experiments in which SR was confounded inextricably with stocking method (Anderson, 1988). Only in 17% of experiments was an advantage for rotational grazing recorded; this advantage was in the range of c. 4–15%, which would not pay for the additional fencing, provision of water and shade, and labour for stock management entailed in a rotational system. Stocking method had no influence in 32% of experiments, and the adjustment of SR on continuously grazed pastures is confirmed as the key management decision which farmers make. The advantage of continuous grazing in Table 10.1 operated for liveweight gain, conception rate and milk production, for cattle, sheep and goats, for tropical and subtropical pastures in semi-arid, subhumid to humid regions on different continents, and for natural and planted pastures.

There are three main misconceptions associated with rotational grazing systems.

(ii) 'Non-selective grazing' has been regarded as necessary to maintain the more palatable species in the sward, which should therefore increase animal production (Acocks, 1966), and to avoid wastage of pasture.

In fact, systems of rotational grazing have not been very successful in containing the spread of undesirable, unpalatable plants, which is primarily decided by SR rather than stocking method. As indicated in Section 4.2, frequency of defoliation at reasonable SR is similar under continuous and rotational grazing systems. Grazing method may have no effect on botanical composition (for example, Denny & Barnes, 1977; Clatworthy, 1984 for Zimbabwe) or continuous grazing may even lead to fewer weeds (Harrington & Pratchett, 1974). The concepts of rotational grazing are inappropriate and inadequate for application to the control of botanical composition in unpredictable environments (Hoffman, 1988), as illustrated in Section 5.2.8. *Medicago sativa* is an isolated example of a legume which will not persist under continuous grazing (Leach, 1983), and this plant does not have great relevance for pastures of the tropics and subtropics.

The emphasis on 'non-selective grazing' has mini-

Table 10.1. *Comparison of effects of stocking method on individual performance of animals grazing tropical and subtropical pastures (% experiments)*

Proposition	Yes	No	No difference
Continuous grazing increases animal performance relative to rotational grazing	51[1]	17[2]	32[3]
Long rest interval is detrimental to animal performance relative to short rest interval	33[4]	17	50[5]
Long duration of grazing in the rotation cycle is detrimental to animal performance relative to short duration of grazing	45[6]	9	45[7]
Long rotation cycle is detrimental to animal performance relative to short rotational cycle	14	0	86[8]

Sources:

[1] Chopping *et al.* (1978); CIAT (1972); Clatworthy & Muyotcha (1983); Davison *et al.* (1981); Denny *et al.* (1977); Eguiarte *et al.* (1984); Harrington & Pratchett (1974); Humphreys (1978); Jackson (1972); McKay (1968); Mannetje (1976); Nava & Gomez (1981); O'Donovan *et al.* (1978); Ottosen *et al.* (1976); Paladines & Leal (1979); Pfister *et al.* (1984); Prajapati (1970); Rodel (1971); Roe & Allen (1945); Stobbs (1969b); Wahab, H. (pers. comm.); Walker & Scott (1968); Winks, L. (pers. comm.).

[2] French *et al.* (1988c); Grof & Harding (1970); Gutierrez & Simon (1974); Jones & Jones (1979); Irulegui *et al.* (1984); Kornelius *et al.* (1979); Vicente-Chandler *et al.* (1953); Walker (1968).

[3] Aguirre Hernandez *et al.* (1984); Ahuja *et al.* (1974); Barnes (1977); Bogdan & Kidner (1967); Johnson & Leatch (1975); Lima *et al.* (1968); Lucci *et al.* (1983); McKay (1971); Norman (1960b); O'Rourke (1978); Pratchett & Shirvel (1978); Sheldrick & Goldson (1978); Soldevila (1980); Thorton & Harrington (1971); Tierney & Taylor (1983).

[4] Includes: Abramides *et al.* (1985); Creek & Nestel (1965).

[5] Includes: Caro-Costas & Vincente-Chandler (1981); Herrera (1978); Joblin (1963).

[6] Includes: Hernández *et al.* (1985); Senra *et al.* (1981); Vasquez & Lao (1978).

[7] Denny & Steyn (1977); Gutierrez *et al.* (1978).

[8] Includes: Delgado & Alfonso (1974); Hernández & Rosete (1985); Stobbs (1969a).

mised the significance for animal performance of the low nutritive value of C_4 grasses, and the need for animals to select young leaf and seed in preference to stem and senescent or dead material (Sections 6.1, 6.3) if satisfactory liveweight gain, reproduction or milk flow is to occur.

Strip grazing reduces selectivity and is practised in intensive dairy systems but it is difficult to locate any hard evidence of its benefit to production from tropical pastures; it is convenient when determining the forage allowance. A 10-strip grazing system with a back fence was compared at the same SR with continuous grazing

of *P. maximum* var. *trichoglume*/*Neonotonia wightii* cv. Tinaroo pastures at Atherton, Queensland (Ottosen, Brown & Maraske, 1976). Milk yield (Table 10.2) from Friesian cows in an advanced stage of lactation was marginally superior under continuous grazing in the first measurement period, when forage on offer averaged 2440 kg ha^{-1}. This difference disappeared as forage allowance decreased to 1550 kg ha^{-1} in period 3 and the opportunity for selective grazing was perhaps reduced. Percentage solids-not-fat was independent of treatment, and % butter fat (Table 10.2) was slightly higher under continuous grazing.

Table 10.2. *Effect of strip grazing on milk yield and fat content of milk*

	Strip grazing	Continuous grazing
Milk yield (kg cow^{-1} d^{-1})		
Period 1	11.5	12.3
Period 2	10.6	10.9
Period 3	10.6	10.6
Mean	10.9	11.4
Fat content (%)		
Period 1	3.8	3.8
Period 2	3.6	3.7
Period 3	3.4	3.8
Mean	3.6	3.7

Source: Ottosen, Brown & Maraske (1976).

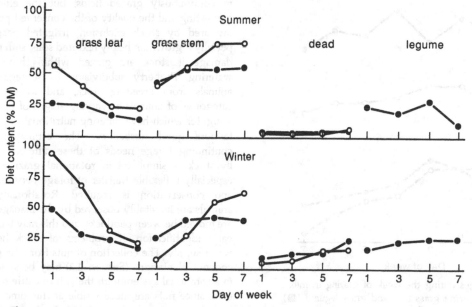

Fig. 10.1. Changes during the week of grazing in the leaf, stem, dead material and legume content of cows grazing grass (○) and grass–legume (●) pastures. (From Cowan et al., 1986.)

Decreased animal production as rotationally grazed animals remain in the paddock is a common feature; production against time gives a 'saw toothed' graph relative to the more even production expected under continuous grazing. Fig. 10.1. shows how the dietary constituents change after animals are introduced to a new paddock. These data come from rotationally grazed pastures (1 week on, 3 weeks off) of pure *P. maximum* cv. Gatton or grass–legume pastures also containing *N. wightii* and *D. intortum* (Cowan, Davison & Shephard, 1986). There was a steep decline in the % grass leaf ingested from day 1 to day 7 in both summer and winter, and an increase in the proportion of grass stem eaten; the legume composed *c.* 20% of the diet in the mixed pasture. Milk yield in winter (Fig. 10.2) decreased *c.* 2 kg cow^{-1} d^{-1} from day 3 to day 7. In summer milk yield was well maintained during the whole week when the cows were in the same paddock; the digestibility of stem material was higher in summer than in winter and the digestibility of the diet decreased less sharply with time after cows entering the paddock.

Rapid rotational grazing, in which animals spend one to three days only in a fresh pasture, is more akin to continuous grazing in providing better opportunity for selection than occurs when a high density of animals is retained in one paddock for an extended period. Table 10.1 provides some support for this view.

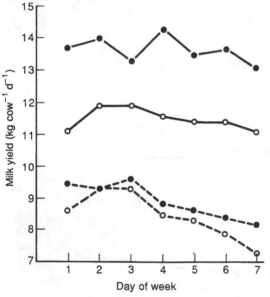

Fig. 10.2. Changes during the week of grazing in milk yield of cows grazing grass (○) and grass–legume (●) pastures in summer (–) and in winter (- - -). (From Cowan *et al.*, 1986.)

(iii) The second false idea is that spelling the pasture would enable animals to be introduced when the sward provides optimum nutritive value. However, digestibility, intake, nitrogen concentration and the concentration of most minerals decrease with age, as clearly demonstrated in Section 6.1; it is the presentation of young rather than of aged pasture to the animal which maximises output. This factor is less influential in legume dominant pastures, where the rate of decrease in pasture quality with time is less.

(iv) Pastures require resting from defoliation in order to replenish carbohydrate reserves in the roots and crown, to seed down and to allow seedling regeneration. This proposition ignores the reduced level of pasture utilisation which occurs in field practice under continuous grazing during the main growing season when growth greatly exceeds consumption, except in severely overgrazed pastures. It also overvalues the significance of carbohydrate reserves in determining growth (Section 4.5.2) and neglects the usual need for heavy grazing to promote seedling regeneration (Sections 5.2.5, 5.2.7). Systems of spelling usually imply stock concentration on another part of the farm, and spelling may not have had much effect on subsequent pasture vigour (Barnes, 1989).

There are three considerations which may justify rotational or deferred grazing.

(v) Convenience of farm operation may not be served by continuous grazing. It is possible to make hay or silage in continuously grazed fields, but the efficiency of harvesting and the quality of the conserved product are favoured by stock exclusion. Irrigated pastures, especially if grown on heavy textured soils, suffer pugging damage if stock are grazed within three days of watering. Property subdivision to segregate young animals from breeding stock, and to separate the categories of animals at varying stages of finishing for slaughter which have differing nutritional requirements, favours good husbandry. The estimation of the continuing forage needs of these different classes of livestock is simplified in rotational grazing systems, especially if flexible transfer of forage between grazing and conservation is required. Rotationally grazed animals are inevitably observed by the manager as they are moved between paddocks, and this may facilitate the early identification of problems of stock health and water supply. The collection of nuts from the pasture in coconut plantations (Section 2.2) may be synchronised in a rotational system with the exit of cattle to the next field, since nuts are more visible at this time.

(vi) Systems of rotational grazing may be designed to meet specific animals needs. The deferment of grazing of special purpose pastures to meet seasonal feed shortfalls has been canvassed in Section 8.2. A long rotation cycle has the effect of reserving forage for later use; the rate of feed deterioration as it stands in the field has to be less than the animal's efficiency in accumulating and using bodily energetic reserves if net gains are to occur relative to a system where the animal consumes the feed earlier. Stobbs (1969*b*) gives an example where continuous grazing in Uganda usually increased animal production during the dry season, except in an abnormally dry year when more animals died. The alternative view is that maximum opportunity for selection, as offered by continuous grazing, is at a premium during the dry season when feed quality is low and separate provision should be made for unusual droughts.

Some control of parasites is achieved by spelling the pastures and denying the infective larvae a host animal on which the life cycle may be continued. The success of this practice depends upon clean mustering and the absence of alternative hosts, and upon a sufficient duration of rest to ensure the death of the infective organism. Spelling for three months in summer and for five months in winter may be necessary in Queensland to kill the larvae of the cattle tick *Boophilus microplus* (Wilkinson, 1964; Dunwell, 1967). Similarly, for internal parasites such as the barber's pole worm (*Haemonchus contortus*) a two-month rest has been advocated (Hall, 1977).

The disadvantage of these spelling systems is that the negative effect of spelling on pasture nutritive value and on the opportunity for selective grazing may counterbalance the beneficial effects of parasite control. For example, at Townsville, Queensland, strategic dipping of continuously grazed cattle with acaricide led to control of ticks and gains of 45 kg hd^{-1} more than undipped continuously grazed cattle. However, the latter showed the same depressed weight gain as cattle in which rotational grazing controlled ticks (Johnson & Leatch, 1975). This rotation employed five paddock movements in the year; a modified system involves the prediction of the seasonal explosion in tick population and the control of this by having a paddock free of ticks into which animals may be mustered; at Townsville this stock concentration in a clean paddock should cover the 10-week period from the end of February, after which animals might be returned to other paddocks which would be relatively tick free.

(vii) The reclamation of a degraded landscape requires stock exclusion or sufficient reduction of SR to enable species to seed down and to increase in basal cover (Barnes, 1982). The period of stock exclusion may be for a whole year, or for longer. A '3 herd to 4 camps' system has been advocated in which each paddock is rested for one year in four. This may be combined with a burning policy directed to the control of shrubs, as discussed in Section 10.3. The manipulation of botanical composition through grazing and cutting management was treated in Chapter 5, where variation in seasonal grazing pressure directed to the regeneration of particular species (and especially legumes) was described.

Some of the detrimental effects of rotational grazing which are so evident in Table 10.1 arise from the application of experimental procedures on a regular calendar basis. Observation of the phenology of the plants to be encouraged and consequential adjustment of animal density are needed for the successful development of grazing systems.

10.2
Mixed grazing

Mixed grazing, or the use of different species of animals in the same grazing group, is widely practised in the tropics and especially in extensive production systems. This may provide (1) more efficient use of the available diet, (2) beneficial effects on botanical composition, and (3) diversification and sometimes optimisation of producer income. This topic was introduced in Section 2.5 when the main features of nomadic pastoral systems were mentioned.

(i) Particular animal species exhibit patterns of dietary preference which differ from other animal species. Preference varies both between plant organs and between plant species. It is therefore argued that complementary grazing of the vegetation components by different animal species should lead to more uniform utilisation; it also makes a higher level of utilisation feasible. The object of management is to use a knowledge of species preferences to devise a balance of different animal species and an appropriate SR which will lead to effective utilisation of vegetation and which will moderate competition between different animal species.

The relative preference of sheep and cattle for particular grasses was studied at Mazoe, Zimbabwe (17° S, 910 mm rainfall), in a cafeteria grazing trial using Afrikander cows and Dorper × Blackhead Persian ewes (Mills, 1977). Experimental plots of different cultivars were grazed on seven occasions from December to April. The preference index for each cultivar was calculated from the weight of herbage present before and after grazing, and the consumption of one cultivar compared with the average of all cultivars; a preference index of 1.0 indicates a neutral acceptance for that cultivar. There was no correlation between the preferences of sheep and cattle (Table 10.3). Cattle preferred the tufted *Eragrostis curvula* and the stoloniferous *C. aethiopicus* cv. Ngorongoro, whilst sheep ate more of *C. aethiopicus* cv. No. 2 and the rhizomatous *P. notatum* than did cattle.

A second study illustrates differences in acceptance of both plant species and plant parts. Maasai haired sheep (*Ovis aries*), Maasai goats (*Capra hircus*), Thomson's gazelle (*Gazella thomsoni*), Grant's gazelle (*Gazella granti*) and impala (*Aepycerus melampus*) grazed a mixed range on the Athi-Kapiti plains in Kenya Maasailand (Hoppe, Qvortrup & Woodford, 1977*b*). Marked differences in preferences for plant species were shown by these animals (Table 10.4). Sheep were overwhelmingly grazers, gaining 92% of their diet from grasses, which were principally *Themeda triandra* and *Cynodon* spp. Goats depended more upon grass than upon browse in this association at this sampling occasion; Thomson's gazelle and impala also had a mixed grass–browse diet.

Grant's gazelle depended upon browse, and consumed seed pods of *Acacia stuhlmannii*, fresh shoots of the shrub *Barleria eranthemoides*, and leguminous herbs. It may be expected that the combination of all animals gave more effective use of the diversity of plant species present than would have occurred if a single animal species had been grazed.

The two gazelles selected a higher ratio of green to dead grass leaf and sheath material (3.0–3.2) than that of goats (1.0) or sheep (0.9) and this led to higher N content in the diet of Thompson's gazelle. Reference was made earlier in Section 2.5 and Section 6.3 to the anatomical modifications which vary the capacity of animals to graze small plant organs selectively.

The browse intake of goats in the last example was low relative to other studies. Fig. 10.3 (Lightfoot & Posselt, 1977) illustrates a more usual pattern in which browse represented about half the goat's diet. This study was undertaken in the southern lowveld of Zimbabwe in *Colophospermum mopane* savanna. This also

Table 10.3. *Preferences indices (utilisation relative to availability) of grass cultivars offered to sheep and cattle in Zimbabwe*

Cultivar	Sheep	Cattle
Cynodon aethiopicus cv. Lake Manyara	0.91	1.07
C. aethiopicus cv. No. 2	1.09	0.83
C. aethiopicus cv. Ngorongoro	1.10	1.27
Eragrostis curvula cv. Ermelo	0.68	1.26
Paspalum notatum cv. Paraquay	0.91	0.74

Source: Mills (1977).

Table 10.4. *Rumen composition (%) of five animal species grazing rangeland in Kenya*

Plant part	Sheep	Goat	Thomson's gazelle	Grant's gazelle	Impala
Grasses					
Leaf	54	40	55	13	38
Sheath	27	22	17	8	18
Stem	12	11	7	3	8
Subtotal	93	73	79	24	64
Other monocotyledons	3	3	2	9	11
Dicotyledons					
Leaf	2	6	5	11	4
Fruit and flower	0	8	2	24	7
Stem	3	10	14	33	14
Subtotal	5	24	21	68	25

Source: Hoppe *et al.* (1977*b*).

confirms the dependence of cattle upon herbage, and only a small intake of browse was evident, predominantly in the dry season winter months. Fig. 10.3 shows the use of browse by the eland in this association. Eland (*Taurotragus oryx*) is potentially a domesticated animal, and contributes significantly to the meat harvest in some African ranch situations. This large antelope consumed only 1–11% of its diet as grass, and ate some forbs during the early summer wet season. Its main food was *Colophospermum mopane*, *Combretum apiculatum* and *Grewia* spp., whilst the evergreen *Euclea divinorum* was eaten in the dry season when the deciduous plants had shed their leaves or declined in nutritive value. The dietary pattern varies with the vegetation association, and on savanna grassland with scattered *Acacia tortilis* and *Commiphora riparia* trees at Kiboko, Kenya, eland grazed as much as they browsed (Nge'the & Box, 1976).

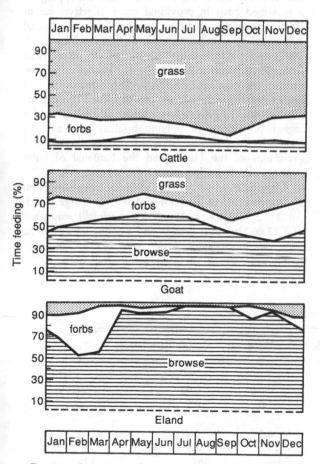

Fig. 10.3. Percentage of time feeding on woody browse, forbs and grass by cattle, goat and eland in *Colophospermum mopane* savanna in Zimbabwe. (From Lightfoot & Posselt, 1977.)

(ii) These examples show the diversity of the intake by different animal species when offered the same mixed vegetation. The factors contributing to the successful conversion of these diets need consideration. The development of profiles of seasonal energy requirements for the component animal species assists management. The accessibility of the diet to small ruminants may be improved by the presence of large ruminants. Merino sheep in western Queensland are loath to graze tall pastures of *Astrebla lappacea* and of *Cenchrus ciliaris*, and surplus forage occurs in years of abnormally high summer rainfall. In these seasons the purchase of store cattle for fattening on holdings predominantly grazing sheep is believed to advantage the performance of sheep, since cattle make tracks through the pasture and knock down tall material so that it is more available to sheep (Purcell & Stubbs, 1965).

Conversely, in seasons of abnormal drought the presence of a mixed animal population which varies in its capacity for drought survival will provide more sustained grazing pressure and continued product output than would occur if a single drought susceptible species were used. Sheep and especially cattle are more vulnerable to drought than goats or camels (Cossins, 1983), and some wild ungulates appear not to depend upon surface water (Reul, 1979).

It is possible that parasite burdens may be lessened through mixed grazing, and that this will then lead to more efficient conversion of dietary intake. This effect would operate in two ways if animals differ in their susceptibility to a particular internal parasite. The parasite population would be lessened by the dilution effect of a reduced SR of the susceptible ruminant, in so far as the infective population is related to domestic SR. It may also be less accessible; lambs in a forward grazing system consume the top of the sward where larval density is less than in the lower layers, which the animals following consume (Lambert & Guerin, 1990). Secondly, the resistant animal would eat and destroy infective larvae, thereby reducing the level of infection. In particular, sheep tend to graze more readily over areas fouled by faeces than do cattle. This topic has been reviewed by Nolan & Connolly (1977) for temperate pasture species in relation to *Nematodirus* spp. but critical studies of this factor in tropical conditions have not been sighted. The higher resistance of wild ungulates to parasites relative to that of sheep and cattle is claimed as an argument for game ranching (Reul, 1979).

Complete complementarity of diet is usually unattainable, and animal species compete for some components of the available feed, especially as forage allowance decreases and selectivity also diminishes with increasing SR. Sheep graze more closely to the ground than cattle and nutritional stress is first evident in the latter. Some studies with temperate pastures show a beneficial effect of cattle on sheep performance, but the deleterious effect on the growth of cattle following a seasonal shortfall in forage availability under conditions of mixed grazing may disappear as compensatory weight gain occurs (Bennett *et al.*, 1970). A narrow ratio of sheep to cattle number is considered beneficial to cattle, but critical studies from the tropics are lacking. Ebersohn (1966) grazed sheep and cattle in ratios varying from 16 to 0.5:1 in *Tarchonanthus* shrub bushveld at Koopmansfontein, South Africa (lat. 28° S, 370 mm rainfall). There was no effect of sheep:cattle ratio on either cattle or sheep performance, but feed supply was usually non-limiting. Hildyard (1970) advocated a narrow ratio of 2–5:1 of sheep:cattle in the Dohne sourveld in South Africa. The concept of selecting a key plant species upon which to base the calculation of grazing capacity (Smith, 1965) is rejected because of its subjectivity and inappropriate conceptual basis.

The variation in the efficiency of use of different diets is reflected in different patterns of rumen function. Another study in the Athi-Kapiti plains in Kenya (Hoppe, Qvortrup & Woodford, 1977a) showed that cattle ingested a diet lower in N% than that of wildebeest, Coke's hartebeest and topi, and slightly higher in fibre than the last two species (Table 10.5). This was associated with a slower rumen fermentation rate in cattle; volatile fatty acid (VFA) production was especially high in the topi. Similarly the domestic sheep and goats whose dietary intake is described in Table 10.4 had a lower rumen fermentation rate than that of Thomson's gazelle, Grant's gazelle and impala; the last species had a particularly high rate of VFA production. These wild animals have a smaller ratio of rumen weight to body weight, which assists their mobility; the high rate of gas production in the rumen is linked to a high

rate of VFA removal associated with the absorptive area of the rumen being enlarged in the antelope by papillae (Hoppe *et al.*, 1977a). The slower rate of digestion in domestic sheep and cattle is in part compensated by their larger rumen capacity.

Rumen fermentation rate and VFA production did not differ significantly between sheep and goats in the Hoppe *et al.* (1977b) study, but there are other reports, reviewed by Devendra (1978), which suggest that goats exhibit higher digestibility of fibre than sheep or cattle, which will increase the ME value to them of high fibre diets. This appears to be associated with differences in rumination and salivary secretion, as well as in rumen fermentation and rate of passage. Certainly goats survive in degraded landscapes inimical to the success of sheep and cattle.

Finally, attention is drawn to the comparative success of game animals in pastoral environments relative to that of sheep and cattle. The role of game cropping on a sustained basis in providing more effective use of tropical vegetation is being canvassed for its potential either to supplement or to replace cattle ranching (Reul, 1979), and is in tune with current emphases on the need to conserve animal resources. Wild ungulates may show greater reproductive efficiency, faster maturity, and higher extraction rate than that of cattle. The carcass meat of wild ungulates has a lower fat content and a higher proportion of polyunsaturates; the capybara (*Hydrochoerus hydrochaeris*) which grazes the flooded grasslands of the Llanos and the Pantanal of Latin America also has a low carcass fat content.

A comparative study in the southern *Colophospermum mopane* lowveld of Zimbabwe at Buffalo Range Ranch (lat. 21° S, 400 m.a.s.l., 250–550 mm rainfall), was made of a 12 000 ha cattle section and a contiguous 8000 ha game section (Taylor & Walker, 1978). Ten large herbivorous animals occurred on the latter, and provided

Table 10.5. *Rumen characteristics of four animal species grazing rangeland in Kenya*

Feature	Cattle	Wildebeest	Coke's hartebeest	Topi
Washed rumen contents				
N (%)	0.53	0.64	0.62	0.99
Crude fibre (%)	51.0	51.0	49.5	44.1
Rumen function data				
pH	6.7	6.5	6.4	6.1
NH_3-N (mg 100 ml^{-1})	6.1	13.9	3.9	7.0
Fermentation rate (1 μmol gas g DM^{-1} h^{-1})	147	210	261	272
Volatile fatty acids (meq l^{-1})	120	139	138	279

Source: Hoppe *et al.* (1977a).

an estimated standing biomass of 45–59 kg LW ha^{-1}. This was similar to the total biomass on the cattle section, which was 59 kg ha^{-1}, of which 5 kg ha^{-1} were contributed by impala, wart hog and sable. A higher per cent utilisation (especially of browse) occurred in the game section, but grazing in the game section was more area specific. Both plant cover and soil erosion were reduced in the game section, which represents an apparent anomaly in the data. Problems of management with respect to control of animal density, harvesting and meat hygiene are clearly acute in ranches devoted wholly to game, but the presence and supplementary harvesting of game animals on cattle ranches has beneficial effects on feed conversion, botanical composition and farm income.

(iii) Some studies of temperate pastures show that mixed grazing of sheep and cattle favours a desirable herbage botanical composition with respect to the proportion of legumes and of perennial grasses (Bennett *et al.*, 1970). In tropical savannas the control of woody species is the declared botanical rationale for mixed grazing. The effect of animal avoidance in increasing the contribution of the uneaten species to overall botanical composition was discussed in Section 5.3.2. Accessibility of foliage, flowers and seed affects consumption. The narrow muzzle, mobile upper lip and biting action of the goat enable it to extract leaf from thorny bushes, whilst its capacity to stand on its hind legs while browsing (Taylor, 1985), and to climb on the lower horizontal branches of trees, increases the height to which feed is available. In Africa some farmers favour the presence of giraffe, since their browsing up to 6 m above ground may reduce the canopy in the upper strata of shrubs and small trees (Skovlin, 1971).

The acceptability of accessible feed to particular animal species, as influenced by chemical composition,

odour, detachability and structure then becomes the overriding factor for control. The general thesis which has been developed that browse is eaten by goats and wild ungulates has its exceptions.

At Makoholi, Zimbabwe, various management practices were tested in *Brachystegia spiciformis/ Julbernardia globiflora* woodland which had *Burkea africana* as a common associate (Ward & Cleghorn, 1970). Initial tree density was *c.* 370 stems ha^{-1}. A mixed grazing comparison was applied following stumping to 30 cm soil depth, and replicated paddocks were grazed intermittently by Mashona cattle, and by indigenous sheep and goats on the basis that 7.5 small ruminants = 1 steer of 2–3 yr age. Table 10.5 shows the situation after eight years of grazing cattle, goats, or goats, sheep and cattle in the ratio 1.2:2.3:1 respectively. Mixed grazing had a beneficial effect in containing tree density, but Table 10.6 indicates that goats alone failed to control *J. globiflora* and *B. africana*; other species were reduced by goats alone to densities lower than or equal to their occurrence in the other two treatments. The rejection by goats of *J. globiflora*, the most common species, led to reduced grass growth and a lower level of animal productivity in the goats alone treatment.

This study illustrates the benefits of mixed grazing and suggests that goats should not be regarded merely as agents of bush clearing, since control of woody regrowth requires the additional use of fire or mechanical or chemical measures. Since goats eat grass as well as browse, a high SR designed to apply pressure to the woody species may also result in land degradation.

(iv) Subsistence farming places a high premium on the avoidance of risk, and diversification of production through the use of more than one animal species reduces the chance of total family disaster occurring through

Table 10.6. *Effect of animal species on woodland density (stems ha^{-1}) in Zimbabwe*

Woodland species	Cattle	Goats	Goats, sheep and cattle
Julbernardia globiflora	270	1410	210
Brachystegia spiciformis	100	30	40
Terminalia sericea	100	0	3
Burkea africana	40	330	70
Cassia singueana	180	180	0
Maytenus senegalansis	150	10	20
Other species	280	230	100
Total	1110	2190	440

Source: Ward & Cleghorn (1970).

disease, predation, or abiotic catastrophes. Entrepreneurial producers may also favour mixed grazing as an insurance against sudden shifts in income; the collapse of a boom in wool prices may be cushioned by the stability of beef prices, and vice versa. The production of cashmere or of angora may meet a specific market niche. Labour requirements may be spread more evenly through the year. It is also hoped that the overall level of animal production may be increased by mixed grazing, if more efficient conversion of vegetation to animal product and a more productive botanical composition result from its adoption.

Mixed grazing entails additional investment on ranch properties. The conversion of a sheep enterprise to one which embraces cattle requires additional fencing and cattle handling facilities; diversification of a cattle enterprise requires the establishment or leasing of a wool shed and perhaps the installation of additional stock watering points. A higher order of skills is needed to manage a mixed animal enterprise successfully. Connolly (1974) describes a scheme of linear programming which optimises profit according to the varying margins for the product output of the animal species concerned.

Mixed grazing offers the producer opportunities for flexible responses to changing market and climatic conditions. It is also a tool for moderating botanical changes and directing these towards the maintenance of a desirable balance of plant species which may best sustain animal production, especially in situations with high forage allowance and diverse plant species composition and topography.

10.3
Fire and pasture management

Burning pastures is controversial and generates heat between protagonists and those opposed to the use of fire, which remains a traditional, inexpensive and widely used tool of management in extensive production systems. It may be the sole management intervention of hunting societies, and these include some tropical cattle ranches in Australia and Latin America. A clear distinction needs to be made between forest fires, in which the great fuel load results in considerable environmental damage and the sterilisation of soil to some depth, and grassland fires, whose effects on soil are superficial.

(i)   The rationale for burning has many facets, whose significance varies according to the type of production enterprise, the climate and the pasture composition. A primary objective is to increase the accessibility of young green forage to the grazing animal. The removal of dry standing pasture by fire facilitates the prehension

of young growth. This can have strong positive effects on cattle growth in tall tropical savanna; in the Llanos Orientales at Carimagua, Colombia, cattle grazing unburnt native pastures dominated by *Trachypogon vestitus* at 0.2 AU ha^{-1} gained only 28 kg hd^{-1} yr^{-1}, whilst burning at the beginning of the next dry season led to gains of 92 kg hd^{-1} yr^{-1} (Paladines & Leal, 1979). In Florida burning *Aristida* pastures increased calf production per cow from 82 to 106 kg (Kirk & Hodges, 1971).

The proportion of green leaf in the pasture is increased by burning, and this leads to a greater ingestion of this high quality material. At Gayndah, Queensland (lat. 26° S, 730 mm rainfall), Ash *et al.* (1982) burnt a native pasture dominated by *Heteropogon contortus* and *Bothriochloa bladhii* in mid-October. The proportion of green leaf increased to a maximum of 68% in the burnt pasture in January (Fig. 10.4a), and the dietary intake contained a higher proportion of green leaf throughout. This led to better animal gains in the three months following burning (Fig. 10.4c), but not thereafter; earlier flowering and maturity occurred in the burnt pastures.

The effects of fire on pasture growth *per se* are problematical, and some reports of growth stimulation simply reflect the greater visibility of green material in a burnt pasture. In the present example the presentation yield of green leaf (Fig. 10.4b) in the burnt treatment was initially less than and subsequently similar to the yield in the unburnt treatment. Burning *C. ciliaris/M. atropurpureum* pastures in a comparable environment (Narayen, lat. 26° S, 710 mm rainfall) in August gave a similar result with respect to yield of green material (Mannetje, Cook & Wildin, 1983), but did not affect liveweight gain.

Young material has a higher nutritive value than aged material, as described in Section 6.1.1. A positive response in animal performance to burning implies some deficiency in the unburnt pasture; alternatively a strong, positive effect of burning may require the rectification of an overriding mineral deficiency. On infertile red earths at Manbulloo in northern Australia (lat. 15° S, 930 mm rainfall), Winter (1987) considered the effects of burning half an open woodland pasture dominated by the perennial grasses *Themeda triandra*, *Sehima nervosum*, *Sorghum plumosum* and *Chrysopogon fallax* each year in June/July, feeding salt (Na Cl) plus the trace elements Zn, Cu, I, Co and Se, or feeding an additional mineral supplement containing N, P and S. Cattle liveweight gain (Fig. 10.5) in the early wet season (November through January) showed a strong positive response to fire, but only if salt was also fed. This suggests that although burning increased N% in this period, low levels of Na exerted control. In the late wet period

(February through to May) when the pasture was maturing and N, P, and S concentration decreasing, a further LWG response only occurred when the NPS supplement was also fed.

(ii) A second objective is to reduce the content of plant species unacceptable to the grazing animal. Firing may give a temporary increase in the herbage content of the vegetation by equalising the competition between a fire-resistant woody species such as *Acacia harpophylla* and grasses such as *C. ciliaris* in central Queensland; firing reduced suckers of *A. harpophylla* to ground level and regrowth of all species occurs on more similar terms. Alternatively, fire-susceptible woody species may be more permanently controlled if adequate fuel for a hot fire is accumulated. The woody plants *Carissa ovata*, *Eremophila gilesii*, *Terminalia oblongata* and the poisonous

Myoporum desertii are susceptible to hot fires in western Queensland (Pressland, 1982). Similarly, brush control in *C. ciliaris* pastures in southern Texas is favoured by burning in the cool season every third year (Hamilton & Seifres, 1982).

A combination of burning and some other management intervention may be needed to effect control of woody species. Coppicing of *Brachystegia spiciformis* and *Julbernardia globiflora*, which are fire susceptible, could only be eliminated in the long term in the Marandera sandveld in Zimbabwe by a combination of mattocking and burning (Barnes, 1965). In south-east Queensland regrowth of *Eucalyptus maculata*, *E. intermedia* and *Acacia cunninghamii* is primarily controlled by grazing, but fire has a secondary effect in suppressing regrowth (Tothill, 1971). Chemical control in association with burning is a further option.

Tropical savanna has evolved under conditions of intermittent firing, whose frequency has been increased by humans. The dominant grass species are usually fire resistant, and burning practice rarely causes substantial changes in the balance of grass species (for example, Winter, 1987). Nearly all the planted tropical grass species are fire resistant (Humphreys, 1981b), but the susceptibility of many planted tropical legumes to fire gives a contraindication for burning grass/legume pastures. Plants of *Stylosanthes guianensis* var. *guianensis* cv. Schofield fail to survive the coolest fire (Gardener, 1980), and *S. humilis*, *S. hamata* and *S. viscosa* are readily killed (Mott, 1982). Fire resistance in *Stylosanthes* depends upon the capacity to produce new shoots from below ground level, and in *S. scabra* these develop 14–38 mm below the surface. The creeping *S. guianensis* var. *intermedia* cv. Oxley increases its density under regular firing (Bowen & Rickert, 1979). *S. capitata* in the Llanos Orientales of Colombia showed 74% plant survival after burning, and plants regrew from the crown, but not the roots (Alejo, Rodriguez & Fisher, 1988). Fire stimulates legume seedling regeneration, as discussed in Section 5.2.5. This arises from the 'gap' created in the pasture, and from some diminution of hardseededness; surface seed may be killed and a decrease in the soil seed bank usually follows burning (Mannetje *et al.*, 1983; Alejo *et al.* 1988).

(iii) A third rationale relates to the ease and convenience of property management. Animals are attracted to recently burnt areas where they are more readily mustered. Riding or driving over the pasture can be done more safely, since fallen logs, holes and ditches or obstacles are more visible after a fire. Burning particular areas in alternate years has the effect of resting the unburnt areas, since cattle graze preferentially on the burnt area throughout the year (Andrew, 1986). The risk

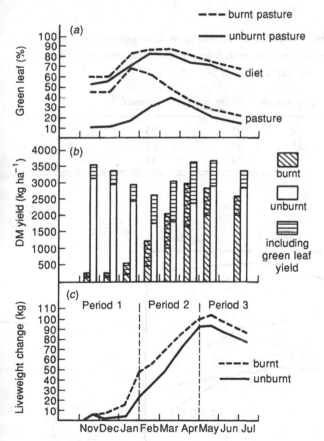

Fig. 10.4. (*a*) Percentage of green leaf in the diet and in the available pasture of burnt and unburnt native pasture; (*b*) yield of standing pasture of burnt and unburnt native pasture, including yield of green leaf; (*c*) liveweight change (kg hd⁻¹) of cattle grazing burnt and unburnt pasture. (From Ash *et al.*, 1982.)

of fire escaping from a neighbouring property to burn the landholder's own area at an inappropriate and unscheduled time is reduced if the area has already been burnt or a firebreak developed.

Fire also meets specific management objectives in more intensive production systems. Fire is used as an alternative to mowing or slashing in order to favour the synchrony of flowering of seed crops of grasses such as *B. decumbens* (Stür & Humphreys, 1987, 1988a,b). Fire has the effect of disinfestation for some pests and diseases, as mentioned in Section 3.2.2. Land preparation for pasture or crop planting is facilitated by fire, and establishment of pastures by aerial seeding is more effective after burning. This has especial force where an early wet-season burn damages the existing native grasses, and promotes the success of legume establishment (Stocker & Sturtz, 1966).

(iv) There are many negative effects of burning. A grazing property is less susceptible to the effects of drought in the absence of burning, since previously unconsumed standing pasture is available to meet a seasonal shortfall in forage supply. Burning is less fashionable amongst managers who have come to recognise the energy value of dry roughage if animals are supplemented from a leguminous protein bank (Section 8.2.2) or with some other nitrogen source, such as urea or cottonseed meal (Section 9.3.1). Planted pastures, especially if they contain legumes, will have greater feeding value as standing forage than native savanna.

Burning is damaging to the environment in several senses. Soil erosion is aggravated by loss of cover (Section 3.1.2), especially if rains of high intensity occur soon after the fire and before ground cover is restored.

Table 10.7. *Liveweight changes (kg head^{-1}) of steers grazing native savanna of Carimagua under two systems of burning*

Stocking rate (beasts ha^{-1})	Total burning			Sequential burning		
	Dry season	Rainy season	Annual	Dry season	Rainy season	Annual
0.20	15	60	75	4	91	95
0.35	−7	73	66	−25	87	62
0.50	−24	55	31	−29	64	35

Source: Paladines & Leal (1979).

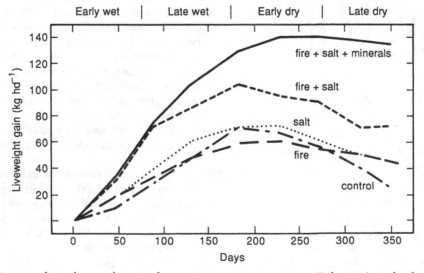

Fig. 10.5. Mean cumulative liveweight gain of steers grazing native pastures at Katherine Australia showing effects of fire and supplements. (From Winter, 1987.)

Runoff is increased and burnt catchment areas exhibit a higher incidence of flash floods than unburnt catchments. Some scientists emphasise the loss of nutrients consumed in the fire. Native pastures usually have low nutrient concentrations at the time of burning; losses from grass shoot material of 3 t ha^{-1} containing *c.* 0.4% N and 0.08% S would represent only *c.* 12 kg N and 2 kg S, whilst most of the remaining nutrients would be available in ash. Even frequent burning had no deleterious effect on soil N content at Katherine in northern Australia (Norman & Wetselaar, 1960). Improved pastures of higher nutrient content would exhibit greater losses (Mannetje *et al.*, 1983). Soil is a good insulator, and grassland fires have little influence on the immediate soil temperature *c.* 1 cm below the soil surface. Some damage to fauna may occur and smoke constitutes a considerable environmental nuisance if burning occurs near to population centres.

(v) Management policy may mitigate some of these negative effects through (1) the timing, (2) the frequency of burning, and (3) the control of stock. The rapid restoration of pasture cover and the provision of forage for animals are ensured if burning is restricted to occasions when adequate soil moisture is present. Granier & Cabanis (1976) recommend for the Sudanese savanna that burning should take place after about 75% of the annual rainfall for the wet season has been registered. Paladines & Leal (1979) recommend for the Llanos Orientales a system of sequential burning which provides an area of recently burned pasture on several occasions through the year. Table 10.7 contrasts animal production at three stocking rates with the system where the total area is burnt at the end of the rainy season. The sequential burning system disadvantaged cattle during the dry season, since less young feed was produced throughout this period, but a subsequent advantage to LWG during the rainy season gave a net annual superiority at the lowest SR of 0.2 beasts ha^{-1}. Annual burning provides less cushioning of unfavourable seasons than burning at a lesser frequency and the latter would also contribute less environment damage.

Forage allowance following burning may be controlled by the extent of the area burnt; firing a small area promotes great stock concentration. Forage allowance is also increased by restricting animal access immediately after the burn. Alternatively, SR may be reduced before the burn to generate sufficient fuel to give a fire hot enough to kill woody species.

Livestock management in the tropics is evolving in the direction of reduced burning as the area of planted pastures and leguminous protein banks increase. Managers are becoming more aware of the potential value of standing feed and of the environmental damage arising from fires.

11

Conclusion

11.1
The balance of management objectives

Farmers accord differing roles to pasture in a farming system, depending upon their goals and skills, the nature of the production enterprise, the biological constraints to the successful use of pasture, and the changing economic incentives available. There are also community pressures which influence the manager's action, according to the societal benefits or damage which flow from the type of pasture utilisation adopted and its integration with whatever regional plans are in place.

(i)   Sustainability of the pasture ecosystem is the primary long term goal, and this may conflict with short-term objectives. A sustainable system is conventionally defined as one which meets the needs of the present generation without prejudicing the capacity of future generations to meet their own needs.

Exploitation of non-renewable resources at a level which leads to a decreased trend in long-term productivity and an increased incidence of environmental damage (for example, stream sedimentation) indicates a lack of sustainability. Fluctuations in climate make it difficult to assess the direction of change, and a succession of dry years will bring out the alarmists in full force and these often undervalue the resilience of tropical vegetation in recovering from stress. On the other hand the evidence of soil erosion and the changed hydrological characteristics of a watershed are usually unequivocal, once bench-marks have been established. The degree to which it is acceptable to utilise fossil fuels in intensifying pasture utilisation and to clear forests for animal production are more controversial, but it is expected that these practices exacerbate the 'greenhouse effect' (World Commission on Environment and Development, 1987).

Technologies are available for much of the subhumid and humid zones of the tropics and subtropics which provide opportunities for further intensification of pasture production in systems which are sustainable. These opportunities will not necessarily be introduced if social and economic incentives are insufficient. In the semi-arid and arid zones the situation is entirely different. Few technical inputs are available for introduction. Any attempt to make the ecosystem sustainable is usually contingent upon reducing SR. The reduction of the present animal SR in those sectors of these regions dominated by a subsistence or near-subsistence economy is only feasible if the human population dependent upon the vegetation is also reduced.

(ii)   There may be conflict between the objectives of productivity and of environmental protection. In some developed communities the emphasis has changed from one of enhancing productivity to one in which the maintenance of a lower input pasture system without damage to the environment has become the key question. In most tropical regions farmers are trying to grow more feed in order to care for their families, to create capital needed for further development, or to pay for imports. The sustainability of the marginal lands which have in consequence been brought into production is both technically and socially difficult to achieve and in many of these countries the thrust to increased productivity means that the plough has become 'the tool of the desert'.

These questions overlap the sustainability issue raised at the outset of this section, but there are questions of environmental degradation which are specific to community well-being and which arise in ways which are distinct from their effects on individual farm production. These include the pollution of rivers with nitrate and phosphate from intensive systems of animal production, the methane generated by ruminants leading to depletion of the ozone layer, the nuisance to urban communities of smoke from grassland fires, and the loss of potential for urban recreation and tourism in rural areas for other reasons.

(iii)   Pasture managers are faced with requirements for the welfare of their livestock which may conflict with requirements for the persistence and sustained productivity of the pastures. This dichotomy commonly arises in times of drought when SR is insufficiently flexible and a SR which is satisfactory for a normal season becomes overstocking for this particular season. The pastoral concern of the manager for the immediate welfare of his animals usually overrides his long term concern for the maintenance of the pasture resource and its productivity; this may involve the loss of pasture legumes in the sward. This bifurcation of interests is less acute if resilient pastures are grown which tolerate a wide range of SR so that the manager's concern for the animals can operate without detriment to the pasture.

Ease of reseeding or of regeneration of the seed bank is a further objective, and the degree of damage from overstocking is less if this occurs when the pasture is quiescent in drought. The interests of the pasture manager in manipulating botanical composition may require plants to seed down at an advanced stage of low nutritive value and to regenerate as seedlings at very low forage allowance; both practices may be inimical to animal production in the short term.

(iv)   The choice of the degree of intensification of livestock production depends upon the manager's acceptance of risk, the availability of investment inputs, and the changing economic incentives as well as the potential environmental damage associated with intensification.

In the tropics many livestock enterprises are already labour intensive, and these achieve greater animal output per unit area than ranching systems based on high output per labour unit in regions such as the Llanos, the Cerrado and the northern Australian savannas. It should be emphasised that in many societies entrepreneurial objectives are secondary to the primary objective of ensuring that the family is well fed; the adoption of risky technology which increases average farm income is not in synchrony with the goals of the farmer if adoption brings a concomitant incidence of some years with lower income. Similarly farmers who own land may avoid technology which involves them in an increased debt burden which places their tenure of the farm in jeopardy, especially if the policies of the lending institutions are land acquisitive. This places a premium on the robustness of the technology scientists offer the farmer, and the farmer has good grounds for scepticism about the claims of scientists for short-term research results obtained in the sheltered research station environment. Managers of capital-intensive operations may also prefer to invest in strategies of risk avoidance, such as security of stock water supplies, rather than to invest in pasture improvement and its attendant increase in SR.

Intensification of livestock production may evolve gradually from a steady increase in the adoption of new technology: the planting of legume hedgerows, the overseeding of communal grazing areas and paddy bunds with herbage legumes, the production of pasture seed or pasture nurseries on the farm for expanded plantings, the introduction of fodder legumes in a cropping cycle, or the control of woody regrowth. Substantial increases in income emerge quite slowly, since the concomitant increase in SR may require that animal sales be foregone in order to increase breeder numbers, unless more working capital is readily available. The incentives for intensification rest also in confidence about the market for livestock products, and the attractiveness of markets for alternative products which may be emphasised to a varying degree in a mixed farm. The adoption of new crops may require that the feeding value of their residues is at least equivalent to the crops they replace. The export strength of the local economy may determine whether it is legitimate to expand farming based on the import of the fossil fuels required, for example, for N fertiliser use.

The concluding note I wish to strike is that the balance of management objectives is best determined at the individual farm level. The reservoir of experience and skills developed by the pasture manager in the husbandry of his animals may be complemented by the advice from the scientist. The work of the latter is to provide a hard data base indicating the probable performance of animals and pastures in the various forage development options available, so that the pasture manager may choose from amongst these the pattern of pasture utilisation most appropriate to his own goals.

11.2
Priorities for research

Pasture research is spread unevenly through tropical and subtropical zones. The performance of pasture systems requires long-term assessment, but even in the more developed countries governments and institutions are loth to provide the long-term research support which is needed not only to generate innovations but to adapt and to test them in the farm situation (Humphreys, 1989). Granting bodies wish to retain a flexibility of funding which is inimical to the sustained critical thought and experimentation needed to develop new insights. Pasture science runs out of innovations if it neglects the basic research upon which their generation depends. Institutions are also unwilling to put in place the organisational infrastructure and the long-term professional development of scientists required for good quality research and extension outputs, upon which depend sustained pasture productivity and environmental repair.

Research needs vary greatly between regions and the lateral transfer of technology to other regions and other societies is fraught with difficulty. The following six themes are singled out as basic to the improved utilisation of tropical pastures.

(i)   Programmes of plant improvement represent the investment most likely to lead to increased ruminant production in the tropics, and the incorporation of elite seed or planting material in farming systems is the single innovation most effective in contributing to sustained gains in productivity and in utilisation. It is the genetically controlled deficiencies in growth, persistence and nutritive value which can be most simply overcome through modern techniques of plant introduction, selection, breeding and evaluation. The priority is to find pasture legumes whose introduction into farming systems is readily facilitated, since on the one hand this is the key to increased utilisation through the higher dietary value of the legume and the longer seasonal availability of edible feed, and on the other hand the N accretion to the ecosystem leads to increased utilisation of grass, as discussed in Chapters 6, 7 and 8.

It would be foolish to claim that, despite the success of pasture legumes on research stations, pasture legumes are widely used in tropical farm practice, except in some districts of northern Australia, pockets of northern Thailand, eastern Indonesia and the Amazon, and in plantation agriculture. Legumes are successful which (1) seed freely, so that both seed crops and plant replacement from seed are satisfactory, (2) are resistant to heavy grazing, (3) are resistant to pests and disease, (4) have promiscuous *Bradyrhizobium* affinities, and (5) are adapted to acid, low fertility soils. The vine type legumes, such as *M. atropurpureum* and *N. wightii* require too low a SR to survive in intensive systems, but many shrub legumes, and low growing herbage legumes such as *S. hamata*, *Arachis pintoi*, *Cassia rotundifolia*, *Aeschynomene falcata*, and *Vigna parkeri* which are resistant to heavy grazing offer great promise. The mineral nutrition of these plants, in so far as it controls growth, N fixation and dietary value, is part of the assessment requirement. The challenge to science is to produce more robust technologies using legumes than has been achieved in the past.

(ii)   Reference was made in Chapter 6 to the intransigent problem of the low nutritive value of the C_4 grasses, and the contrasting nutritive value of *Cenchrus* lines associated with structural variability was illustrated. The low digestibility of bundle sheath cells, and the protection of cell contents from digestion by cell wall material are associated with lignin and the ways in which lignin is laid down. Genetic variability in these characteristic is available; there is also the possibility that gene manipulation might interfere with lignin synthesis. Plant improvement programmes have often targeted increased DM production as a desirable and readily measurable attribute; advances in grass utilisation will arise when the attributes sought are high digestibility and intake protected by satisfactory plant persistence.

(iii)   Plant improvement programmes might also be directed to the grossly neglected area of the nutritive value of crop residues, as mentioned in Section 9.1.4. Breeders have concentrated on increasing grain production. The increases in grain yield which have resulted from a decrease in grain/straw ratio or from the production of stiffer straws of higher silica content have been to the detriment of animal performance in mixed farming systems where crop residues may form the main constituents of animal diet. This has been especially disadvantageous in annual cropping systems dependent for their success on animal draught power, and other aspects, such as the transfer and cycling of nutrients, are also involved.

(iv)   Some gaps in our knowledge of pasture plant physiology have been mentioned in this book; a key question is the ways in which the growth and persistence of the sward are influenced by defoliation (Chapter 4). Despite several decades of cutting trials and a lesser period of studies of plant response to grazing, it is still not possible to be predictive about sward response to defoliation with any degree of precision, or to define the modifying factors to that response in quantitative terms. Studies in the tropics of tillering, leaf growth and senescence, light relations in pastures, carbon allocation, the interaction of mineral nutrition and moisture relations with defoliation and the effects of hormonal balance have been insufficiently critical to provide a good basis for modelling pasture growth and utilisation.

(v)   Studies at the plant-animal interface are notoriously difficult, because of the inherent variability in the grazing behaviour of animals and the problems of describing the processes occurring in a selectively grazed sward. However, a knowledge of sward characteristics, forage allowance, diet selection, grazing behaviour and their interrelations with animal performance is essential to the development of sustainable grazing systems. We know little of what regulates the capacity and willingness of the grazing animal to eat particular plant materials, which is relevant to the definition of objectives in plant improvement as well as being directly applicable in the management of pasture utilisation, as discussed in Chapters 6 and 7.

(iv) Physical information about the interrelations of plant growth, senescence, consumption, persistence, nutritive value and nutrient flows with environmental factors, farm inputs and animal performance are basic to the development of descriptive models of the grassland ecosystem (Humphreys, 1986b).

Computer science has contributed greatly to our understanding of grassland ecosystems. Immense data sets may be handled painlessly to establish patterns of performance of animals grazing different combinations of pastures at various stocking rates in a varying climatic environment, using rigorous mathematical techniques. The effects of rainfall on pasture productivity can be simulated for 100 individual years. The building of holistic models leads to the discovery of particular weaknesses in the understanding of linked processes, which helps to define research needs; we also use these techniques to generate hypotheses.

Brougham (1983) has mentioned the dangers of overplaying a dependence upon simulation modelling, especially when dealing with a complex livestock–forage system. The modelling of tropical pasture systems to this point still contains much noise and some remarkable hypotheses are incorporated when scientists become removed from the field situation, or simply lack the type of data which need to be incorporated in the model (for example, Sullivan *et al.*, 1980). There is no substitute for direct, detailed field observations obtained as scientists spend a good part of their lives on their knees in the pasture, living with the plants and observing directly the effects of and on the grazing animal.

It must be recognised that although describing the grassland ecosystem in quantitative terms may lead to better management and to sustained productivity, the basic advances only come from the removal or mitigation of constraints to component processes. Much grassland research is and will always be based on reductionist science and the analysis of component processes; on the other hand the application of these research outputs is interdisciplinary. The challenge to pasture scientists is to ensure that communication and co-operation between grassland workers is sufficiently good to ensure that research outputs at the component process level take account of their interactions with other processes and are adapted and modified, so that they may be validated in the optimisation of the new grassland ecosystems which evolve, and may help the farmers' situation.

The primary objective of tropical pasture utilisation is to effect a synchrony between forage allowance and animal requirement in terms which sustain the basic resources of vegetation, soils and animals. The success of pasture technology and of farm practice is, in the final analysis, judged by whether the long-term trend lines in pasture growth and cover and in animal performance are positive, neutral, or negative. A good deal of tropical pasture technology has generated a negative trend line as the legumes have disappeared, due to overgrazing or the ravages of diseases or pests. We look for the trend exemplified in Table 7.3 (Shaw, 1978) where individual animal performance and the appropriate stocking rate increase as soil fertility and pasture production gradually rise. Tropical pastures are grown in environments which periodically impose considerable stresses on plants and animals and the systems of utilisation devised need to exhibit a singular resilience.

References

Abramides, P. L. G., Meirelles, N. M. F. & Bianchini, D. (1985). Effect of three grazing systems and two pasture types on performance of dairy breed bullocks. *Zootecnia*, **23**, 69–83.

Acocks, J. P. H. (1953). *Veld Types of South Africa*. Pretoria: Government Printer.

Acocks, J. P. H. (1966). Non-selective grazing as a means of veld reclamation. In *Proceedings of the Grassland Society of Southern Africa*, **1**, 33–9.

Adams, J. E., Arkin, G. F. & Ritchie J. T. (1976). Influence of row spacing and straw mulch on first stage drying. *Soil Science Society of America Journal*, **40**, 436–42.

Addison, K. B. (1970). Management systems on spear grass country. In *Proceedings of the XIth International Grassland Congress*, pp. 789–93. St Lucia: University of Queensland Press.

Addison, K. B., Cameron, D. G. & Blight, G. W. (1984). Biuret, sorghum and cottonseed meal as supplements for weaner cattle grazing native pastures in sub coastal south-east Queensland. *Tropical Grasslands*, **18**, 113–20.

Addy, B. L. & Thomas, D. (1977). Intensive fattening of beef cattle on Rhodes grass pastures on the Lilongwe plain, Malawi, *Tropical Animal Health and Production*, **9**, 99–106.

Adebowale, E. A. (1983). The performance of West African Dwarf goats and sheep fed silage and silage plus concentrate. *World Review of Animal Production*, **19**, 15–20.

Adegbola, A. (1966). Preliminary observations on the reserve carbohydrate of and regrowth potential of tropical grasses. In *Proceedings of the Xth International Grassland Congress*, pp. 933–6. Helsinki, Finland.

Adegbola, A. A. & McKell, C. M. (1966). Effect of nitrogen fertilization on the carbohydrate content of coastal bermuda grass. *Agronomy Journal*, **58**, 60–4.

Ademosun, A. A. (1973) Nutritive evaluation of Nigerian forages. 4. The effect of stage of maturity on the nutritive value of *Panicum maximum* (Guinea grass). *Nigerian Agricultural Journal*, **10**, 170–7.

Adjei, M. B., Mislevy, P. & Ward, C. Y. (1980). Response of tropical grasses to stocking rate. *Agronomy Journal*, **72**, 863–8.

Adjei, M. B., Mislevy, P. & West, R. L. (1988). Effect of stocking rate on the location of storage carbohydrates in the stubble of tropical grasses. *Tropical Grasslands*, **22**, 50–6.

Agamuthu, P. & Broughton, W. J. (1985). Nutrient cycling with the developing oil palm–legume ecosystem in Malaysia. *Agricultural Ecosystems & Environment*, **13**, 111–23.

Agamuthu, P., Chan, Y. K., Broughton, W. J., Jesinger R. & Khoo, K. M. (1981). Effect of differently managed legume on the early development of oil palms (*Elaeis guineensis* Jacq). *Agro-ecosystems*, **6**, 315–23.

Agata, W. (1985a). Studies on dry matter production of bahiagrass (*Paspalum notatum*) sward. 1. Characteristics of dry matter production during the regrowth period. In *Proceedings of the XVth International Grassland Congress*, pp. 1235–6. Nishi-nasuno, Japan: The Japanese Society of Grassland Science.

Agata, W. (1985b). Studies on dry matter production of bahiagrass (*Paspalum notatum*) sward. II. Characteristics of CO_2 balance and solar energy during the regrowth period. In *Proceedings of the XVth International Grassland Congress*, pp. 1237–8. Nishi-nasuno, Japan: The Japanese Society of Grassland Science.

Aguirre Hernández A., Eguiarte Vázquez, J. A., Carrete Carréon, F., Rodríguez Preciado, C. G., Garza Trevino, R. (1984). Use of two grazing systems on Para and Pangola grass pastures in the dry tropical conditions of the North Pacific coast. *Técnica Pecuaria en México*, **46**, 79–84.

Ahuja, L. D., Vishwanatham, M. K., Vyas, K. K. & Lal, K., (1974). Growth of sheep of Chokla breed under different systems of grazing on rangelands in the arid zone of western Rajasthan. *Annals of Arid Zone*, **13**, 259–65.

Aii, T. & Stobbs, T. H. (1980). Solubility of the protein of tropical pastures species and the rate of its digestion in the rumen. *Animal Feed Science Technology*, **5**, 183–92.

Alfonso, A., Hernández, C. A. & Batista, J. (1986). Alternatives for meat production from Guinea grass cv. Likoni pastures with different fertilizer levels. Final fattening stage *Pastos y Forrajes*, **9**, 177–84.

Akin, D. E., Willemse, M. T. M., & Barton, F. E. (1985). Histochemical reactions, autofluorescence, and rumen microbial degradation of tissues in untreated and delignified bermuda-grass stems. *Crop Science*, **25**, 901–5.

Akin, D. E., Wilson, J. R., & Windham, W. R. (1983). Site and rate of tissue digestion in leaves of C_3, C_4, and C_3/C_4 intermediate *Panicum* species. *Crop Science*, **23**, 147–55.

Akinola, J. O., Mackenzie, J. A. & Chheda, H. R. (1971). Effects of cutting frequency and level of applied nitrogen on productivity, chemical composition, growth components and regrowth potential of three *Cynodon* strains. 3. Regrowth potential. *West African Journal of Biology and Applied Chemistry*, **14**, 7–12.

Albrecht, K. A. & Boote, K. J. (1985). Photosynthetic recovery of *Aeschynomene americana* and *Hemarthria altissima* swards. In *Proceedings of the XVth International Grassland Congress*, pp. 415–6. Nishi-nasuno, Japan: The Japanese Society of Grassland Science.

Alejo, A. P., Rodríguez, A. G. & Fisher, M. J. (1988). Tolerencia de *Stylosanthes capitata* a la guerra en los Llanos Orientales de Colombia. *Pasturas Tropicales*, **10**, 2–7.

Alfonso, A., Hernández, C. A. & Batista, J. (1986). Alternatives for meat production from guinea grass cv. Likoni pastures with different fertilizer levels. Final fattening stage. *Pastos y Forrajes*, **9**, 177–84.

Alfonso, A., Valdés, L. R. & Batista, J. (1985). Effect of supplementation on yearlings grazing pangola grass (*Digitaria decumbens* Stent) with different stocking rates, segregation and fertilizer application. *Pastos y Forrajes*, **8**, 307–20.

Alli, I., Fairbairn, R., Noroozi, E. & Baker, B. E. (1984). The effects of molasses on the fermentation of chopped whole-plant leucaena. *Journal of the Science of Food and Agriculture*, **35**, 285–9.

Anderson, D. M. (1988). Seasonal stocking of tobosa managed under continuous and rotation grazing. *Journal of Range Management*, **41**, 78–83.

Andrew, M. H. (1986). Use of fire for spelling monsoon tallgrass pasture grazed by cattle. *Tropical Grasslands*, **20**, 69–78.

Anon. (1957). Feeding hay and silage to cattle. *Queensland Agricultural Journal*, **83**, 565–73.

Antoni Padilla, M., Fernández Van Cleve, J., Arroyo Aguilú, J. A. & Quinones Torres, R. (1983). Performance of Holstein cows grazing on intensively managed tropical grass pastures at three stocking rates. *Journal of Agriculture of the University of Puerto Rico*, **67**, 317–27.

Araújo, E. C. de & Languidey, P. H. (1982). Chemical composition, voluntary intake and apparent digestibility of hay from aerial parts of cassava. *Pesquisa Agropecuária Brasileira*, **17**, 1679–84.

Archer, S. G. & Bunch, C. E. (1953). *The American Grass Book*. Norman, USA: University of Oklahoma Press.

Arias, I., López, G. & Aurrecoechea, P. (1980). Weight gain of cattle continuously grazing sorghum stubble and regrowth in eastern Guárico. *Agronomía Tropical*, **30**, 269–78.

Arias, J. F., Cardozo, R., Casal, J. R., Mejía, E. A., Rodríguez, F. & Venegas, M. (1984). Mineral deficiencies as a factor limiting cattle production in the plains of Venezuela. 2. Response to mineral supplementation. *Tropical Animal Production*, **9**, 95–102.

Arias, P., Luciani, J. & Novoa, L. (1983). Sweet potato (*Ipomoea batata* Lamb.) vs. legumes as pasture plants. *Informe Anual 1982*, Maracay, 88–9.

Arnold, G. W. (1981). Grazing behaviour. In *Grazing Animals* ed. F. H. W. Morley, pp. 79–104. Amsterdam: Elsevier.

Arnold, G. W. & Dudzinski, M. L. (1978). *Ethology of Free Ranging Domestric Animals*. Amsterdam: Elsevier.

Ash, A. J., Prinsen, J. H., Myles, D. J. & Hendricksen, R. E. (1982). Short-term effects of burning native pasture in spring on herbage and animal production in south-east Queensland. *Proceedings of the Australian Society of Animal Production*, **14**, 377–80.

Awad, A. S., Edwards, D. G. & Huett, D. O. (1979). Seasonal changes in chemical composition of heavily fertilized Kikuyu pasture and their potential effects on the mineral nutrition of grazing cattle. *Australian Journal of Experimental Agriculture and Animal Husbandry*, **19**, 183–91.

Bahnisch, L. M. & Humphreys, L. R. (1977). Urea application and time of harvest effects on seed production of *Setaria anceps* cv. Narok. *Australian Journal of Experimental Agriculture and Animal Husbandry*, **17**, 621–8.

Barcellos, J. M., Echeverria, L. C. R., Pimentel, D. M., Soares, W. V. & Valle, L. S. (1979). Beef cattle production on low-fertiilty soils of Brazil: study of two production systems in Mato Grosso do Sul, using the model simulation method. In *Pasture Production in Acid Soils of the Tropics*, ed. P. A. Sánchez & L. E. Tergas, pp. 301–25. Cali, Colombia: CIAT.

Barker, S. J. & Levitt, M. S. (1969). Studies on ensilage of green panic/glycine mixed pasture. *Queensland Journal of Agriculture and Animal Science*, **26**, 541–52.

Barnes, D. L. (1965). The effects of frequency of burning and mattocking on the control of coppice in the Marandellas sandveld. *Rhodesian Journal of Agricultural Research*, **3**, 55–68.

Barnes, D. L. (1977). An analysis of rotational grazing on veld. *Rhodesia Agricultural Journal*, **74**, 147–51.

Barnes, D. L. (1982). Management strategies for the utilisation of southern African savanna. In *Ecology of Tropical Savannas*, ed. B. J. Huntley & B. H. Walker, pp. 626–56, Ecological Studies 42. New York: Springer-Verlag.

Barnes, D. L. (1989). Reaction of three veld grasses to different schedules of grazing and resting. 2. Residual effects on vigour. *South African Journal of Plant and Soil*, **6**, 8–13.

Barnes, D. L. & Hava, K. (1963). Effects of cutting on seasonal changes in the roots of Sabi panicum (*Panicum maximum* Jacq.) *Rhodesian Journal of Agricultural Research*, **1**, 107–10.

Barrow, N. J. (1987). Return of nutrients by animals. In *Managed Grasslands. Analytical Studies*, ed. R. W. Snaydon, pp. 181–6. Amsterdam: Elsevier.

Barry, G. A. (1984). Cobalt concentration in pasture species grown in several cattle grazing areas of Queensland. *Queensland Journal of Agricultural and Animal Sciences*, **41**, 73–82.

Barry, T. N. & Blaney, B. J. (1987). Secondary compounds of forages. In *The Nutrition of Herbivores*, ed J. B. Hacker & J. H. Ternouth, pp. 91–119. Sydney: Academic Press.

Bayer, W. & Otchere, E. O. (1985). Effect of livestock-crop integration on grazing time of cattle in a subhumid African savanna. In *Ecology and Management of the World's Savannas*, ed J. C. Tothill & J. J. Mott, pp. 256–9. Canberra, Australian Academy of Science.

Beaty, E. R., Sampaio, E. V. S. B., Ashley, D. A. & Brown, R. H. (1974). Partitioning and translocation of ^{14}C photosynthate by Bahiagrass (*Paspalum notatum* Flugge). In *Proceedings of the XIIth International Grassland Congress*, **1**, 259–67. Moscow.

Beaty, E. R., Tan, K. H., McCreery, R. A. & Powell, J. D. (1980). Yield and N content of closely clipped Bahiagrass as affected by N treatments. *Agronomy Journal*, **72**, 56–60.

Belyuchenko, I. S. (1976). Rhythmicity in the development of tropical and subtropical fodder crops. *Beiträge zur Tropischen Landwirtschaft und Vetinärmedizin*, **14**, 223–40.

Belyuchenko, I. S. (1980). Features of regrowth of paniculate and eragrostoid perennial grasses. In *Proceedings of the XIIIth*

International Grassland Congress, pp. 193–6. Berlin: Akademie-Verlag.

Benacchio, S. S., Baumgardner, M. F. & Mott, G. O. (1970). Residual effect of grain-pasture feeding systems on the fertility of the soil under a pasture sward. *Proceedings of Soil Science Society of America*, **34**, 621–4.

Ben-Ghedalia, D., Yosef, E., Solomon, R., Rivera Villarreal, E., & Ellis, W. C. (1988). Effects of ammonia treatment and stage of maturity of coastal Bermuda grass on monosaccharide residue composition and digestibility by steers. *Journal of the Science of Food and Agriculture*, **45**, 1–8.

Benitez, J. G., Bravo, M. H. & Gómez, O. R. (1984). Intake, digestibility and nitrogen balance of sheep supplemented with sodium hydroxide treated maize stover. *Revista Chapingo*, **9**, 167–71.

Bennett, D., Morley, F. H. W., Clark, K. W. & Dudzinski, M. L. (1970). The effect of grazing cattle and sheep together. *Australian Journal of Experimental Agriculture & Animal Husbandry*, **10**, 696–709.

Bianchine, D., Abramides, P. L. G., Meirelles, N. M. F., Werner, J. C. & Alcantara, P. B. (1987). Behaviour of four different mixed pastures in the South Part of the State of Sao Paulo, Brazil. II. Animal production. *Boletim Indústria Animal*, **44**, 173–93.

Billore, S. K., & Mall, L. P. (1976). Nutrient composition and inventory in a tropical grassland. *Plant and Soil*, **45**, 509–20.

Bishop, A. H. & Birrell, H. A. (1973). Efficiency of grazing – fodder conservation systems. *Proceedings of World Conference of Animal Production*, **2(b)**, 9–20.

Bisset, W. J. & Marlowe, G. W. C. (1974). Productivity and dynamics of two Siratro based pastures in the Burnett coastal foothills of south-east Queensland. *Tropical Grasslands*, **8**, 17–24.

Blunt, C. G. (1978). Production from steers grazing nitrogen fertilized irrigated pangola grass in the Ord Valley. *Tropical Grasslands*, **12**, 90–6.

Blunt, C. G. & Fisher, M. J. (1973). Production and utilization of fodder and grain sorghum as forage for cattle in the Ord River valley, Western Australia. *Australian Journal of Experimental Agriculture and Animal Husbandry*, **13**, 234–7.

Blunt, C. G. & Fisher, M. J. (1976). Production and utilization of oats as forage for cattle in the Ord River Valley, Western Australia. *Australian Journal of Experimental Agriculture and Animal Husbandry*, **16**, 88–93.

Blunt, C. G. & Humphreys, L. R. (1970). Phosphate response of mixed swards at Mt. Cotton, south-eastern Queensland. *Australian Journal of Experimental Agriculture and Animal Husbandry*, **10**, 31–41.

Bogdan, A. V. (1977). *Tropical Pasture and Fodder Plants*. London: Longman.

Bogdan, A. V. & Kidner, E. M. (1967). Grazing natural grassland in western Kenya. *East African Agriculture & Forestry Journal*, **33**, 31.

Böhnert, E., Lascano, C. & Weniger, J. H. (1985). Botanical and chemical composition of the diet selected by fistulated steers under grazing on improved grass-legume pastures in the tropical savannas of Colombia. I. Botanical composition of forage available and selected. *Zeitschrift für Tierzüchtung und Züchtungsbiologie*, **102**, 385–94.

Böhnert, E., Lascano, C. & Weniger, J. H. (1986). Botanical and chemical composition of the diet selected by fistulated steers under grazing on improved grass-legume pastures in the tropical savannas of Colombia. II. Chemical composition of forage available and selected. *Zeitschrift für Tierzüchtung und Züchtungsbiologie*, **103**, 69–79.

Boonman, J. G. (1971). Experimental studies on seed production of tropical grasses in Kenya. 2. Tillering and heading in seed crops of eight grasses. *Netherlands Journal of Agricultural Science*, **19**, 237–49.

Boulière, F. (ed.) (1983). *Tropical Savannas. Ecosystems of the World*, **13**. Amsterdam: Elsevier.

Bowen, G. D. (1959). Field studies on nodulation and growth of *Centrosema pubescens* Benth. *Queensland Journal of Agricultural Science*, **16**, 253–81.

Bowen, E. J. & Rickert, K. G. (1979). Beef production from native pastures sown to fine-stem stylo in the Burnett region of south-eastern Queensland. *Australian Journal of Experimental Agriculture and Animal Husbandry*, **19**, 140–9.

Braithewaite, B. M., Jane, A., & Swain, F. G. (1958). *Amnemus quadrituberculatus* (Boh.), a weevil pest of clover pastures on the north coast of N.S.W. *Journal of the Australian Institute of Agricultural Science*, **24**, 146–54.

Bray, R. A. & Sands, D. P. A. (1987). Arrival of the *Leucaena* psyllid in Australia: impact, dispersal and natural enemies. *Leucaena Research Reports*, **7(2)**, 61–5.

Breinholt, K. A., Gowen, F. A. & Nwosu, C. C. (1981). Influence of environmental and animal factors on day and night grazing activity of imported Holstein-Friesian cows in the humid lowland tropics of Nigeria. *Tropical Animal Production*, **6**, 300–7.

Britton, C. M., Dodd, J. D. & Wiechert, A. T. (1978). Net aerial primary production of an *Andropogon–Paspalum* grassland ecosystem. *Journal of Range Management*, **31**, 381–6.

Brougham, R. W. (1956). Effect of intensity of defoliation on regrowth of pasture. *Australian Journal of Agricultural Research*, **7**, 337–87.

Brougham, R. W. (1983). Practical livestock-forage systems: model to manager. *Proceedings of the XIV International Grassland Congress*, pp. 48–55. Boulder, Colorado: Westview.

Broughton, W. J. (1977). Effect of various covers on soil fertility under *Hevea brasiliensis* Muell. Arg. and on growth of the tree. *Agro-ecosystems*, **3**, 147–70.

Brown, R. F. (1982). Tiller development as a possible factor in the survival of the two grasses, *Aristida armata* and *Thyridolepis mitchelliana*. *Australian Rangeland Journal*, **4**, 34–8.

Bruce, R. C., & Ebersohn, J. P. (1982). Litter measurements in two grazed pastures in south-east Queensland. *Tropical Grasslands*, **16**, 180–5.

Burrows, W. (1985). Woodland management in south-east Queensland. *Tropical Grasslands*, **19**, 186–9.

Burton, G. W. & Jackson, J. E. (1962). A method for measuring sod reserves. *Agronomy Journal*, **54**, 53–5.

Butterworth, M. H. (1985). *Beef Cattle Nutrition and Tropical Pastures*. Harlow, UK: Longman.

Cameron, D. F. & Ludlow, M. M. (1977). Variation in the reaction of *Stylosanthes guianensis* lines to radiation frosts in controlled environments. *Australian Journal of Agricultural Research*, **28**, 795–806.

Carew, G. W. (1976). Stocking rate as a factor determining profitability of beef production. *Rhodesian Agricultural Journal*, **73**, 111–5.

Caro-Costas, R. & Vicente-Chandler, J. (1981). Effect of three grazing intervals on carrying capacity and weight gains produced by star grass pastures. *Journal of Agriculture of the University of Puerto Rico*, **65**, 14–20.

Carr, D. J. & Eng Kok Ng (1956). Experimental induction of flower formation in kikuyu grass (*Pennisetum clandestinum* Hochst. ex Chiov.). *Australian Journal of Agricultural Research*, **7**, 1–6.

Catchpoole, V. R. (1966). Laboratory ensilage of *Setaria sphacelata* (Nandi) with molasses. *Australian Journal of Experimental Agriculture and Animal Husbandry*, **6**, 76–81.

Catchpoole, V. R. (1969). Preliminary studies on curing and storing Nandi setaria hay. *Tropical Grasslands*, **3**, 65–74.

Catchpoole, V. R. (1970). Laboratory ensilage of three tropical pasture legumes – *Phaseolus atropurpureus*, *Desmodium intortum*, and *Lotononis bainesii*. *Australian Journal of Experimental Agriculture and Animal Husbandry*, **10**: 568–76.

Catchpoole, V. R. (1972a). Time of harvest, composition and silage characteristics of *Sorghum almum*. *Tropical Grasslands*, **6**, 171–6.

Catchpoole, V. R. (1972b) Laboratory ensilage of *Setaria sphacelata* cv. Nandi and *Chloris gayana* cv. Pioneer at a range of dry matter contents. *Australian Journal of Experimental Agriculture and Animal Husbandry*, **12**, 269–73.

Catchpoole, V. R. & Henzell, E. F. (1971). Silage and silage-making from tropical herbage species. *Herbage Abstracts*, **41**, 213–21.

Cernuda, C. F., Smith, R. M. & Vicente-Chandler, J. (1954). Influence of initial soil moisture condition on resistance of macroaggregates to slaking and to water drop impact. *Soil Science*, **77**, 19–28.

Chacon, E. & Stobbs, T. H. (1976). Influence of progressive defoliation of a grass sward on the eating behaviour of cattle. *Australian Journal of Agricultural Research*, **27**, 709–29.

Chacon, E. A., Stobbs, T. H. & Dale, M. B. (1978). Influence of sward characteristics on grazing behaviour and growth of Hereford steers grazing tropical grass pastures. *Australian Journal of Agricultural Research*, **29**, 89–102.

Chacon, E., Stobbs, T. H. & Sandland, R. I. (1976). Estimation of herbage consumption by grazing cattle using measurements of eating behaviour. *Journal of British Grassland Society*, **31**, 81–7.

Chadhokar, P. A. (1982). *Gliricidia maculata*. A promising legume fodder plant. *World Animal Review*, **44**, 36–43.

Chadhokar, P. A. & Humphreys, L. R. (1973). Effect of tiller age and time of nitrogen deficiency on seed production of *Paspalum plicatulum*. *Journal of Agricultural Science*, **81**, 219–29.

Chadhokar, P. A. & Kantharaju, H. R. (1980). Effect of *Gliricidia maculata* on growth and breeding of Bannur ewes. *Tropical Grasslands*, **14**, 78–82.

Chadhokar, P. A. & Lecamwasam, A. (1982). Effect of feeding *Gliricidia maculata* to milking cows: a preliminary report. *Tropical Grasslands*, **16**, 46–8.

Chantkam, S. (1978). Effects of phosphorus supply and defoliation on growth and phosphorus nutrition of tropical pasture legumes with special reference to centro (*Centrosema pubescens* Benth.). Ph.D. Thesis, University of Queensland.

Chantkam, S. (1982). Growth and phosphorus concentrations of *Centrosema pubescens* Benth. in flowing solution culture under differing defoliation regimes. *The Kasetsart Journal. Natural Sciences*, **16**, 1–11.

Chapman, H. D., Marchant, W. H., Utley, P. R., Hellwig, R. E. & Monson, W. G. (1972). Performance of steers on Pensacola Bahiagrass, Coastal Bermudagrass and Coastcross – 1 Bermudagrass pastures and pellets. *Journal of Animal Science*, **34**, 373–8.

Chen, C. P. & Othman, A. (1984). Performance of tropical forages under the closed canopy of the oil palm. 2. Legumes. *MARDI Research Bulletin*, **12**, 21–37.

Chen, M-C., Chen, C. P. & Chung, P. (1972). The nutritive value of pineapple by-products for ruminants. *Journal of the Agricultural Association of China*, **79**, 22–6.

Chik, A. B., Hassan, W. E. W. & Idris, M. S. H. (1984). Effect of concentrate supplementation on voluntary forage intake and growth response of young dairy Jersey cattle. *MARDI Research Bulletin*, **12**, 55–60.

Chopping, G. D., Deans, H. D., Sibbick, R., Thurbon, P. N. & Stokoe, J. (1976). Milk production from irrigated nitrogen fertilized Pangola grass. *Proceedings of the Australian Society of Animal Production*, **11**, 481–4.

Chopping, G. D., Lowe, K. J. & Clarke, L. G. (1983). Irrigation Systems. In *Dairy Management in the 80s. Focus on Feeding Seminar*, pp. 109–20. Brisbane: Queensland Department of Primary Industries.

Chopping, G. D., Moss, R. J., Goodchild, I. K. & O'Rourke, P. K. (1978). The effects of grazing systems and nitrogen fertilizer regimes on milk production from irrigated pangola-couch pastures. *Proceedings of the Australian Society of Animal Production*, **12**, 229.

Chopping, G. D., Thurbon, P. N., Moss, R. J. & Stephenson, H. (1982). Winter-spring milk production responses from the annual autumn oversowing of irrigated tropical pastures with ryegrass and clovers. *Animal Production in Australia*, **17**, 421–4.

Christiansen, S., Ruelke, O. C. & Lynch, R. O. (1981). Regrowth in darkness as influenced by previous cutting treatment of four limpograss genotypes. *Proceedings of Soil and Crop Science Society of Florida*, **40**, 156–9.

Christie, E. K. (1978). Herbage condition assessment of an infertile range grassland based on site production potential. *Australian Rangelands Journal*, **1**, 87–94.

Christie, E. K. (1979). Ecosystem processes in semiarid grasslands. II. Litter production, decomposition and nutrient dynamics. *Australian Journal of Agricultural Research*, **30**, 29–42.

CIAT (1972). In *Tropical Pastures Program Annual Report 1972*, p. 27. Cali, Colombia: Centro Internacional de Agricultura Tropical.

CIAT (1976). Pasture utilization. In *Annual Report 1976*. pp. C–33–34. Cali, Colombia: Centro Internacional de Agricultura Tropical.

CIAT (1982). Animal excreta as nutrient pools. In *Tropical Pastures Program Annual Report 1981*, pp. 186–91. Cali, Colombia: Centro Internacional de Agricultura Tropical.

CIAT (1983). Effects of cutting and deferred grazing on seed production in *Andropogon*. *Tropical Pastures Program Annual Report 1983*, pp. 157–8. Cali, Colombia: Centro Internacional de Agricultura Tropical.

CIAT (1985). *Tropical Pastures Program. Annual Report 1984*. Cali, Colombia: Centro Internacional de Agricultura Tropical.

CIAT (1986). Grazing of *B. dictyoneura + D. ovalifolium*. *Tropical Pastures Program Annual Report 1985*, pp. 291–61. Cali, Colombia: Centro Internacional de Agricultura Tropical.

CIAT (1988). *Tropical Pastures Program. Annual Report 1987*. Cali, Colombia: Centro Internacional de Agricultura Tropical.

Ciesiolka, C. (1987). *Catchment Management in the Nogoa Watershed*, Australian Water Resources Council Research Project Report 80/128. Canberra: Department of Resources and Energy.

Clapp, J. G. & Chamblee, D. S. (1970). Influence of different defoliation systems on the regrowth of pearl millet, hybrid sudangrass, and two sorghum-sudangrass hybrids from terminal, axillary and basal buds. *Crop Science*, **10**, 345–9.

Clapp, J. G., Chamblee, D. S. & Gross, H. D. (1965). Interrelationships between defoliation systems, morphological characteristics, and growth of "coastal" bermudagrass. *Crop Science*, **5**, 468–71.

Clark, N. A., Hemken, R. W. & Vandersall, J. H. (1965). A comparison of pearl millet, sundangrass and sorghum–sundangrass hybrid as pasture for lactating dairy cows. *Agronomy Journal*, **57**, 266–9.

Clatworthy, J. N. (1984). The Charter Estate Grazing Trial. Results of the botanical analysis. *Zimbabwe Agricultural Journal*, **81**, 49–52.

Clatworthy, J. N. & Muyotcha, M. J. (1983). Effect of stocking rate and grazing procedure on animal production and botanical composition of a Silverleaf desmodium/star grass pasture. In *Annual Report 1980–81, Division of Livestock and Pastures*, pp. 171–3. Marondera, Zimbabwe: Department of Research and Specialist Services, Grasslands Research Station.

Clements, F. E. (1920). *Plant Indicators. The Relation of Plant Communities to Process and Practice*. Washington: Carnegie Institution of Washington.

Clements, R. J. (1985). The patterns of grazing of Siratro plants by steers in a Siratro/Setaria pasture. In *CSIRO Division of Tropical Pasture Annual Report 1984–85*, pp. 91–4. Brisbane: CSIRO.

Clements, R. J. (1986). Rate of destruction of growing points of pasture legumes by grazing cattle. *CSIRO Division of Tropical Crops and Pastures Annual Report 1985–86*, pp. 73–4. Brisbane, Australia: CSIRO.

Clements, R. J. & Ludlow, M. M. (1977). Frost avoidance and frost resistance in *Centrosema virginianum*. *Journal of Applied Ecology*, **14**, 551–66.

Clifford, P. E. (1977). Tiller bud suppression in reproductive plants of *Lolium multiflorum* Lam. cv. Westerwoldicum. *Annals of Botany*, **41**, 605–15.

Cobbina, J. (1988). Vertisols of Ghana: uses and potential for improved management using cattle. In *Management of Vertisols in Sub-Saharan Africa*, ed. S. C. Jutzi, I. Haque, J. McIntire & J. E. S. Stares, pp. 359–78. Addis Ababa: ILCA.

Coleman, S. W., Neri-Flores, O., Allen, R. J. Jr. & Moore, J. E. (1978). Effect of pelleting and of forage maturity on quality of 2 sub-tropical grasses. *Journal of Animal Science*, **46**, 1103–12.

Colman, R. L. & Holder, J. M. (1968). Effect of stocking rate on butter fat production of dairy cows grazing kikuyu grass pastures fertilized with nitrogen. *Proceedings of the Australian Society of Animal Production*, **7**, 129–32.

Colman, R. L. & Kaiser, A. G. (1974). The effect of stocking rate on milk production from kikuyu grass pastures fertilized with nitrogen. *Australian Journal of Experimental Agriculture and Animal Husbandry*, **14**, 155–60.

Combellas, J., Baker, R. D. & Hodgson, J. (1979). Concentrate supplementation, and the herbage intake and milk production of heifers grazing *Cenchrus ciliaris*. *Grass and Forage Science*, **34**, 303–10.

Combellas, J., Centeno, A. & Mazzani, B. (1971). Utilization of the aerial parts of the groundnut. 1. Yield, chemical composition and digestibility *in vitro*. *Agronomia Tropical*, **21**, 533–7.

Conniffe, D., Browne, D. & Walshe, M. J. (1970). Experimental design for grazing trials. *Journal of Agricultural Science*, **74**, 339–42.

Connolly, J. (1974). Linear programming and the optimum carrying capacity of range under common use. *Journal of Agricultural Science*, **83**, 259–66.

Connolly, J. (1976). Some comments on the shape of the gain-stocking rate curve. *Journal of Agricultural Science*, **86**, 103–9.

Cooksley, D. G. (1983). A physical model of beef cattle production using inputs of native pasture and *Leucaena leucocephala* (Leucaena). *Animals Production in Australia*, **15**, 11–13.

Cooksley, D. G., Prinsen, J. H. & Paton, C. J. (1988). *Leucaena leucocephala* production in subcoastal south-east Queensland. *Tropical Grasslands*, **22**, 21–16.

Coombe, J. B. (1980). Utilization of low-quality residues. In *Grazing Animals*, ed. F. H. W. Morley, pp. 319–34. Amsterdam: Elsevier.

Cossins, N. J. (1983). Production strategies and pastoral man. In *Pastoral Systems Research in Sub-Saharan Africa*, pp. 213–31. Addis Ababa: ILCA.

Cossins, N. J. & Upton, M. (1987). The Borana pastoral system of southern Ethiopia. *Agricultural Systems*, **25**, 199–218.

Cossins, N. J. & Upton, M. (1988). Options for the improvement of the Borana pastoral system. *Agricultural Systems*, **27**, 251–78.

Costa, J. L. & Gomide, J. A. (1989). Haymaking from tropical grasses. In *Proceedings XVI International Grassland Congress*, **2**, 997–8. Versailles: Association Francaise pour la Production Fourragère.

Coupland, R. T. (ed.). (1979). *Grassland Ecosystems of the World: Analysis of Grasslands and Their Uses*. Cambridge: Cambridge University Press.

Cowan, R. T. (1975). Grazing time and pattern of grazing of Friesian cows on a tropical grass–legume pasture. *Australian Journal of Experimental Agriculture and Animal Husbandry*, **15**, 32–7.

Cowan, R. T., Byford, I. J. R. & Stobbs, T. H. (1975). Effects of stocking rate and energy supplementation on milk production from tropical grass–legume pasture. *Australian Journal of Experimental Agriculture and Animal Husbandry*, **15**, 740–6.

Cowan, R. T. & Davison, T. M. (1978). Milk yields of cows fed maize and molasses supplements on tropical pastures at two stocking rates. *Australian Journal of Experimental Agricultural and Animal Husbandry*, **18**, 12–15.

Cowan, R. T., Davison, T. M. & O'Grady, P. (1977). Influence of level of concentrate feeding on milk production and pasture utilization by Friesian cows grazing tropical grass-legume pasture. *Australian Journal of Experimental Agriculture and Animal Husbandry*, **17**, 373–9.

Cowan, R. T., Davison, T. M. & Shephard, R. K. (1986). Observations on the diet selected by Friesian cows grazing tropical grass and grass–legume pastures. *Tropical Grasslands*, **20**, 183–92.

Cowan, R. T., O'Grady, P., Moss, R. J. & Byford, I. J. R. (1974).

Milk and fat yields of Jersey and Friesian cows grazing tropical grass—legume pastures. *Tropical Grasslands*, **8**, 117–20.

Cowan, R. T. & Stobbs, T. H. (1976). Effects of nitrogen fertilizer applied in autumn and winter on milk production from a tropical grass—legume grazed at four stocking rates. *Australian Journal of Experimental Agriculture and Animal Husbandry*, **16**, 829–37.

Creek, M. J. & Nestel, B. L. (1965). The effect of grazing cycle duration on liveweight output and chemical composition of pangola grass (*Digitaria decumbens* Stent.) in Jamaica. *Proceedings IX International Grassland Congress*, **2**, 1613–18.

Crowder, L. V. & Chheda, H. R. (1982). *Tropical Grassland Husbandry*. London: Longman.

Cruz, L. C. & Calub, A. D. (1981). Performance of grade weanling calves in confinement or in native grassland (*Themeda triandra* FORSK) with and without supplementation. *Philippines Journal of Veterinary Animal Science*, **324**, 123–33.

Cubillos, G., Vohnout, K. & Jiménez, C. (1975). Intensive cattle feeding under grazing systems. In *Potential to Increase Beef Production in Tropical America*, Series CE (10), pp. 131–47. Cali, Colombia: CIAT.

Cumming, B. G. (1963). The dependence of germination on photoperiod, light quality, and temperature in *Chenopodium* spp. *Canadian Journal of Botany*, **41**, 1211–33.

Davidson, J. L. & Donald, C. M. (1958). The growth of swards of subterranean clover with particular reference to leaf area. *Australian Journal of Agricultural Research*, **9**, 53–72.

Davidson, J. L. & Milthorpe, F. L. (1966). Leaf growth in *Dactylis glomerata* following defoliation. *Annals of Botany*, **30**, 173–84.

Davidson, R. L. (1962). The influence of edaphic factors on the species composition of early stages of the subsere. *Journal of Ecology*, **50**, 401–10.

Davies, A. (1965). Carbohydrate levels and regrowth in perennial rye-grass. *Journal of Agricultural Science*, **65**, 213–21.

Davies, T. (1963). Fodder conservation in Northern Rhodesia. *Journal of Agricultural Science*, **61**, 309–28.

Davis, F. E. & Norton, B. W. (1978). The effects of rust (*Puccinia oahuensis*) on lamb production from pangola grass (*Digitaria decumbens* Stent.). *Proceedings of the Australian Society of Animal Production*, p. 283. St Lucia, Queensland: Department of Agriculture, University of Queensland.

Davison, T. M., Cowan, R. T., & O'Rourke, P. K. (1981). Management practices for tropical grasses and their effects on pasture and milk production. *Australian Journal of Experimental Agriculture and Animal Husbandry*, **21**, 196–202.

Davison, T. M., Cowan, R. T. & Shepherd, R. K. (1985). Milk production from cows grazing on tropical grass pastures. 2. Effects of stocking rate and level of nitrogen fertilizer on milk yield and pasture—milk yield relationship. *Australian Journal of Experimental Agriculture*, **25**, 515–32.

Davison, T. M., Jarrett, W. D. & Martin, P. (1985). A comparison of four patterns of allocating maize during lactation to Friesian cows grazing tropical pastures. *Australian Journal of Experimental Agriculture*, **25**, 241–8.

Davison, T. M., Marschke, R. J. & Brown, G. W. (1982). Milk yields from feeding maize silage and meat-and-bone meal to Friesian cows grazing on tropical grass and legume pasture. *Australian Journal of Experimental Agriculture and Animal Husbandry*, **22**, 147–54.

Davison, T. M., Murphy, G. M., Maroske, M. M. & Arnold, G.

(1980). Milk yield response following sodium chloride supplementation of cows grazing a tropical grass-legume pasture. *Australian Journal of Experimental Agriculture and Animal Husbandry*, **20**, 543–6.

Davison, T. M., Orr, W. N., Silver, B. A. & Duncalfe, F. (1989). Phosphorus fertiliser and the long term productivity of nitrogen fertiliser dairy pastures. In *Proceedings XVI International Grassland Congress*, **2**, 1133–4. Versailles: Association Française pour la Production Fourragère.

Davison, T. M., Orr, W. N., Silver, B. A. & Kerr, D. V. (1988). The effect of level of nitrogen fertilizer and season on milk yield and composition in cows grazing tropical grass pastures. *Proceedings of the Australian Society of Animal Production*, **17**, 170–3.

Deans, H. D., Chopping, G. D., Sibbick, R., Thurbon, P. N., Copeman, D. B. & Stokoe, J. (1976). Effect of stocking rate, breed, grain supplementation, nitrogen fertilizer level and anthelmintic treatment on growth rate of dairy weaners grazing irrigated Pangola grass. *Proceedings of the Australian Society of Animal Production*, **11**, 449–52.

De Boer, A. J. (1973). Selected measures of bovine performance in three Thai villages. *Thai Journal of Agricultural Science*, **6**, 177–90.

Delgado, A. & Alfonso, F. (1974). Effect of grazing systems and stocking rate on beef fattening on pangola grass. *Cuban Journal of Agricultural Science*, **8**, 129–35.

Denny, R. P. & Barnes, D. L. (1977). Trials of multi-paddock grazing systems on veld. 3. A comparison of six grazing procedures at two stocking rates. *Rhodesian Journal of Agricultural Research*, **15**, 129–42.

Denny, R. P., Barnes, D. L. & Kennan, T. C. D. (1977). Trials of multi-paddock grazing systems on veld. 1. An exploratory trial of systems involving twelve paddocks and one herd. *Rhodesian Journal of Agricultural Research*, **15**, 11–23.

Denny, R. P. & Steyn, J. S. H. (1977). Trials of multi-paddock grazing systems on veld. 2. A comparison of a 16-paddocks-to-one-herd system with a four-paddocks-to-one-herd system using breeding cows. *Rhodesian Journal of Agricultural Research*, **15**, 119–27.

Deregibus, V. A., Sanchez, R. A., Casal, J. J. & Trlica, M. J. (1985). Tillering responses to enrichment of red light beneath the canopy in a humid natural grassland. *Journal of Applied Ecology*, **22**, 199–206.

Devendra, C. (1978). The digestive efficiency of goats. *World Review of Animal Production*, **14**, 9–22.

Devendra, C. (1979a). Chemical treatment of rice straw in Malaysia. 1. The effect on digestibility of treatment with high levels of sodium and calcium hydroxide. *MARDI Research Bulletin*, **7**, 75–88.

Devendra, C. (1979b). The nutritive value of cassava (*Manihot esculenta* Crantz) leaves as a source of protein for ruminants in Malaysia. *MARDI Research Bulletin*, **7**, 112–17.

Devendra, C. (1981). Non-conventional feed resources in the S.E. Asian Region. *World Review of Animal Production*, **17**, 65–80.

Devendra, C. (1983). The nutritive value of sugar cane (*Saccharum officinarum*) tops. *MARDI Research Bulletin*, **11**, 389–94.

Devendra, C. (1987). Herbivores in the arid and wet tropics. In *The Nutrition of Herbivores*, ed. J. B. Hacker & J. H. Ternouth, pp. 23–46. Sydney: Academic Press.

Dixon, R. M. (ed.). (1987). *Ruminant Feeding Systems Utilizing Fibrous Agricultural Residues – 1986.* Canberra: IDP.

Doak, B. W. (1952). Some chemical changes in the nitrogenous constituents of urine when voided on pasture. *Journal of Agricultural Science,* **42,** 162–71.

Donald, C. M. (1946). *Pastures and Pasture Research.* Sydney: University of Sydney.

Donald, C. M. (1963). Competition among crop and pasture plants. *Advances in Agronomy,* **15,** 1–118.

Dovrat, A., Dayan, E. & van Keulen, H. (1980). Regrowth potential of shoot and of roots of Rhodes grass (*Chloris gayana* Kunth) after defoliation. *Netherlands Journal of Agricultural Science,* **28,** 185–99.

Dovrat, A., Deinum, B. & Dirven, J. G. P. (1972). The influence of defoliation and nitrogen on the regrowth of Rhodes grass. 2. Etiolated growth and non-structural carbohydrate content. *Netherlands Journal of Agricultural Science,* **20,** 97–103.

Doyle, P. T. (1987). Supplements other than forage. In *The Nutrition of Herbivores,* ed. J. B. Hacker & J. H. Ternouth, pp. 429–64. Sydney: Academic Press.

Doyle, P. T., Devendra, C. & Pearce, G. R. (1986). *Rice Straw as a Feed for Ruminants.* Canberra: IDP.

Dunwell, G. H. (1967). Paddock rotation reduces dippings. *Queensland Agricultural Journal,* **93,** 577–9.

Ebersohn, J. P. (1966). Effects of stocking rate, grazing method and ratio of cattle to sheep on animal liveweight gains in a semi-arid environment. *Proceedings of the X International Grassland Congress,* pp. 453–8. Helsinki: Finnish Grassland Association.

Ebersohn, J. P., Evans, J. & Limpus, J. F. (1983). Grazing time and its diurnal variation in beef steers in coastal south-east Queensland. *Tropical Grasslands,* **17,** 76–81.

Ebersohn, J. P. & Moir, K. W. (1984). Effect of pasture growth rate on live-weight gain of grazing beef cattle. *Journal of Agricultural Science,* **102,** 265–8.

Ebersohn, J. P., Moir, K. W. & Duncalfe, F. (1985). Inter-relationships between pasture growth and senescence and their effects on liveweight gain of grazing beef cattle. *Journal of Agricultural Science,* **104,** 299–301.

Edye, L. A. & Gillard, P. (1985). Pasture improvement in semi-arid tropical savannas: a practical example in northern Queensland. In *Ecology and Management of the World's Savannas,* ed. J. C. Tothill and J. J. Mott, pp. 303–9. Canberra: Australian Academy of Science.

Edye, L. A., Ritson, J. B. & Haydock, K. P. (1972). Calf production of Droughtmaster cows grazing a Townsville stylo-spear grass pasture. *Australian Journal of Experimental Agriculture and Animal Husbandry,* **12,** 7–12.

Edye, L. A., Ritson, J. B., Haydock, K. P. & Davies, the late J. Griffiths, (1971). Fertility and seasonal changes in liveweight of Droughtmaster cows grazing a Townsville stylo-spear grass pasture. *Australian Journal of Agricultural Research,* **22,** 963–77.

Edye, L. A., Williams, W. T. & Winter, W. H. (1978). Seasonal relations between animal gain, pasture production and stocking rate on two tropical grass–legume pastures. *Australian Journal of Agricultural Research,* **29,** 103–13.

Egan, A. R., Frederick, F. & Dixon, R. M. (1987). Improving efficiency of use of supplements by manipulation of management procedures. In *Ruminant Feeding Systems Utilising Fibrous Agricultural Residues – 1986,* ed. R. M. Dixon, pp. 69–81. Canberra: IDP.

Egan, A. R., Wanapat, M., Doyle, P. T., Dixon, R. M. & Pearce, G. R. (1986). Production limitations of intake, digestibility and rate of passage. In *Forages in Southeast Asian and South Pacific Agriculture,* ed. G. J. Blair, D. A. Ivory & T. R. Evans, pp. 104–10. Canberra: Australian Centre for International Agricultural Research.

Eguiarte V. J. A., Garza T. R., Lagunes, L. J., Rodríguez, P. C. G., Carrete C. F. O. & Sánchez, A. R. (1984). Beef production from African star grass under two grazing systems and two rates of fertilizer application. *Técnica Pecuaria en México,* **47,** 60–5.

Ehara, K., Maeno, N. & Yamada, Y. (1966). Physiological and ecological studies on the regrowth of herbage plants. 4. The evidence of utilization of food reserves during the early stage of regrowth in bahiagrass (*Paspalum notatum* Flugge) with $^{14}CO_2$. *Journal of Japanese Society of Grassland Science,* **12,** 1–3.

Eiten, G. (1972). The Cerrado vegetation of Brazil. *Botanical Review,* **38,** 201–341.

Elliott, R. C. & Fokkema, K. (1960). Protein digestibility relationships in ruminants. *Rhodesian Agricultural Journal,* **57,** 301.

Elwell, H. A. & Stocking, M. A. (1974). Rainfall parameters and a cover model to predict runoff and soil loss from grazing trials in the Rhodesia sandveld. *Proceedings of Grassland Society of Southern Africa,* **9,** 157–63.

Elwell, H. A. & Stocking, M. A. (1976). Vegetal cover to estimate soil erosion hazard in Rhodesia. *Geoderma,* **15,** 61–70.

Eng, P. K., Kerridge, P. C. & Mannetje, L.'t. (1978). Effects of phosphorus and stocking rate on pasture and animal production from a guinea grass–legume pasture in Johore, Malaysia. 1. Dry matter yields, botanical and chemical composition. *Tropical Grasslands,* **12,** 188–97.

Eng, P. K. & Mannetje, L.'t & Chen, C. P. (1978). Effects of phosphorus and stocking rate on pasture and animal production from a Guinea grass–legume pasture in Johore, Malaysia. 2. Animal liveweight change. *Tropical Grasslands,* **12,** 198–207.

Entwistle, K. W. & Knights, G. (1974). The use of urea–molasses supplements for sheep grazing semi-arid tropical pastures. *Australian Journal of Experimental Agriculture and Animal Husbandry,* **14,** 17–22.

Eriksen, F. I. & Whitney, A. S. (1982). Growth and nitrogen fixation of some tropical forage legumes as influenced by solar radiation regimes. *Agronomy Journal,* **74,** 703–9.

Ernst, A. J., Limpus, J. F. & O'Rourke, P. K. (1975). Effect of supplements of molasses and urea on intake and digestibility of native pasture hay by steers. *Australian Journal of Experimental Agriculture and Animal Husbandry,* **15,** 451–5.

Esperance, M. (1984). Studies on the improvement of systems of segregation of areas for conservation in milk production. *Pastos y Forrajes,* **7,** 95–109.

Esperance, M., Caceres, O., Ojeda, F. & Perdomo, A. (1980). Fermentation characteristics, nutritive value and production of milk from Pangola grass ensiled at two stages. *Pastos y Forrajes,* **3,** 147–61.

Evans, T. R. & Bryan, W. W. (1973). Effects of soils, fertilizers and stocking rates on pastures and beef production on the Wallum of south-eastern Queensland. 2. Liveweight change and beef production. *Australian Journal of Experimental Agriculture and Animal Husbandry,* **13,** 530–6.

Falvey, J. L. (1983). The response of cattle in the Thai highlands to a supplement containing sodium and phosphorus. *Tropical Animal Production*, **8**, 45–9.

Falvey, J. L. & Hengmichai, P. (1979). Invasion of *Imperata cylindrica* (L) Beauv. by *Eupatorium* species. *Journal of Range Management*, **32**, 340–4.

FAO (1988). *1987 FAO Production Yearbook, 41*. Rome: Food and Agriculture Organisation of the United Nations.

Farias, I. & Gomide, J. A. (1973). Effect of wilting and the addition of cassava meal on the characteristics of silage from elephant grass cut at various DM contents. *Experientiae*, **16**, 131–49.

Ferdinandez, D. E. F. (1972). Intercropping with coconut. *Ceylon Coconut Quarterly*, **23**, 51–3.

Ferguson, K. A., Hemsley, J. A. & Reis, P. J. (1967). Nutrition and wool growth. The effect of protecting dietary protein from microbial degradation in the rumen. *Australian Journal of Science*, **30**, 215–17.

Ferreira, J. J., Silva, J. F. C. da & Gomide, J. A. (1974). Effect of growth stage, wilting and the addition of cassava scrapings on the nutritive value of elephant grass silage (*Pennisetum purpureum* Shum). *Experientiae*, **17**, 85–108.

Ferreiro, H. M., Preston, T. R. & Herrera, F. (1979). Sisal by-products as cattle feed; effect of supplementing ensiled pulp with rice polishings and ramon (*Brosimum alicastrum*) on growth rate, digestibility and glucose entry rate by cattle. *Tropical Animal Production*, **4**, 73.

Ferreiro, H. M., Preston, T. R. & Sutherland, T. M. (1977). Digestibility of stalk and tops of mature and immature sugar cane. *Tropical Animal Production*, **2**, 100–4.

Ffoulkes, D., Espejo, S., Marie, D., Delpeche, M. & Preston, T. R. (1978). The banana plant as cattle feed; composition and biomass production. *Tropical Animal Production*, **1**, 45–50.

Ffoulkes, D., Hovell, F. D. DeB., & Preston, T. R. (1978). Sweet potato forage as cattle feed: voluntary intake and digestibility of mixtures of sweet potato forage and sugar cane. *Tropical Animal Production*, **3**, 140–4.

Ffoulkes, D. & Preston, T. R. (1978). The banana plant as cattle feed: digestibility and voluntary intake of different proportions of leaf and pseudostem. *Tropical Animal Production*, **3**, 114–17.

Ffoulkes, D. & Preston, T. R. (1979). Fattening cattle with molasses/urea and cassava or sweet potato forage. *Tropical Animal Production*, **4**, 97–8.

Filho, A. B. & Lopez, J. (1979). Evaluation of the quality of pearl millet (*Pennisetum americanum* (L.) Leeke) silage with N or energy supplementation. *Revista da Sociedade Brasileira de Zootecnia*, **8**, 316–31.

Filho, L. C. P. M. & Mühlbach, P. R. F. (1986). Effect of wilting on the quality of chemically evaluated elephant grass cv. Cameron (*Pennisetum purpureum* Schumach.) and pearl millet (*Pennisetum americanum* (L.) Leeke) silages. *Revista da Sociedade Brasileira de Zootecnia*, **15**, 224–33.

Fisher, M. J. (1973). Effect of times, height and frequency of defoliation on growth and development of Townsville stylo in pure ungrazed swards at Katherine, N. T. *Australian Journal of Experimental Agriculture and Animal Husbandry*, **13**, 389–97.

Floate, M. J. S. (1987). Nitrogen cycling in managed grasslands. In *Managed Grasslands. Analytical Studies*, ed. R. W. Snaydon, pp. 163–72. Amsterdam: Elsevier.

Flores, J. F., Stobbs, T. H. & Minson, D. J. (1979). The influence of the legume *Leucaena leucocephala* and formal-casein on the production and composition of milk from grazing cows. *Journal of Agricultural Science*, **92**, 351–8.

Fonseca, D. M. da & Escuder, C. J. (1983). Stocking rate and productivity on buffel grass pastures. *Revista da Sociedade Brasileira de Zootecnia*, **12**, 11–24.

Ford, C. W., Morrison, I. M. & Wilson, J. R. (1979). Temperature effects on lignin, hemicellulose and cellulose in tropical and temperate grasses. *Australian Journal of Agricultural Research*, **30**, 621–33.

French, A. V., O'Rourke, P. K. & Cameron, D. G. (1988a). Beef production from forage crops in the brigalow regions of central Queensland. 1. Forage sorghums. *Tropical Grasslands*, **22**, 79–84.

French, A. V., O'Rourke, P. K. & Cameron, D. G. (1988b). Beef production from forage crops in the brigalow regions of central Queensland. 2. Winter forage crops. *Tropical Grasslands*, **22**, 85–90.

French, A. V., O'Rourke, P. K. & Cameron, D. G. (1988c). Rotational and continuous grazing of Zulu forage sorghum (*Sorghum* spp. hybrid) by beef cattle grazed at 3 stocking rates. *Tropical Grasslands*, **22**, 91–3.

Fresco, L. O. & Westphal, E. (1988). A hierarchical classification of farm systems. *Experimental Agriculture*, **24**, 399–419.

Fribourg, H. A., Overton, J. R. & Mullins, J. A. (1975). Wheel traffic on regrowth and production of summer annual grasses. *Agronomy Journal*, **67**, 423–6.

Gallaher, R. N. & Brown, R. H. (1977). Starch storage in C_4 vs. C_3 grass leaf cells as related to nitrogen deficiency. *Crop Science*, **17**, 85–8.

Gammon, D. M. & Roberts, B. R. (1978). Patterns of defoliation during continuous and rotational grazing of the Matopos sandveld of Rhodesia. 3. Frequency of defoliation. *Rhodesian Journal of Agricultural Research*, **16**, 147–64.

Gardener, C. J. (1975). Mechanisms regulating germination in seeds of *Stylosanthes*. *Australian Journal of Agricultural Research*, **26**, 281–94.

Gardener, C. J. (1980). Tolerance of perennating *Stylosanthes* plants to fire. *Australian Journal of Experimental Agriculture and Animal Husbandry*, **20**, 587–93.

Gardener, C. J. (1982). Population dynamics and stability of *Stylosanthes hamatu* cv. Verano in grazed pastures. *Australian Journal of Agricultural Research*, **33**, 63–74.

Gardener, C. J., Freire, L. C. L. & Murray, R. M. (1988). Effect of superphosphate application on the nutritive value of *Stylosanthes* spp.–native grass pasture for cattle. 1. Composition of the diet selected. *Proceedings of the Australian Society of Animal Production*, **17**, 190–3.

Gardener, C. J., Megarrity, R. G. & McLeod, M. N. (1982). Seasonal changes in the proportion and quality of plant parts of nine *Stylosanthes* lines. *Australian Journal of Experimental Agriculture and Animal Husbandry*, **22**, 391–401.

Gartner, R. J. W. & Anson, R. J. (1966). Vitamin A reserves of sheep maintained on mulga (*Acacia aneura*). *Australian Journal of Experimental Agriculture & Animal Husbandry*, **6**, 321–5.

Gartner, R. J. W., McLean, R. W., Little, D. A. & Winks, L. (1980). Mineral deficiencies limiting production of ruminants grazing tropical pastures in Australia. *Tropical Grasslands*, **14**, 266–72.

Geoffroy, F. & Despois, P. (1978). Banana leaves and stems as a forage resource. 2. Animal utilization: intake. *Nouvelles Agronomiques des Antilles et de la Guyana*, **4**, 81–5.

Geoffroy, F. & Barreto-Velez, F. (1983a). Review of cassava (*Manihot esculenta* Crantz) in ruminant nutrition. I. Chemical composition feeding value, toxicity and processing. *Turrialba*, **33**, 231–41.

Geoffroy, F. & Barreto-Velez, F. (1983b). Review of manioc (*Manihot esculenta*) in the feeding of ruminants. 2. Utilization by ruminants. *Turrialba*, **33**, 245–56.

Geoffroy, F., Fabert, V., Calif, E., Saminadin, G. & Varo, H. (1978). Potential of banana leaves and stems as forage. 1. Availability and nutritive value. *Nouvelles Agronomiques de Antilles et de la Guyana*, **4**, 1–9.

Gibson, T. (1987). Northeast Thailand. A ley farming system using dairy cattle in the infertile uplands. *World Animal Review*, **61**, 36–43.

Gihad, E. A. (1979). Intake, digestibility and nutrient utilization by sheep of sodium hydroxide-treated tropical grass supplemented with soybean or urea. *Journal of Animal Science*, **48**, 1172–6.

Gildersleeve, R., Ocumpaugh, W. R., Quesenberry, K. H. & Moore, J. E. (1987). Mob-grazing of morphologically different *Aeschynomene* species. *Tropical Grasslands*, **21**, 123–32.

Gillard, P. (1967). Coprophagous beetles in pasture ecosystems. *Journal of Australian Institute of Agricultural Science*, **33**, 30–4.

Gillard, P. (1969). The effect of stocking rate on botanical composition and soils in natural grassland in South Africa. *Journal of Applied Ecology*, **6**, 489–97.

Gillingham, A. G. (1987). Phosphorus cycling in managed grasslands. In *Managed Grasslands. Analytical Studies*, ed. R. W. Snaydon, pp. 173–80. Amsterdam: Elsevier.

Golding, E. J., Moore, J. E., Franke, D. E. & Ruelke, O. C. (1976). Formulation of hay–grain diets for ruminants. 2. Depression in voluntary intake of different quality forages by limited grain in sheep. *Journal of Animal Science*, **42**, 717–23.

Gomide, J. A. & da Cruz, M. E. (1986). Haymaking from surplus pasture during the grazing season. *Revista da Sociedade Brasileira de Zootecnia*, **15**, 85–93.

Goncalves de Assis, A. (1984). Feeding of dairy cows in Zona da Mata, Minas Gerais. II. Effect of stocking rate on the relationship between herbage supply and demand. *Pesquisa Agropecuária Brasileira*, **19**, 1145–56.

Graham, T. G. & Mayer, B. G. (1972). Effect of method of establishment of Townsville stylo and application of superphosphate on the growth of steers. *Queensland Journal of Agriculture and Animal Sciences*, **29**, 289–96.

Graham, T. W. G., Wood, S. J., Knight, J. L. & Blight, G. W. (1983). Urea and molasses as a winter supplement for weaner steers grazing improved pasture in central Queensland. *Tropical Grasslands*, **17**, 11–20.

Granier, P. & Cabanis, Y. (1976). Burning and animal production on the Sudanese savanna. *Revue d'Elevage et de Médecine Vétérinaire des Pays Tropicaux*, **29**, 267–75.

Grant, S. A., Banthram, G. T., Torvell, L., King, J. & Smith, H. K. (1983). Sward management, lamina turnover and tiller population density in continuously stocked *Lolium perenne*-dominated swards. *Grass and Forage Science*, **38**, 333–44.

Gray, S. G. (1966). Inheritance of growth habit and quantitative characters in intervarietal crosses in *Leucaena leucocephala* (Lam.) De Wit. *Australian Journal of Agricultural Research*, **18**, 63–70.

Grime, J. P. (1979). *Plant Strategies and Vegetation Processes*. Chichester: Wiley.

Grof, B. & Harding, W. A. T. (1970). Dry matter yields and animal production of Guinea grass (*Panicum maximum*) on the humid tropical coast of North Queensland. *Tropical Grasslands*, **4**, 85–95.

Grotheer, M. D., Cross, D. L., Grimes, L. W., Caldwell, W. J. & Johnson, L. J. (1985). Effect of moisture level and injection of ammonia on nutrient quality and preservation of Coastal Bermudagrass hay. *Journal of Animal Science*, **61**, 1370–77.

Grünwaldt, E. G., Escuder, C. J., Rodriguez, N. M. & Vasconcelos, A. C. (1981). Effect of stocking rate and seasonal changes on pastures and cattle diet. I. Chemical composition. *Arquivos da Escola de Veterinária da Universidade Federal de Minas Gerais*, **33**, 519–27.

Gryseels, G. & Asamenew, G. (1985). Links between livestock and crop production in the Ethiopian highlands. *ILCA Newsletter*, **4(2)**, 5–6.

Guerrero, J. N., Conrad, B. E., Holt, E. C. & Wu, H. (1984). Prediction of animal performance on bermuda grass pasture from available forage. *Agronomy Journal*, **76**, 577–80.

Gutierrez, A. & Simon, L. (1974). Effect of rotation and stocking rate of grazing calves on liveweight gains and incidence of parasites. *Serie Técnico Científica, Estación Experimental de Pastos y Forrajes Indio Hatuey*, **3**, 14–20.

Gutierrez, A., Simon, L., Perdomo, A. & Cruz, R. (1978). Effect of two management systems on the growth of calves. *Pastos y Forrajes*, **1**, 155–61.

Gutteridge, R. C. (1982). The productivity and pathways of persistence of legumes sown in grazed native pasture situations in northeast Thailand. Ph.D. thesis, University of Queensland.

Gutteridge, R. C. (1983). Productivity of forage legumes on rice-paddy walls in north-east Thailand. *Proceedings XIV International Grassland Congress*, 226–9.

Gutteridge, R. C. (1985). Survival and regeneration of four legumes oversown into native grasslands in northeast Thailand. *Journal of Applied Ecology*, **22**, 885–94.

Gutteridge, R. C., Shelton, H. M., Wilaipon, B. & Humphreys, L. R. (1983). Productivity of pastures and responses to salt supplements by beef cattle on native pasture in northeast Thailand. *Tropical Grasslands*, **17**, 105–14.

Hacker, J. B. (1974). Variation in oxalate, major cations, and dry matter digestibility of 47 introductions of the tropical grass setaria. *Tropical Grasslands*, **8**, 145–54.

Hacker, J. B. (ed.) (1982). *Nutritional Limits to Animal Production from Pastures*. Farnham Royal, UK: Commonwealth Agricultural Bureaux.

Hacker, J. B., Forde, B. J. & Gow, J. M. (1974). Simulated frosting of tropical grasses. *Australian Journal of Agricultural Research*, **25**, 45–58.

Hacker, J. B. & Minson, D. J. (1972). Cultivar differences in *in vitro* dry matter digestibility in *Setaria*, and the effects of site, age and season. *Australian Journal of Agricultural Research*, **23**, 959–67.

Hacker, J. B., Strickland, R. W. & Basford, K. E. (1985). Genetic variation in sodium and potassium concentration in herbage of *Digitaria milanjiana*, and its relation to site of origin. *Australian Journal of Agricultural Research*, **36**, 201–12.

Hacker, J. B. & Ternouth, J. H. (eds.) (1987). *The Nutrition of Herbivores*. Sydney: Academic Press.

Haggar, R. J. (1965). The production of seed from *Andropogon gayanus. Proceedings of the International Seed Testing Association,* **31**, 251–9.

Haggar, R. J. (1971). The production and management of *Stylosanthes gracilis* at Shika, Nigeria. II. In savanna grassland. *Journal of Agricultural Science,* **77**, 347–44.

Haggar, R. J. & Ahmed, M. B. (1970). Seasonal production of *Andropogon gayanus.* 2. Seasonal changes in digestibility and feed intake. *Journal of Agricultural Science,* **75**, 369–73.

Hall, H. T. B. (1977). *Diseases and Parasites of Livestock in the Tropics,* pp. 202–3. London: Longman.

Hall, R. L. (1978). The analysis and significance of competitive and non-competitive interference between species. In *Plant Relations in Pastures,* ed. J. R. Wilson, pp. 163–74. Melbourne: CSIRO.

Hamilton, R. I., Catchpole, V. R., Lambourne, L. J. & Korr, J. D. (1978). The preservation of a Nandi *Setaria* silage and its feeding value for dairy cows. *Australian Journal of Experimental Agriculture and Animal Husbandry,* **18**, 16–24.

Hamilton, W. T. & Seifres, C. J. (1982). Prescribed burning during winter for maintenance of buffelgrass. *Journal of Range Management,* **35**, 9–12.

Harker, K. W. (1960). Defaecating habits of a herd of Zebu cattle. *Tropical Agriculture,* **37**, 193–200.

Harlan, J. R. (1958). Generalised curves for gain per head and gain per acre in rates of grazing studies. *Journal of Range Management,* **11**, 140–7.

Harrington, G. N. & Pratchett, D. (1974). Stocking rate trials in Ankole, Uganda. 1. Weight gain of Ankole steers at intermediate and heavy stocking rates under different managements. *Journal of Agricultural Science,* **82**, 497–506.

Harvey, J. M., Beames, R. M., Hegarty, A. & O'Bryan, M. S. (1963). Influence of grazing management and copper supplementation on the growth rate of Hereford cattle in south-eastern Queensland. *Queensland Journal of Agricultural Science,* **20**, 137–59.

Hegarty, M. P. (1982). Deleterious factors in forages affecting animal production. In *Nutritional Limits to Animal Production from Pastures,* ed J. B. Hacker, pp. 133–50. Farnham Royal, UK: Commonwealth Agricultural Bureaux.

Hegarty, M. P., Lee, C. P., Christie, G. S., Court, R. D. & Haydock, K. P. (1979). The goitrogen 3-hydroxy-4 (IH)-pyridone, a ruminal metabolite from *Leucaena leucocephala:* Effects in mice and rats. *Australian Journal of Biological Science,* **32**, 27–40.

Hendricksen, R. E. & Minson, D. J. (1980). The feed intake and grazing behaviour of cattle grazing a crop of *Lablab purpureus* cv. Rongai. *Journal of Agricultural Science,* **95**, 547–54.

Hendricksen, R. E. & Minson, D. J. (1985). Growth, canopy structure and chemical composition of *Lablab purpureus* cv. Rongai at Samford, S.E. Queensland. *Tropical Grasslands,* **19**, 81–7.

Hendricksen, R. E., Poppi, D. P. & Minson, D. J. (1981). The voluntary intake, digestibility and retention time by cattle and sheep of stem and leaf fractions of a tropical legume (*Lablab purpureus*). *Australian Journal of Agricultural Research,* **32**, 389–98.

Hennessy, D. W. (1980). Protein nutrition of ruminants in tropical areas of Australia. *Tropical Grasslands,* **14**, 260–5.

Hennessy, D. W. & Williamson, P. J. (1976). The nutritive value of Kikuyu grass (*Pennisetum clandestinum*) leaf and the use of pellet leaf in rations high or low in energy. *Australian*

Journal of Experimental Agriculture and Animal Husbandry, **16**, 729–34.

Henzell, E. F. & Ross, P. J. (1973). The nitrogen cycle of pasture ecosystems. In *Chemistry and Biochemistry of Herbage,* vol. 2, eds G. W. Butler & R. W. Bailey, pp. 227–45. London: Academic Press.

Hermans, C. (1986). Pattern of seasonality in livestock production cattle rations in Bangladesh. *World Review of Animal Production,* **23**, 43–50.

Hernández, C. A., Alfonso, A. & Duquesne, P. (1986). Producción de carne basada en pastos naturales mejorados con reguminosas arbustivas y herbaceas. 1. Ceba inicial. *Pastos y Forrajes,* **9**, 79–88.

Hernández, C. A., Alfonso, A. & Duquesne, P. (1988). Banco de proteina de *Neonotonia wightii* y *Macroptilium atropurpureum* corro complemento al pasto natural en la cabe de bovinos. *Pastos y Forrajes,* **11**, 74–81.

Hernández, D. & Rosete, A. (1985). Milk production with *Cynodon dactylon.* Integral analysis of the rotational cycle and the rest period. *Pastos y Forrajes,* **8**, 423–34.

Hernández, D., Rosete, A. & Robles, F. (1985). A rotational grazing system for milk production with *C. dactylon.* II. Effect of grazing time on each paddock. *Pastos y Forrajes,* **8**, 279–95.

Herrera, J. (1978). Effect of interval of rotation on milk production and on the pasture. In *Primer Seminario Científico Técnico, Provincia de Las Tunas,* pp. 58–60. Havana, Cuba: Estación Central de Pastos.

Herrera, R. S., Martinez, R. O., Ruiz, R. & Hernandez, Y. (1986). Milk production of cows grazing coast cross/bermuda grass (*Cynodon dactylon*). 4. Vertical distribution of structural carbohydrates and pasture digestibility. *Cuban Journal of Agricultural Science.* **20**, 183–90.

Hildyard, P. (1970). The utilisation of certain native pastures composed of grasses of varying palatability. *Proceedings of the XI Grassland Congress,* pp. 41–5. St Lucia: University of Queensland Press.

Hirakawa, M., Okubo, T. & Kayama, R. (1985). Seasonal dry matter production in grazed pasture of *Paspalum notatum* and cumulative effect of defoliation. In *Proceedings of the XVth International Grassland Congress,* pp. 598–600. Nishi-nasuno, Japan: The Japanese Society of Grassland Science.

Hirata, M., Sugimoto, Y. & Ueno, M. (1986). Energy and matter flows in bahiagrass pasture. II. Net primary production and efficiency for solar energy utilisation. *Journal of Japanese Society of Grassland Science,* **31**, 387–96.

Hodgkinson, K. C., Mott, J. J. & Ludlow, M. M. (1985). Coping with grazing: a comparison of two savanna grasses differing in tolerance to defoliation. In *Proceedings of the XVth International Grassland Congress,* pp. 1089–91. Nishi-nasuno, Japan: The Japanese Society of Grassland Science.

Hoffman, M. T. (1988). The rationale for karoo grazing systems: criticisms and research implications. *South African Journal of Science,* **84**, 556–9.

Hogan, J. P. (1982). Digestion and utilization of proteins. In *Nutritional Limits to Animal Production from Pastures,* ed. J. B. Hacker, pp. 245–57. Farnham Royal, UK: Commonwealth Agricultural Bureaux.

Holling, C. S. (1973). Resilience and stability of ecological systems. *Annual Review of Ecology and Systems,* **4**, 1–23.

Holm, J. (1972). The treatment of rice straw with sodium

hydroxide and its economic limitations in northern Thailand. *Thai Journal of Agricultural Science*, **5**, 89–100.

Holm, J. (1974). Nutritive value and acid contents of silages made from tropical forages at Chiang Mai, Thailand. *Thai Journal of Agricultural Science*, **7**, 11–21.

Holroyd, R. G., Allan, P. J. & O'Rourke, P. K. (1977). Effect of pasture type and supplementary feeding on the reproductive performance of cattle in the dry topics of north Queensland. *Australian Journal of Experimental Agriculture and Animal Husbandry*, **17**, 197–206.

Holroyd, R. G., O'Rourke, P. K., Clarke, M. R. & Loxton, I. D. (1983). Influence of pasture type and supplement on fertility and liveweight of cows, and progeny growth rate in the dry tropics of northern Queensland. *Australian Journal of Experimental Agriculture and Animal Husbandry*, **23**, 4–13.

Holt, J. H. & Easey, J. F. (1984). Biomass of mound-building termites in a red and yellow earth landscape, north Queensland. *Proceedings National Soils Conference*, p. 363. Brisbane, Australia: Australian Society Soil Science Incorporated.

Hong, A. (1978). Evaluation on the use of vegetative covers for soil conservation in FELDA. *Malaysian Agricultural Journal*, **51**, 335–42.

Hongyantarachai, S., Nithichai, G., Wongsuwan, N., Prasanpanich, S., Siwichai, S., Pratumsuwan, S., Tasapanon, T. & Watkin, B. R. (1989). The effects of grazing versus indoor feeding during the day on milk production in Thailand. *Tropical Grasslands*, **23**, 8–14.

Hoppe, P. P., Qvortrup, S. A. & Woodford, M. H. (1977a). Rumen fermentation and food selection in East African Zebu cattle, wildebeest, Coke's hartebeest and topi. *Journal of Zoology*, **181**, 1–9.

Hoppe, P. P., Qvortrup, S. A. & Woodford, M. H. (1977b). Rumen fermentation and food selection in East African sheep, goats, Thomson's gazelle, Grant's gazelle and impala. *Journal of Agricultural Science*, **89**, 129–35.

Hudson, N. W. (1957). Erosion control research. *Rhodesian Agricultural Journal*, **54**, 297–323.

Humphreys, L. R. (1966a). Sub-tropical grass growth. II. Effects of variation in leaf area index in the field. *Queensland Journal of Agricultural and Animal Sciences*, **23**, 337–58.

Humphreys, L. R. (1966b). Sub-tropical grass growth. III. Effects of stage of defoliation and infloresence removal. *Queensland Journal of Agricultural and Animal Sciences*, **23**, 499–531.

Humphreys, L. R. (1978). *Tropical Pastures and Fodder Crops*, p. 117. London: Longman.

Humphreys, L. R. (1981a). Humid and sub-humid tropical range lands. In *Potential of the World's Forages for Ruminant Animal Production*, ed. R. D. Child & E. K. Byington, pp 29–48. Morrilton, Arkansas: Winrock International Livestock Research and Training Center.

Humphreys, L. R. (1981b). *Environmental Adaptation of Tropical Pasture Plants*. London: Macmillan.

Humphreys, L. R. (1986a). The improved integration of forage production with rice culture in South-East Asia. *International Rice Commission Newsletter*, **34(2)**, 275–96.

Humphreys, L. R. (1986b). Perspectives on pasture management and improvement in Australia. In *Science for Agriculture: The Way Ahead*, ed. L. W. Martinelli, pp. 96–103. Melbourne: Australian Institute of Agricultural Science, Melbourne.

Humphreys, L. R. (1987). *Tropical Pastures and Fodder Crops*, 2nd edn. Harlow, UK: Longman.

Humphreys, L. R. (1989). Future directions in grassland science and its applications. In *Proceedings of the XVI International Grassland Congress*, **3**, 1705–10. Versailles: Association Française pour la Production Fourragère.

Humphreys, L. R. & Riveros, F. (1986). *Tropical Pasture Seed Production*. FAO Plant Production and Protection Paper 8, pp. 111–21. Rome: Food and Agriculture Organization of United Nations.

Humphreys, L. R. & Robinson, A. R. (1966). Subtropical grass growth. 1. Relationship between carbohydrate accumulation and leaf area in growth. *Queensland Journal of Agriculture and Animal Science*, **23**, 211–59.

Hunt, H. W. (1977). A simulation model for decomposition in grasslands. *Ecology*, **58**, 469–84.

Hunt, W. F. (1979). Effects of treating and defoliation height on the growth of *Paspalum dilatatum* Poir. *New Zealand Journal of Agricultural Research*, **22**, 69–75.

Hunter, R. A. & Siebert, B. D. (1985). Utilization of low-quality roughage by *Bos taurus* and *Bos indicus* cattle. 2. The effect of rumen-degradable nitrogen and sulphur on voluntary food intake and rumen characteristics. *British Journal of Nutrition*, **53**, 649–56.

Hunter, R. A. & Siebert, B. D. (1987). The effect of supplements of rumen-degradable protein and formaldehyde-treated casein on the intake of low-nitrogen roughages by *Bos taurus* and *Bos indicus* steers at different stages of maturity. *Austalian Journal of Agricultural Research*, **38**, 209–18.

Hunter, R. A., Siebert, B. D. & Webb, C. D. (1979). The positive response of cattle to sulphur and sodium supplementation while grazing *Stylosanthes guianensis* in north Queensland. *Australian Journal of Experimental Agriculture and Animal Husbandry*, **19**, 517–21.

Huntley, B. J. & Walker, B. H. (eds) (1982). *Ecology of Tropical Savannas*. Berlin: Springer Verlag.

Husz, G.St. (1977). Agro-ecosystems in South America. *Agro-Ecosystems*, **4** (special issue), 244–76.

Hutchinson, K. J. (1966). A note on wool production responses to fodder conservation in pastoral systems. *Journal of British Grassland Society*, **21**, 303–4.

Hutchinson, K. J. (1971). Productivity and energy flow in grazing/fodder conservation systems. *Herbage Abstracts*, **41**, 2–8.

Hutton, E. M. (1970). Legume 'little leaf' resistance and susceptibility in pasture legumes adapted to the Australian tropics. *SABRO Newsetter*, **2**, 151–3.

ILCA (1987). Field Programmes. In *Annual Report 1986/87*, pp. 1–28. Addis Ababa, Ethiopia: International Livestock Centre for Africa.

Imrie, B. C. (1971). The effects of severity of defoliation and soil moisture stress on *Desmodium intortum*. *Australian Journal of Experimental Agriculture and Animal Husbandry*, **11**, 521–4.

Irulegui, G. S. de., Maraschin, G. E. & Riboldi, J. (1984). Yield of a subtropical pasture mixture under continuous and rotational grazing. *Pesquisa Agropecuária Brasileira*, **19**, 101–7.

Isaraseenee, A., Shelton, H. M., Jones, R. M. & Bunch, G. A. (1984). Accumulation of edible forage of *Leucaena leucocephala* čv. Peru over late summer and autumn for use as dry season feed. *Leucaena Research Reports*, **5**, 3.

Ivory, D. A. & Whiteman, P. C. (1978). Effect of temperature on growth of five subtropical grasses. 1. Effect of day and night temperature on growth and morphological development. *Australian Journal of Plant Physiology*, **5**, 131–48.

Izac, A-M. N., Anaman, K. A. & Jones, R. J. (1990). Biologic and economic optima in a tropical grazing ecosystem in Australia. *Agriculture, Ecosystems and the Environment*, **16**, (in press).

Jackson, I. J. (1989). *Climate, Water and Agriculture in the Tropics.* Burnt Mill, UK: Longman.

Jackson, J. J. (1972). Some observations on the comparative effects of short duration grazing systems and continuous grazing systems on the reproductive performance of ranch cows. *Rhodesian Agricultural Journal*, **69**, 95–102.

Jarrige, R., Demarquilly, C. & Dulphy, J. P. (1982). Forage conservation. In *Nutritional Limits to Animal Production from Pastures*, ed. J. B. Hacker, pp. 363–87. Farnham Royal, UK: Commonwealth Agricultural Bureaux.

Jayasuriya, M. C. N. & Panditharatne, S. (1978). Spent tea leaf as a ruminant feed. *Animal Feed Science and Technology*, **3**, 219–26.

Jewiss, O. R. (1972). Tillering in grasses – its significance and control. *Journal of British Grassland Society*, **27**, 65–82.

Jobin, M. & Shelton, H. M. (1980). Studies on *Crotalaria juncea* L. (Sunnhemp). II. Influence of stage of growth on pyrrolizidine alkaloid concentration in plant parts. In *Khon Kaen Pasture Improvement Project Annual Report*, pp. 19–22. Khon Kaen, Thailand: Khon Kaen University.

Joblin, A. D. H. (1963). Strip grazing versus paddock grazing under tropical conditions. *Journal of British Grassland Society*, **18**, 69–73.

Johnson, L. A. Y. & Leatch, G. (1975). Effect of different tick control techniques on tick populations and cattle productivity. *CSIRO Division of Animal Health Annual Report 1975*, pp. 60–1.

Jones, C. A. & Carabaly, A. (1981). Some characteristics of the regrowth of 12 tropical grasses. *Tropical Agriculture Trinidad*, **58**, 37–44.

Jones, R. J. (1967). Losses of dry matter and nitrogen from autumn-saved improved pastures during the winter at Samford, south-eastern Queensland. *Australian Journal of Experimental Agriculture and Animal Husbandry*, **7**, 72–7.

Jones, R. J. (1974a). The relation of animal and pasture production to stocking rate on legume based and nitrogen fertilized subtropical pastures. *Proceedings of Australian Society of Animal Production*, **10**, 340–3.

Jones, R. J. (1974b). Effect of previous cutting interval and of leaf area remaining after cutting on the regrowth of *Macroptilium atropurpureum* cv. Siratro. *Australian Journal of Experimental Agriculture and Animal Husbandry*, **14**, 344–8.

Jones, R. J. (1976). Grass species, fodder conservation and stocking rate effects on nitrogen fertilized sub-tropical pastures. *Proceedings of the Australian Society of Animal Production*, **11**, 445–8.

Jones, R. J. (1988a). The effect of pasture management on grass and animal production following frosting of nitrogen fertilized sub-tropical grass pastures. *Tropical Grasslands*, **22**, 57–62.

Jones, R. J. (1988b). The future for the grazing herbivore. *Tropical Grasslands*, **22**, 97–115.

Jones, R. J., Blunt, C. G. & Holmes, J. H. G. (1976). Enlarged thyroid glands in cattle grazing *Leucaena* pastures. *Tropical Grasslands*, **10**, 113–6.

Jones, R. J. & Ford, C. W. (1972). Some factors affecting the oxalate content of the tropical grass *Setaria sphacelata*. *Australian Journal of Experimental Agriculture and Animal Husbandry*, **128**, 400–6.

Jones, R. J. & Jones, R. M. (1979). Beef production from nitrogen fertilized grasses. *CSIRO Tropical Crops and Pastures Divisional Report 1978–1979*, p. 6.

Jones, R. J. & Megarrity, R. G. (1986). Successful transfer of DHP-degrading bacteria from Hawaiian goats to Australian ruminants. *Australian Veterinary Journal*, **63**, 259–62.

Jones, R. J. & Sandland, R. L. (1974). The relation between animal gain and stocking rate. Derivation of the relation from the results of grazing trials. *Journal of Agricultural Science*, **83**, 335–42.

Jones, R. M. (1979). Effect of stocking rate and grazing frequency on a Siratro (*Macroptilium atropurpureum*)/*Setaria anceps* cv. Nandi pasture. *Australian Journal of Experimental Agriculture and Animal Husbandry*, **19**, 318–24.

Jones, R. M. (1980). Survival of seedlings and primary taproots of white clover (*Trifolium repens*) in sub-tropical pasture in south-east Queensland. *Tropical Grasslands*, **14**, 19–22.

Jones, R. M. (1981). Studies on the population dynamics of Siratro: the fate of Siratro seeds following oversowing into sub-tropical pastures. *Tropical Grasslands*, **15**, 95–101.

Jones, R. M. (1984). White clover (*Trifolium repens*) in subtropical south-east Queensland. III. Increasing clover and animal production by use of lime and flexible stocking rates. *Tropical Grasslands*, **18**, 186–93.

Jones, R. M. (1988). The effect of stocking rate on the population dyamics of Siratro in Siratro (*Macroptilium atropurpureum*)–Setaria (*Setaria spachelata*) pastures in south-east Queensland. 3. Effects of spelling on restoration of Siratro in overgrazed pastures. *Tropical Grasslands*, **22**, 5–11.

Jones, R. M. (1989a). Productivity and population dynamics of silverleaf desmodium (*Desmodium uncinatum*), greenleaf desmodium (*D. intortum*) and two *D.intortum* × *D. sandwicense* hybrids in coastal south-east Queensland. *Tropical Grasslands*, **23**, 43–55.

Jones, R. M. & Bunch, G. A. (1987a). The effect of stocking rate on the population dynamics of Siratro in Siratro (*Macroptilium atropurpureum*)–setaria (*Setaria sphacelata*) pastures in south-east Queensland. 1. Survival of plants and stolons. *Australian Journal of Agricultural Research*, **39**, 209–19.

Jones, R. M. & Bunch, G. A. (1987b). The effect of stocking rate on the population dynamics of Siratro in Siratro (*Macroptilium atropurpureum*)–setaria (*Setaria sphacelata*), pastures in south-east Queensland. 2. Seed set, soil seed reserves, seedling recruitment and seedling survival. *Australian Journal of Agricultural Research*, **39**, 221–34.

Jones, R. M. & Evans, T. R. (1977). Soil seed levels of *Lotononis bainesii*, *Desmodium intortum* and *Trifolium repens* in subtropical pastures. *Journal of Australian Institute of Agricultural Science*, **43**, 164–6.

Jones, R. M. & Harrison, R. E. (1980). Note on the survival of individual plants of *Leucaena leucocephala* in grazed stands. *Tropical Agriculture*, **57**, 265–6.

Jones, R. M., Kerridge, P. C. & McLean, R. W. (1987). Seed of *Stylosanthes scabra* in faeces excreted by grazing cattle. *CSIRO Division of Tropical Crops and Pastures Annual Report, 1986–87*, pp. 84–5. Brisbane, Australia: CSIRO.

Jones, R. M. & Ratcliff, D. (1983). Patchy grazing and its relation to deposition of cattle dung pats in pastures in coastal sub-tropical Queensland. *Journal of the Australian Institute of Agricultural Science*, **49**, 109–11.

Kanbe, M., Inami, S., Fujimoto, F., Yamashita, K. & Seki, M. (1984). Suitability of tropical grasses for haymaking with reference to their growth stage. *Research Bulletin of the Aichiken Agricultural Research Center*, **16**, 85–94.

Kang, B. T., Grimme, H. & Lawson, T. L. (1985). Alley cropping sequentially cropped maize and cowpea with leucaena on a sandy soil in southern Nigeria. *Plant and Soil*, **85**, 267–77.

Kayongo-Male, H., Karue, C. N. & Mutinga, E. R. (1977). The effect of supplementation on the growth of dairy heifers grazed on medium quality pasture under East African conditions. *East African Agriculture & Forestry Journal*, **42**, 435–40.

Kellems, R., Wayman, O., Nguyen, A. H., Nolan, J. C. Jr., Campbell, C. M., Carpenter, J. R. & Ho-a, E. B. (1979). Post-harvest pineapple plant forage as a potential feedstuff for beef cattle: evaluated by laboratory analyses, *in vitro* and *in vivo* digestibility and feedlot trials. *Journal of Animal Science*, **48**, 1040–8.

Kellman, M. (1980). Longevity and susceptibility to fire of *Paspalum virgatum* L. seed. *Tropical Agriculture*, **57**, 301–4.

Kennedy, P. M. & Siebert, B. D. (1973). The utilization of spear grass (*Heteropogon contortus*). 3. The influence of the level of dietary sulphur on the utilization of spear grass by sheep. *Australian Journal of Agricultural Research*, **24**, 143–52.

Kerr, D., Bird, A. C. & Buchanan, I. K. (1985). Heifer liveweight influences lifetime production. *Queensland Agricultural Journal*, **111**, 32.

Kessler, C. D. J. & Shelton, H. M. (1980). Dry-season legume forages to follow paddy rice in N. E. Thailand. III. Influence of time and intensity of cutting on *Crotalaria juncea*. *Experimental Agriculture*, **16**, 207–14.

King, K. L., & Hutchinson, K. J. (1980). Effects of superphosphate and stocking intensity on grassland microarthropods. *Journal of Applied Ecology*, **17**, 581–91.

Kipnis, T., Dovrat, A. & Lavee, S. (1977). *Morphological and physiological aspects of regrowth of Rhodes Grass* (*Chloris gayana* Kunth.) *after cutting*. Volcani Center Pamphlet 173. Bet Dagan, Israel.

Kirk, W. G. & Hodges, E. M. (1971). Effect of controlled burning on production of cows on native range. *Proceedings of Soil and Crop Science Society of Florida*, pp. 341–3.

Kirk, W. G., Hodges, E. M., Carpenter, J. W., Peacock, F. M. & Martin, F. G. (1974). Supplemental feeding of steers on *Pangola digitgrass* and *Pensacola* Bahiagrass pastures. *Proceedings, Soil and Crop Science Society of Florida*, **33**, 53–5.

Knox, J. P. & Wareing, P. E. (1984). Apical dominance in *Phaseolus vulgaris* L. The possible roles of abscisic and indole-3-acetic acid. *Journal of Experimental Botany*, **35**, 239–44.

Kobayashi, T., Koguchi, S. & Nishimura, S. (1980). The effect of cutting stage and additives on the quality of and subtropical grass silage. *Journal of Japanese Society of Grassland Science*, **26**, 81–8.

Kobayashi, T. & Nishimura, S. (1978). Winter hardiness and carbohydrate reserve of some tropical and subtropical grasses as affected by final cutting date in autumn. *Journal of Japanese Society of Grassland Science*, **24**, 27–33.

Kornelius, E., Saueressig, M. G. & Goedert, W. J. (1979). Pastures establishment and management in the cerrado of Brazil. In *Pasture Production in Acid Soils of the Tropics*, ed. P. A. Sanchez & L. E. Tergas, pp. 147–66. Cali, Colombia: CIAT.

Kowithayakorn, L. & Humphreys, L. R. (1987). Influence of withholding irrigation and trellis culture on seed production of *Macroptilium atropurpureum* cv. Siratro. *Tropical Grasslands*, **21**, 107–16.

Lal, R., Kang, B. T., Moorman, F. R., Juo, A. S. R. & Moomaw, J. C. (1975). Soil management problems and possible solutions in Western Nigeria. In *Soil Management in Tropical America*, ed. E. Bornemisza & A. Alvarado, pp. 372–408. Raleigh, N.C.: Soil Science Department, North Carolina State University.

Lambert, M. G. & Guerin, H. (1990). Competitive and complementary effects with different species of herbivores in their utilisation of pastures. In *Proceedings of the XVI International Grassland Congress*, **3**, pp. 1785–9. Versailles: Association Française pour la Production Fourragère.

Lamela, L., García-Trujillo, R. & Cáceres, O. (1980). Milk production, intake and digestibility of Guinea grass IH-127. *Pastos y Forrages*, **3**, 275–85.

Lamotte, M. (1982). Consumption and decomposition in tropical grassland ecosystems at Lamto, Ivory Coast. In *Ecology of Tropical Savannas*, ed. B. J. Huntley & B. H. Walker, pp. 415–29. New York: Springer-Verlag.

Lamotte, M. & Bourlière, F. (1983). Energy flow and nutrient cycling in tropical savannas. In *Ecosystems of the World: 13. Tropical Savannas*, ed. Bourlière, F. pp. 583–603. Amsterdam: Elsevier.

Landon, J. R. (ed.) (1984). *Barker Tropical Soil Manual*. London: Barker Agriculture International.

Lang, R. D. (1979). The effect of ground cover on surface runoff from experimental plots. *Journal of Soil Conservation Service of New South Wales*, **35**, 108–14.

Lange, R. T. (1969). The piosphere, sheep track and dung patterns. *Journal of Range Management*, **22**, 396–400.

Lara, J. G., Parra, R. & Neher, A. (1985). Effect of treating *Paspalum fasciculatum* with anhydrous ammonia for rations for growing sheep. *Informe Anual 1984*, Maracay, 78–9.

Laredo, M. A. & Minson, D. J. (1973). The voluntary intake, digestibility and retention time by sheep of leaf and stem fractions of five grasses. *Australian Journal of Agricultural Research*, **24**, 875–88.

Lascano, C. E. (1987). Canopy structure and composition in legume selectivity. *Proceedings of Workshop on Forage–Livestock Research Needs for the Caribbean Basin*. Tampa, Florida: Carribbean Basin Advisory Group (CBAG).

Lascano, R. J., Bavel, C. H. M. van, Hatfield, J. L. & Upchurch, D. R. (1987). Energy and water balance of a sparse crop: simulated and measured soil and crop evaporation *Soil Science Society of America Journal*, **51**, 1113–21.

Lavezzo, W., Lavezzo, O. E. N. M., Rossi, C. & Bonassi, I. A. (1989). Effects of wilting and formic acid on the chemical composition and nutritive value of elephant grass silage (*Pennisetum purpureum* Schum) purple cultivar. In *Proceedings XVI International Grassland Congress*, **2**, 965–6. Versailles: Association Francaise pour la Production Fourragère.

Lavezzo, W., Lavezzo, O. E. N. M. & Silveira, A. C. (1984). Effects of wilting, formol and formic acid on consumption and digestibility of elephant grass (*Pennisetum purpureum* Schum) silages. *Revista da Sociedade Brasileira de Zootecnia*, **13**, 501–8.

Lazier, J. R. (1981). Effect of cutting height and frequency on dry matter production of *Codariocalyx gyroides* (syn. *Desmodium gyroides*) in Belize, Central America. *Tropical Grasslands*, **15**, 10–16.

Leach, G. J. (1983). Influence of rest interval, grazing duration and mowing on the growth, mineral content and utilization of a lucerne pasture in a subtropical environment. *Journal of Agricultural Science*, **101**, 169–83.

Leaver, J. D. (1982). Grass height as an indicator for supplementary feeding of continuously stocked dairy cows. *Grass and Forage Science*, **37**, 285–90.

Lekchom, C., Witayanuparpyunyong, K., Sukpituksakul, P. & Watkin, B. R. (1989). The use of improved pastures by grazing dairy cows for economic milk production in Thailand. In *Proceedings XVI International Grassland Congress*, **2**, 1163–4. Versailles: Association Française pour la Production Fourragère.

Leng, R. A. (1986). Determining the nutritive value of forage. In *Forages in Southeast Asian and South Pacific Agriculture*, ed. G. J. Blair, D. A. Ivory & T. R. Evans, pp. 111–23. Canberra: Australian Centre for International Agricultural Research.

Lenné, J. M. (1981). Control of anthracnose of the tropical forage legume *Stylosanthes capitata* by burning. *14th International Grassland Congress Summaries of Papers*, 321.

Levitt, M. S., Hegarty, A. & Radel, M. J. (1964). Studies on grass silage from predominantly *Paspalum dilatatum* pastures in south-eastern Queensland. 2. Influence of length of cut on silages with and without molasses. *Queensland Journal of Agricultural Science*, **21**, 181–92.

Levitt, M. S., Taylor, V. J. & Hegarty, A. (1962). Studies on grass silage from predominately *Paspalum dilatatum* pastures in south-eastern Queensland. I. A comparison and evaluation of the additives metabisulphite and molasses. *Queensland Journal of Agricultural Science*, **19**, 153–75.

Lightfoot, C. J., & Posselt, J. (1977). Eland (*Taurotragus oryx*) as a ranching animal complementary to cattle in Rhodesia. 2. Habitat and diet selection. *Rhodesian Agricultural Journal*, **74**, 53–62.

Lima, F. P., Martinelli, D. & Werner, J. C. (1968). Beef production on grass pastures in the 'terra rossa' region. *Boletim Industria Animal*, **25**, 129–37.

Lima, C. R. & Souto, S. M. (1972). Nutritive value of hay cut at various growth stages of a crop of perennial soyabean (*Glycine javanica*). *Pesquisa Agropecuária Brasileira, Zootecnia*, **7**, 59–62.

Lima, C. R., Souto, S. M., Garcia, J. M. R. & Araújo, M. R. (1972). Nutritive value of siratro (*Phaseolus atropurpureus*) hay at various growth stages. *Pesquisa Agropecuária Brasileira, Zootecnia*, **7**, 63–6.

Litscher, T. & Whiteman, P. C. (1982). Light transmission and pasture composition under smallholder coconut plantations in Malaita, Solomon Islands. *Experimental Agriculture*, **18**, 383–91.

Little, D. A. (1976). Assessment of several pasture species, particularly tropical legumes, for oestrogenic activity. *Australian Journal of Agricultural Research*, **27**, 681–6.

Little, D. A. (1980). Observations on the phosphorus requirement of cattle for growth. *Research in Veterinary Science*, **28**, 258–60.

Little, D. A. (1982). Utilization of minerals. In *Nutritional Limits to Animal Production from Pasture*, ed. J. B. Hacker, pp. 259–83. Farnham Royal, UK: Commonwealth Agricultural Bureaux.

Little, D. A. (1987). The influence of sodium supplementation on the voluntary intake and digestibility of low-sodium *Setaria sphacelata* cv. Nandi by cattle. *Journal of Agricultural Science*, **108**, 231–6.

Little, D. A. & Said, A. N. (eds.). (1987). *Utilization of Agricultural By-products as Livestock Feeds in Africa*. Addis Ababa: ILCA.

Llanos, F., Parra, R. & Neher, A. (1985). Effect of treating mixed hay of *Brachiaria mutica* and *Cynodon dactylon* with anhydrous ammonia in rations for sheep. *Informe Anual 1984*, Maracay, 79–80.

Loch, D. S., Hopkinson, J. M. & English, B. H. (1976). Seed production of *Stylosanthes guyanensis*. 2. The consequences of defoliation. *Australian Journal of Experimental Agriculture and Animal Husbandry*, **16**, 226–30.

Loch, D. S. & Humphreys, L. R. (1970). Effects of stage of defoliation on seed production and growth of *Stylosanthes humilis*. *Australian Journal of Experimental Agriculture and Animal Husbandry*, **10**, 577–81.

Lourenco, A. J., Boin, C., Matsui, E. & Abramides, P. L. G. (1984). Utilization of a reserve area of pigeonpeas to supplement a jaragua grass pasture in the dry season. *Zootecnia*, **22**, 83–103.

Lourenco, A. J., Sartini, H. J., Abramides, P. L. G. & Camargo, J. C. de M. (1980). Grazing trial on Guinea grass (*Panicum maximum* Jacq.) mixed with four tropical legumes. *Boletim de Indústria Animal*, **37**, 257–78.

Lourenco, A. J., Sartini, H. J., Santamaraia, M. & Leme da Rocha, G. (1978). A comparison of three nitrogen levels and grass associated with legumes in pasture Napier Elephant grass (*Pennisetum purpureum* Schum.) in determination of stocking rates. *Boletim Indústria Animal, Nova Odessa*, **35**, 69–80.

Lowe, K. F. & Hamilton, B. A. (1986). Dairy pastures in the Australian tropics and subtropics. *Tropical Grassland Society of Australia Occasional Publication*, **3**, 68–79.

Loxton, I. D., Murphy, G. M. & Toleman, M. A. (1983). Effect of superphosphate application on the phosphorus status of breeding cattle grazing Townsville stylo based pasture in northern Queensland. *Australian Journal of Experimental Agriculture and Animal Husbandry*, **23**, 340–7.

Lucci, C. de S., Nogueira Filho, J. C. M. & Borelli, V. (1983). Milk production on grass pastures with and without nitrogen, continuously and rotationally grazed. *Revista da Faculdade de Medicina Veterinária e Zootecnia da Universidade de Sao Paulo*, **20**, 53–6.

Ludlow, M. M. (1978). Light relations of pasture plants. In *Plant Relations in Pastures*, ed. J. R. Wilson, pp. 35–49. Melbourne: CSIRO.

Ludlow, M. M. & Charles-Edwards, D. A. (1980). Analysis of the regrowth of a tropical grass/legume sward subjected to different frequencies and intensities of defoliation. *Australian Journal of Agricultural Research*, **31**, 673–92.

Ludlow, M. M. & Fisher, M. J. (1976). Influence of soil surface litter on frost damage in *Macroptilium atropurpureum*. *Journal of Australian Institute of Agricultural Science*, **42**, 134–6.

Ludlow, M. M., Samarakoon, S. P. & Wilson, J. R. (1988). Influence of light regime and leaf nitrogen concentration on 77K fluorescence in leaves of four tropical grasses: no evidence for photoinhibition. *Australian Journal of Plant Physiology*, **15**, 669–76.

Ludlow, M. M. & Wilson, G. L. (1971a). Photosynthesis of tropical pasture plants. 2. Temperature and illuminance history. *Australian Journal of Biological Science*, **24**, 1065–75.

Ludlow, M. M. & Wilson, G. L. (1971*b*). Photosynthesis of tropical pasture plants. 3. Leaf age. *Australian Journal of Biological Science*, **24**, 1077–87.

Ludlow, M. M. & Wilson, G. L. (1983). The distribution of leaf photosynthetic activity in a mixed grass-legume pasture canopy. *Photosynthesis Research*, **4**, 137–44.

Lufadeju, E. A., Olayiwole, M. B. & Umunna, N. N. (1987). Intake and digestibility of urea-treated gamba (*Andropogon gayanus*) hay by cattle. In *Utilization of Agricultural By-products as Livestock Feeds in Africa*, ed. D. A. Little & A. N. Said, pp. 7–14. Addis Ababa: ILCA.

Lusi, L., Ojeda, F. & Ramírez, M. (1986). Effect of the addition of Lactisil on *Pennisetum purpureum* silage. *Pastos y Forrajes*, **9**, 278–83.

MAFF (1975). *Energy Allowances and Feeding Systems for Ruminants*. MAFF Technical Bulletin NO. 33. London: HMSO.

Mall, L. P. & Billore, S. K. (1974). An indirect estimation of litter disappearances in grassland study. *Current Science*, **43**, 506–7.

Manby, T. C. D. & Shepperson, G. (1975). Increasing the efficiency of grass conservation. *Agricultural Engineer*, **30**, 77–85.

Mannetje, L.'t. (1976). Grazing management on granitic soils. *CSIRO Division of Tropical Pastures Annual Report 1975–76*, pp. 10–11.

Mannetje, L.'t, Cook, S. J. & Wildin, J. H. (1983). The effects of fire on a buffel grass and Siratro pasture. *Tropical Grasslands*, **17**, 30–9.

Mannetje, L.'t & Ebersohn, J. P. (1980). Relations between sward characteristics and animal production. *Tropical Grasslands*, **14**, 273–87

Mannetje, L.'t & Nicholls, D. F. (1974). Beef production from pastures on granitic soils. In *CSIRO Division of Tropical Agronomy Annual Report 1973–74*, pp. 24–5. St Lucia, Australia: CSIRO.

Mannetje, L.'t, Singh Sidu, A. & Murugaiah, M. (1976). Cobalt deficiency in cattle in Johore. Liveweight changes and response to treatments. *MARDI Research Bulletin*, **4**, 90–8.

Maraschin, G. E., Mella, S. C., Iruleghi, G. S. & Riboldi, J. (1983). Performance of a subtropical legume–grass pasture under different grazing management systems. *Proceedings of the XIV International Grassland Congress*, pp. 459–61. Boulder, Colorado: Westview.

Martin, P. C. & Ruiz, R. (1986). Weight gain and consumption of Holstein bulls grazing pangola grass (*Digitaria decumbens* Stent). *Cuban Journal of Agricultural Science*, **20**, 239–43.

Masuda, Y. (1976). [Effect of tillering habits on chemical composition and *in vitro* dry matter digestibility of tropical grasses]. *Science Bulletin Faculty of Agriculture, Kyushu University*, **31**, 107–12.

McCartor, M. M. & Rouquette, F. M. Jr. (1977). Grazing pressures and animal performance with pearl millet. *Agronomy Journal*, **69**, 983–7.

McCown, R. L. (1973). An evaluation of the influence of available soil water storage capacity on growing season length and yield of tropical pastures using simple water balance models. *Agricultural Meteorology*, **11**, 53–63.

McCown, R. L. & Wall, B. H. (1981). The influence of weather on the quality of tropical legume pasture during the dry season in northern Australia. II. Moulding of standing hay in relation to rain and dew. *Australian Journal of Agricultural Research*, **32**, 589–98.

McCown, R. L., Wall, B. H. & Harrison, P. G. (1981). The influence of weather on the quality of tropical legume pasture during the dry season in northern Australia. 1. Trends in sward structure and moulding of standing hay at three locations. *Australian Journal of Agricultural Research*, **32**, 575–87.

McDowell, L. R. (ed.). (1985*a*). *Nutrition of Grazing Ruminants in Warm Climates*. New York: Academic Press.

McDowell, L. R. (1985*b*). Vitamin nutrition of cattle under tropical conditions. *World Review of Animal Production*, **21**, 9–17.

McDowell, L. R., Conrad, J. H., Ellis, G. L. & Loosli, J. K. (1983). *Minerals for Grazing Ruminants in Tropical Regions*. Gainesville, Florida: University of Florida.

McDowell, L. R., Conrad, J. H., Thomas, J. E., Harris, L. E. & Fick, K. R. (1977). Nutritional composition of Latin American forages. *Tropical Animal Production*, **2**, 273–9.

McDowell, L. R., Ellis, G. L. & Conrad, J. H. (1984). Mineral supplementation for grazing cattle in tropical regions. *World Animal Review*, **52**, 2–12.

McDowell, R. E. (1988). Importance of crop residues for feeding livestock in smallholder farming systems. In *Plant Breeding and the Nutritive Value of Crop Residues*, ed. J. D. Reed, B. S. Capper, & P. J. H. Neate, pp. 3–27. Addis Ababa: International Livestock Centre for Africa.

McHan, F. (1986). Cellulase-treated Coastal bermudagrass silage and production of soluble carbohydrates, silage acids, and digestibility. *Journal of Dairy Science*, **69**, 431–8.

McHan, F., Burdick, C. & Wilson, R. (1984). Changes in organic acids and digestibility of Coastal bermudagrass silage pre-treated with monensin. *Journal of Dairy Science*, **67**, 294–8.

McHan, F., Spencer, R., Evans, J. & Burdick, D. (1979). Composition of high and low moisture Coastal Bermudagrass ensiled under laboratory conditions. *Journal of Dairy Science*, **62**, 1606–10.

McIlroy, R. J. (1964). *An Introduction to Tropical Grassland Husbandry*. London: Oxford University Press.

McIvor, J. G. (1978). The effect of cutting interval and associate grass species on the growth of *Stylosanthes* species near Ingham, north Queensland. *Australian Journal of Experimental Agriculture and Animal Husbandry*, **18**, 546–53.

McIvor, J. G. (1984). Leaf growth and senescence in *Urochloa mosambicensis* and *U. oligotricha* in a seasonally dry tropical environment. *Australian Journal of Agricultural Research*, **35**, 177–87.

McKay, A. D. (1968). Rangeland productivity in Botswana. *East African Agriculture and Forestry Journal*, **34**, 178–93.

McKay, A. D. (1971). Seasonal and management effects on the composition and availability of herbage, steer diet and liveweight gains in a *Themeda triandra* grassland in Kenya. *Journal of Agricultural Science*, **76**, 9–26.

McKeon, G. M., Rickert, K. G. & Scattini, W. J. (1986). Tropical pastures in the farming system: Case studies of modelling integration through simulation. In *Proceedings of the Third Australian Conference on Tropical Pastures*, (ed.) G. J. Murtagh & R. M. Jones, Occasional Publication 3, pp. 92–100. St Lucia, Australia: Tropical Grassland Society of Australia.

McLean, R. W., McCown, R. L., Little, D. A., Winter, W. H. &

Dance, R. A. (1983). An analysis of cattle liveweight changes on tropical grass pasture during the dry and early wet seasons in northern Australia. 1. The nature of weight changes. *Journal of Agricultural Science*, **101**, 17–24.

McLeod, C. C. (1974). The performance of beef cattle on improved pastures in north east Thailand. In *Proceedings of the XIIth International Grassland Congress*, pp. 266–73. Moscow, USSR.

Mears, P. T. (1967). Tropical pastures in the Richmond–Tweed region of New South Wales – recent experiences and future development. *Tropical Grasslands*, **1**, 98–105.

Mears, P. T. & Humphreys, L. R. (1974a). Nitrogen response and stocking rate of *Pennisetum clandestinum* pastures. I. Pasture nitrogen requirement and concentration, distribution of dry matter and botanical composition. *Journal of Agricultural Science*, **83**, 451–68.

Mears, P. T. & Humphreys, L. R. (1974b). Nitrogen response and stocking rate of *Pennisetum clandestinum* pastures. II. Cattle growth. *Journal of Agricultural Science*, **83**, 469–78.

Mecelis, N. & Favoretto, V. (1981). Regrowth of Colonial grass (*Panicum maximum* Jacq.) under different nitrogen levels, frequencies, and heights of cuttings. In *Proceedings of the XIVth International Grassland Congress Summaries*, p. 308.

Medina, E. (1982). Nitrogen balance in *Trachypogon* grasslands of central Venezuela. *Plant & Soil*, **67**, 305–14.

Meléndez, F., Pérez, J. & Alvarez, J. (1977). Response to nitrogen fertilizer in African Star grass (*Cynodon plectostachyus*). *Tropical Animal Production*, **2**, 224.

Mellor, W., Hibberd, M. J. & Grof, B. (1973). Beef cattle liveweight gains from mixed pastures of some guinea grasses and legumes on the wet tropical coast of Queensland. *Queensland Journal of Agricultural and Animal Sciences*, **30**, 259–66.

Meyreles, L., MacLeod, N. A. & Preston, T. R. (1977). Cassava forage as a protein supplement in sugar cane diets for cattle: effect of different levels on growth and rumen fermentation. *Tropical Animal Production*, **2**, 73–80.

Michelena, J. B. & Molina, A. (1987). The effect of different levels of formic acid on the quality of king grass silage (*Pennisetum purpureum* × *Pennisetum typhoides*). *Cuban Journal of Agricultural Science*, **21**, 295–300.

Michelena, J. B., Molina, A., Perez, M. & Gonzalez, C. (1988). Effect of different combinations of sulfuric acid-formol on the quality of king grass (*Pennisetum purpureum* × *Pennisetum typhoides*) silage. *Cuban Journal of Agricultural Science*, **22**, 195–203.

Middleton, C. H. & Mellor, W. (1982). Grazing assessment of the tropical legume *Calopogonium caeruleum*. *Tropical Grasslands*, **16**, 213–16.

Milera, M., Martínez, J., Cáceres, O. & Hernández, J. (1986). Influence of herbage allowance on milk production and grazing days of Bermuda grass cv. Coastcross-1. *Pastos y Forrajes*, **9**, 107–76.

Miles, J. W. & Lenné, J. M. (1987). Effect of frequency of defoliation of 40 *Stylosanthes guianensis* genotypes on field reaction to anthracnose caused by *Colletotrichum gloeosporioides*. *Australian Journal of Agricultural Research*, **38**, 309–15.

Milford, R. & Haydock, K. P. (1965). The nutritive value of protein in subtropical pasture species grown in south-east Queensland. *Australian Journal of Experimental Agriculture and Animal Husbandry*, **5**, 13–17.

Miller, C. P. & Van der List, J. T. (1977). Yield, nitrogen uptake, and liveweight gains from irrigated grass–legume pasture on a Queensland tropical highland. *Australian Journal of Experimental Agriculture and Animal Husbandry*, **17**, 949–60

Milligan, K. & de Leeuw, P. (1983). Low altitude aerial surveys in pastoral systems research. In *Pastoral Systems Research in Sub-Saharan Africa*, pp. 81–103. Addis Ababa: ILCA.

Mills, P. F. L. (1977). A comparison of the preferences of cattle and sheep for different grasses. *Rhodesian Agricultural Journal*, **74**, 41–3.

Minson, D. J. (1971a). Influence of lignin and silicon on a summative system for assessing the organic matter digestibility of *Panicum*. *Australian Journal of Agricultural Research*, **22**, 589–98.

Minson, D. J. (1971b). The digestibility and voluntary intake of six *Panicum* varieties. *Australian Journal of Experimental Agriculture and Animal Husbandry*, **11**, 18–25.

Minson, D. J. (1971c). The nutritive value of tropical pastures. *Journal of Australian Institute of Agricultural Science*, **37**, 255–63.

Minson, D. J. (1973). Effect of fertiliser nitrogen on digestibility and voluntary intake of *Chloris gayana*, *Digitaria decumbens* and *Pennisetum clandestinum*. *Australian Journal of Experimental Agriculture and Animal Husbandry*, **13**, 153–7.

Minson, D. J. (1976). Relation between digestibility and composition of feed. *Wageningen Miscellaneous Papers*, **12**, 101–14.

Minson, D. J. (1980). Nutritional differences between tropical and temperate pastures. In *Grazing Animals*, ed. F. H. W. Morley, pp. 143–57. Amsterdam: Elsevier.

Minson, D. J. (1982). Effects of chemical and physical composition of herbage eaten upon intake. In *Nutritional Limits to Animal Production from Pastures*, ed. J. B. Hacker, pp. 167–82. Farnham Royal, UK: Commonwealth Agricultural Bureaux.

Minson, D. J. (1990). *Forage in Ruminant Nutrition*. San Diego: Academic Press.

Minson, D. J. & Bray, R. A. (1986). Voluntary intake and *in vivo* digestibility by sheep of five lines of *Cenchrus ciliaris* selected on the basis of preference rating. *Grass and Forage Science*, **41**, 47–52.

Minson, D. J. & Laredo, M. A. (1972). Influence of leafiness on voluntary intake of tropical grasses by sheep. *Journal of the Australian Institute of Agricultural Science*, **38**, 303–5.

Minson, D. J. & Wilson, J. R. (1980). Comparative digestibility of tropical and temperate forage—a contrast between grasses and legumes. *Journal of the Australian Institute of Agricultural Science*, **46**, 247–9.

Mohammed-Saleem, M. (1982). In *Annual Report, ILCA*, p. 9. Addis Ababa: ILCA.

Mohamed-Saleem, M. A., Suleiman, H. & Otsyina, R. M. (1986). Fodder banks: for pastoralists or farmers. In *Potentials of Forage Legumes in Farming Systems of Sub-Saharan Africa*, ed. I. Haque, S. Jutzi & P. J. H. Neate, pp. 420–37. Addis Ababa: ILCA.

Mohamed-Saleem, M. A. & Von Kaufmann, R. R. (1989). A rapid survey of feeding regimes for draught cattle in Niger State, Nigeria. *ILCA Bulletin*, **33**, 14–7.

Mohr, E. C. J., Van Bahren, F. A. & Van Schuylenborgh, J. (1972). *Tropical Soils. A Comprehensive Study of Their Genesis*. The Hague: Mouton-Ichtiar Baru-Van Hoeve.

Moir, K. W., Ryley, J. W., Pepper, P. M. & Middleton, C. H. (1969). Effect of nitrogen fertilizer on growth rate of Hereford heifers grazing *Paspalum dilatatum* dominant pasture. *Queensland Journal of Agricultural and Animal Sciences*, **26**, 341–52.

Monteiro, L. A., Gardner, A. L. & Chudleigh, P. D. (1981). Beef production in the Cerrado region of Brazil. A bioeconomic analysis of ranch improvement schemes. *World Animal Review*, **37**, 37–44.

Moog, F. A. (1980). Backyard cattle raising and its feeding system in a Batangas barrio. Thesis, University of Philippines, Los Banos.

Moore, J. E., Ruelke, O. C., Rios, C. E. & Franke, D. E. (1971). Nutritive evaluation of Pensacola Bahiagrass hays. *Proceedings of the Soil and Crop Science Society of Florida*, **30**, 211–21.

Moore, J. E., Sollenberger, L. E., Morantes, G. A. & Beede, P. T. (1985). Canopy structure of *Aeschynomene americana–Hemarthria altissima* pastures and ingestive behaviour of cattle. In *Proceedings of the XVth International Grassland Congress*, pp. 1126–8. Nishi-nasuno, Japan: The Japanese Society of Grassland Science.

Moore, L. E., Sollenberger, L. E., Albrecht, K. A., Beede, P. T. & Brown, W. F. (1989). Effect of regrowth interval upon canopy structure and utilization of *Aeschynomene*–limpograss pastures. In *Proceedings XVI International Grassland Congress*, **2**, 1029–30. Versailles: Association Française pour la Production Fourragère.

Morris, J. W., Bezuidenhout, J. J. & Furniss, P. R. (1982). Litter decomposition. In *Ecology of Tropical Savannas*, ed. B. J. Huntley & B. H. Walker, pp. 535–53, Ecological Studies 42. New York: Springer-Verlag.

Mott, G. O. (1961). Grazing pressure and the measurement of pasture production. *Proceedings of the VIII International Grassland Congress*, pp. 606–11. Hurley, UK: British Grassland Society.

Mott, G. O., Quinn, L. R. & Bisschoff, W. V. A. (1970). The retention of nitrogen in a soil–plant–animal system in guinea grass (*Panicum maximum*) pastures in Brazil. In *Proceedings XI International Grassland Congress*, ed. M. J. T. Norman, pp. 414–16. St Lucia, Queensland: University of Queensland Press.

Mott, J. J. (1982). Fire and survival of *Stylosanthes* spp. in the dry savanna woodlands of the Northern Territory. *Australian Journal of Agricultural Research*, **33**, 203–11.

Mott, J. J., McKeon, G. M., Gardener, C. J. & Mannetje, L.'t. (1981). Geographic variation in the reduction of hard seed content of *Stylosanthes* seeds in the tropics and subtropics of northern Australia. *Australian Journal of Agricultural Research*, **32**, 861–9.

Mott, J. J., McKeon, G. M. & Moore, C. J. (1976). Effects of seed bed conditions on the germination of four *Stylosanthes* species in the Northern Territory. *Australian Journal of Agricultural Research*, **27**, 811–23.

Mott, J. J., Williams, J., Andrew, M. H. & Gillison, A. N. (1985). Australian savannah ecosystems. In *Ecology and Management of the World's Savannas*, ed. J. C. Tothill and J. J. Mott, pp. 56–82. Canberra: Australian Academy of Science.

Muldoon, D. K. & Pearson, D. J. (1979a). Primary growth and re-growth of the tropical tallgrass hybrid Pennisetum at different temperatures. *Annals of Botany*, **43**, 709–17.

Muldoon, D. K. & Pearson, C. J. (1979b). Morphology and physiology of regrowth of the tropical tallgrass hybrid Pennisetum. *Annals of Botany*, **43**, 719–28.

Murdoch, J. C. (1980). The conservation of grass. In *Grass. Its Production and Utilization*, ed. W. Holmes, pp. 174–215. Oxford: Blackwell Scientific Publications.

Murphy, H. & Whiteman, P. C. (1981). Yield responses of oats and ryegrass varieties to sowing date in south-east coastal Queensland. *Tropical Grasslands*, **15**, 145–9.

Murray, M., Barry, J. D., Morrison, W. I., Williams, R. O., Hirumi, H. & Rovis, L. (1979). A review of the prospects for vaccination in African *trypanosomiasis*–Part 1. *World Animal Review*, **32**, 9–13.

Murray, R. M., Freire, L. C. L. & Gardener, C. J. (1988). Effect of superphosphate application on the nutritive value of *Stylosanthes* spp.–native grass pasture for cattle. 2. Nutritive value of the diet selected. *Proceedings of the Australian Society of Animal Production*, **17**, 270–3.

Murtagh, G. J. (1987). Estimating the proportion of degradable dry weight of kikuyu from rates of lamina senescence. *Annals of Botany*, **59**, 159–165.

Murtagh, G. J., Kaiser, A. G., Huett, D. O. & Hughes, R. M. (1980). Summer-growing components of a pasture system in a subtropical environment. 1. Pasture growth, carrying capacity and milk production. *Journal of Agricultural Science*, **94**, 645–63.

Murtagh, G. J., Kaiser, A. G. & Huett, D. O. (1980). Summer-growing components of a pasture system in a subtropical environment. 2. Relations between pasture and animal production. *Journal of Agricultural Science*, **94**, 665–74.

Murtagh, G. J. & Moore, K. G. (1987). A pasture budget for dairy farms in the Kyogle district. *Tropical Grasslands*, **21**, 145–53.

Myers, L. F. (1967). Assessment and integration of special purpose pastures. 1. Theoretical. *Australian Journal of Agricultural Research*, **18**, 235–44.

Nada, Y. & Jones, R. M. (1982). Yield and quality of annual forage grasses and of perennial grasses in the year of sowing in south eastern Queensland. *Journal of Japanese Society of Grassland Science*, **28**, 48–58.

Naidoo, G. & Steinke, T. D. (1979). Effect of varying carbohydrate levels on the uptake and translocation of ^{32}P in *Eragrostis curvula* (Schrod.) Nees. *Journal of South African Botany*, **45**, 231–41.

Nascimento, D., Jr & Pinheiro, J. S. (1975). Nutritive value of Jaragua grass at various ages. *Revista da Sociedade Brasileira de Zootecnia*, **4**, 101–13.

Natural, N. C. & Perez, C. B., Jr (1977). Copra meal and liquid molasses–urea as supplements to rice straw for feedlot bulls. *Philippine Agriculturalist*, **61**, 176–85.

Nava, V. G., & Gomez, B. R. (1981). Comparison of the continuous and rotational grazing systems in Guinea grass (*Panicum maximum* Jacq.) under natural conditions in the township of Aldama, Tamps. In *XVIIth Annual Report 1979–1980*. Nuevo León, Mexico: Division of Agricultural and Marine Sciences, Technological Institute of Monterrey.

Nelliat, E. V., Bavappa, K. V. & Nair, P. K. R. (1974). Multi-storeyed cropping. A new dimension in multiple cropping for coconut plantations. *World Crops*, **26**, 262–5.

Nge'the, J. C. & Box, T. W. (1976). Botanical composition of eland and goat diets on an *Acacia*-grassland community in Kenya. *Journal of Range Management*, **29**, 290–3.

Nicol, D. C. & Smith, L. D. (1981). Responses to cobalt therapy in weaner cattle in south-east Queensland. *Australian Journal of Experimental Agriculture and Animal Husbandry*, **21**, 27–31.

Nitis, I. M. (1985). Present state of grassland production and utilisation and future perspectives for grassland farming in humid tropical Asia. *Proceedings XV International Grassland Congress*, pp. 39–44.

Nnadi, L. A. & Haque, I. (1988). Forage legumes in African crop–livestock production systems. *ILCA Bulletin*, **30**, 10–19.

Nojima, H., Oizumi, H. & Takasaki, Y. (1985). Effect of cytokinin on lateral bud development in regrowth of *Sorghum bicolor* M. In *Proceedings of the XVth International Grassland Congress*, pp. 372–3. Nishi-nasuno, Japan: The Japanese Society of Grassland Science.

Nolan, T. & Connolly, J. (1977). Mixed stocking by sheep and steers–a review. *Herbage Abstracts*, **47**, 367–74.

Norman, M. J. T. (1960a). The relationship between competition and defoliation in pasture. *Journal of British Grassland Society*, **15**, 145–9.

Norman, M. J. T. (1960b). *Grazing and feeding trials with beef cattle at Katherine, N.T.* CSIRO Division of Land Research and Regional Survey Technical Paper 12.

Norman, M. J. T. & Phillips, L. J. (1968). The effect of time of grazing on bulrush millet (*Pennisetum typhoides*) at Katherine, N.T. *Australian Journal of Experimental Agriculture and Animal Husbandry*, **8**, 288–93.

Norman, M. J. T. & Phillips, L. J. (1973). Effect of time of cessation of wet season grazing on Townsville stylo–annual grass pasture at Katherine, N.T. *Australian Journal of Experimental Agriculture and Animal Husbandry*, **13**, 544–8.

Norman, M. J. T. & Wetselaar, R. (1960). Losses of nitrogen on burning native pastures at Katherine, N.T. *Journal of the Australian Institute of Agricultural Science*, **26**, 272–3.

Norton, B. W. (1982). Differences between species in forage quality. In *Nutritional Limits to Animal Production from Pastures*, ed. J. B. Hacker, pp. 89–110. Farnham Royal, UK: Commonwealth Agricultural Bureaux.

Norton, B. W. & Deery, M. J. (1985). The productivity of Angora goats grazing improved and native pastures in south-eastern Queensland. *Australian Journal of Experimental Agriculture and Animal Husbandry*, **25**, 35–40.

Norton, B. W. & Gondipon, R. (1984). Effects of alkali treatment on the drying rate and nutritive value of some tropical grasses and legumes. *Journal of the Australian Institute of Agricultural Science*, **50**, 55–8.

Norton, B. W. & Hales, J. W. (1976). A response of sheep to cobalt supplementation in south-eastern Queensland. *Proceedings of the Australian Society of Animal Production*, **11**, 393–6.

Noy-Meir, I. (1980). Structure and function of desert ecosystems. *Israel Journal of Botany*, **28**, 1–19.

Noy-Meir, I. & Walker, B. H. (1986). Stability and resilience in rangelands. In *Rangelands: A Resource Under Siege*, eds. P. J. Joss, P. W. Lynch and O. B. Williams, pp. 21–25. Canberra: Australian Academy of Science.

Obeid, J. A. & Cruz, M. E. (1989). Corn crop grown with and without tropical legumes for silage making. In *Proceedings of the XVI International Grassland Congress*, **2**, 959–60. Versailles: Association Française pour la Production Fourragère.

O'Donovan, P. B., Woldegabriel, A., Taylor, M. S. & Gebrewalde, A. (1978). *Tropical Animal Health and Production*, **10**, 23–9.

Oizumi, H., Takasaki, Y., Nojima, H. & Isono, Y. (1985). Physiology of regrowth in *Sorghum bicolor* Moench: studies on the sequential action of hormones and reserves. In *Proceedings of the XVth International Grassland Congress*, pp. 419–21. Nishi-nasuno, Japan: The Japanese Society of Grassland Science.

Ojeda, F. & Cáceres, O. (1981). Chopping, addition of 4% molasses and wilting on the consumption and digestibility of Guinea grass cv. likoni. *Pastos y Forrages*, **4**, 373–82.

Ojeda, F. & Cáceres, O. (1984). Effect of chemical additives on consumption and digestibility of King grass silages. *Pastos y Forrages*, **7**, 409–19.

Ojeda, F., Varsolomiev, G. (1982). Effect of chemical additives on the quality of Pangola grass silages. *Pastos y Forrages*, **5**, 359–78.

Okubo, T., Hirakawa, M., Okajima, T. Kayama, R., Tano, H. & Kihira, N. (1985). Energy efficiency of primary and secondary production in grazed pasture of *Dactylis glomerata* as compared with those of *Paspalum notatum*. In *Proceedings XV International Grassland Congress*, pp. 736–8. Nishi-nasuno: Japanese Society of Grassland Science.

Oliveira, P. R. P. & Humphreys, L. R. (1986). Influence of level and timing of shading on seed production in *Panicum maximum* cv. Gatton. *Australian Journal of Agricultural Research*, **37**, 417–24.

O'Rourke, J. T. (1978). Grazing rate and system trial over 5 years in a medium-height grassland of northern Tanzania. *Proceedings of First International Rangeland Congress*, pp. 563–6. Denver: Society for Range Management.

Orr, D. M. (1986). Factors affecting the vegetation dynamics of Astrebla grasslands. Ph.D. thesis, University of Queensland.

Orr, D. M., Bawly, P. S. & Evenson, C. J. (1986). Effects of grazing management on the basal area of perennial grasses in *Astrebla* grassland. In *Rangelands: A Resource Under Siege*, ed. P. J. Joss, P. W. Lynch & O. B. Williams, pp. 56–7. Canberra: Australian Academy of Science.

Orskov, E. R. (ed.). (1988). *Feed Science*. Amsterdam: Elsevier Science Publishers B.V.

Othman, W. M. B. W. (1983). The effects of defoliation on carbon assimilation, nitrogen fixation and regrowth of phasey bean (*Macroptilium lathyroides*). Ph.D thesis, University of Queensland.

Othman, W. M. W. & Asher, C. J. (1987). The effects of height and frequency of previous defoliation on nodulation, nitrogen fixation and regrowth of phasey bean. *Pertanika*, **10**, 1–10.

Othman, W. M. W., Asher, C. J. & Wilson, G. L. (1988). ^{14}C-labelled assimilate supply to root nodules and nitrogen fixation of phasey bean plants following defoliation and flower removal. In *Biotechnology of Nitrogen Fixation in the Tropics*, ed. Z. H. Shamsuddin, W. M. W. Othman, M. Marziah & J. Sundram, pp. 217–24. Kuala Lumpur: Universiti Pertanian Malaysia.

Otsyina, R. M. & McKell, C. M. (1985). Africa. Browse in the nutrition of livestock. *World Animal Review*, **53**, 33–9.

Ottosen, E. M., Brown, G. W. & Maraske, M. R. (1976). Strip grazing – advantage or disadvantage. *Queensland Agricultural Journal*, **101**, 569–70.

Owen, J. B. & Ridgman, W. J. (1968). The design and interpretation of experiments to study animal production from grazed pasture. *Journal of Agricultural Science*, **71**, 327–35.

Paladines, O. & Leal, J. A. (1979). Pasture management and

productivity in the Llanos Orientales of Colombia. In *Pasture Production in Acid Soils of the Tropics* ed. P. A. Sánchez & L. E. Tergas, pp. 311–25. Cali, Colombia: CIAT.

Panditharatne, S., Allen, V. G., Fontenot, J. P. & Jayasuriya, M. C. N. (1986). Ensiling characteristics of tropical grasses as influenced by stage of growth, additives and chopping length. *Journal of Animal Science*, **63**, 197–207.

Panditharatne, S., Allen, V. G., Fontenot, J. P. & Jayasuriya, M. C. N. (1988). Effect of stage of growth and chopping length on digestibility and palatability of Guinea-'A' grass silage. *Journal of Animal Science*, **66**, 1005–9.

Parsons, A. J., Johnson, I. R. & Harvey, A. (1988). Use of a model to optimize the interaction between frequency and severity of intermittent defoliation and to provide a fundamental comparison of the continuous and intermittent defoliation of grass. *Grass and Forage Science*, **43**, 49–59.

Parsons, J. J. (1972). Spread of African pasture grasses to the American tropics. *Journal of Range Management*, **25**, 12–17.

Partridge, I. J. (1979a). Improvement of Nadi blue grass (*Dichanthium caricosum*) pastures on hill land in Fiji with superphosphate and Siratro: effects of stocking rate on beef production and botanical composition. *Tropical Grasslands*, **13**, 157–64.

Partridge, I. J. (1979b). Sward structure in grazed Mission grass pastures. *Fiji Agricultural Journal*, **41**, 109–12.

Partridge, I. J. (1986). Effect of stocking rate and superphosphate level on an oversown fire climax grassland of mission grass (*Pennisetum polystachyon*) in Fiji. 2. Animal production. *Tropical Grasslands*, **20**, 174–80.

Partridge, I. J. & Ranacou, E. (1974). The effects of supplemental *Leucaena leucocephala* browse on steers grazing *Dichanthium caricosum* in Fiji. *Tropical Grasslands*, **8**, 107–12.

Pate, J. S., Layzell, D. B. & McNeil, D. L. (1979). Modeling the transport and utilization of carbon and nitrogen in a nodulated legume. *Plant Physiology*, **63**, 730–7.

Paterson, R. T. (ed.) (1987). *Crop Diversification: New Horizons for Agricultural Development. Proceedings of the Forages Session.* St Augustine, Trinidad: CARDI.

Paterson, R. T., Proverbs, G. A. & Keoghan, J. M. (1987). *The Management and Use of Forage Banks.* St Augustine, Trinidad: Caribbean Agricultural Research and Development Institute.

Paterson, R. T., Quiroga, L., Sauma, G. & Samur, C. (1983). Dry season growth of Zebu-Criollo steers with limited access to leucaena. *Tropical Animal Production*, **8**, 138–42.

Paterson, R. T. & Samur, C. (1982). Complementary legume grazing in dry season milk production. *Tropical Animal Production*, **7**, 40–2.

Paterson, R. T., Samur, C. & Sauma, G. (1982). *Leucaena leucocephala* for the complementation of existing pastures. *Tropical Animal Production*, **7**, 9–13.

Paterson, R. T., Sauma, G. & Samur, C. (1979). The growth of young bulls on grass and grass/legume pastures in sub-tropical Bolivia. *Tropical Animal Production*, **4**, 154–61.

Payne, W. J. A. (1976). Systems of beef production in developing countries. In *Proceedings Conference on Beef Cattle Production in Developing Countries*, ed. A. J. Smith. Edinburgh: Centre for Tropical Veterinary Medicine.

Pearson, C. J. & Ison, R. L. (1987). *Agronomy of Grassland Systems.* Cambridge: Cambridge University Press.

Pedreira, J. V. S. (1975). [Tillering habit of colonial grass, *Panicum maximum* Jacq.] *Boletim de Indústria Animal*, **32**, 111–14.

Peducassé, C. A., McDowell, L. R., Parra, L. A., Wilkins, J. V., Martin, F. G., Loosli, J. K. & Conrad, J. H. (1983). Mineral status of grazing beef cattle in the tropics of Bolivia. *Tropical Animal Production*, **8**, 118–30.

Peiris, H. & Ibrahim, M. N. M. (1985). Effect of intensity and frequency of defoliation on the yield and nutritive value of unfertilized Guinea A (*Panicum maximum* ecotype A). *Sri Lanka Veterinary Journal*, **33**, 11–18.

Pemadasa, M. A. (1981). The mineral nutrition of the vegetation of a montane grassland in Sri Lanka. *Journal of Ecology*, **69**, 125–34.

Petersen, R. G., Lucas, H. L. & Mott, G. O. (1965). Relationship between rate of stocking and per animal and per acre performance on pasture. *Agronomy Journal*, **57**, 27–30.

Pfister, J. A., Donart, G. B., Pieper, R. D., Wallace, J. D. & Parker, E. E. (1984). Cattle diets under continuous and four-pasture, one-herd grazing systems in south central New Mexico. *Journal of Range Management*, **37**, 50–4.

Phillips, I. D. J. (1975). Apical dominance. *Annual Review of Plant Physiology*, **26**, 341–67.

Picard, D. (1977). Dynamique racinaire de *Panicum maximum* Jacq. 1. Émission des racines adventives primaires dans un intercoupe en liaison avec le tallage. *Cahiers O.R.S.T.O.M. Biologie*, **12**, 213–26.

Picard, D. (1979). Evaluation of the organic matter supplied to the soil by the decay of the roots of an intensively managed *Panicum maximum* sward. *Plant and Soil*, **51**, 491–501.

Pizarro, E. A., Escuder, C. J. & Andrade, N. de S. (1981). Nutritive value of conserved forage. 2. Perennial soyabean hay and maize silage. *Arquivos da Escola de Veterinária da Universidade Federal de Minas Gerais*, **33**, 545–51.

Pizarro, E. A. & Vera, R. R. (1980). Efficiency of fodder conservation systems: maize silage. In *Fodder Conservation in the 80's*, ed. C. Thomas, pp. 436–41. Occasional Symposium No. 11., Hurley, UK: British Grassland Society.

Playne, M. J. (1970). The sodium concentration in some tropical pasture species with reference to animal requirements. *Australian Journal of Experimental Agriculture and Animal Husbandry*, **108**, 32–5.

Playne, M. J. (1974). The contribution of the seed of the legume *Stylosanthes humilis* to the nutrition of cattle grazing mature tropical pastures. *Proceedings of the XII International Grassland Congress*, **3**, 421–5.

Playne, M. J., McLeod, M. N. & Dekker, R. F. H. (1972). Digestion of dry matter, nitrogen, phosphorus, sulphur, calcium and detergent fibre fractions of the seed and pod of *Stylosanthes humilis* contained in terylene bags in the bovine rumen. *Journal of Science, Food, and Agriculture*, **23**, 925–32.

Plucknett, D. L. (1979). *Managing Pastures and Cattle under Coconuts.* Westview Tropical Agriculture Series 2. Boulder, Colorado: Westview Press.

Pond, K. R., Ellis, W. C., Lascano, C. E. & Akin, D. E. (1987). Fragmentation and flow of grazed coastal Bermudagrass through the digestive tract of cattle., *Journal of Animal Science*, **65**, 609–18.

Pott, A. & Humphreys, L. R. (1983). Persistence and growth of *Lotononis bainesii/Digitaria decumbens* pasture. 1. Sheep stocking rate. *Journal of Agricultural Science*, **101**, 1–7.

Pott, A., Humphreys, L. R. & Hales, J. W. (1983). Persistence and growth of *Lotononis bainesii/Digitaria decumbens* pasture. II. Sheep treading. *Journal of Agricultural Science*, **101**, 9–15.

Powles, S. B. (1984). Photoinhibition of photosynthesis induced by visible light. *Annual Review of Plant Physiology*, **35**, 15–44.

Prajapati, M. C. (1970). Effect of different systems of grazing by cattle on *Lasiurus–Eleusine–Aristida* grassland in arid region of Rajasthan vis-a-vis animal production. *Annals of Arid Zone*, **9**, 114–24.

Pratchett, D. & Shirvel, B. (1978). The testing of grazing systems on semiarid rangeland in Botswana. *Proceedings of the First International Rangeland Congress*, pp. 567–8. Denver: Society for Range Management.

Pressland, A. J. (1982). Fire in the management of grazing lands in Queensland. *Tropical Grasslands*, **16**, 104–12.

Preston, T. R. (1984). New approaches to animal nutrition in the tropics. In *Development of Animal Production Systems*, ed. B. Nestel, pp. 379–96. Amsterdam: Elsevier.

Preston, T. R. & Leng, R. A. (1978). Sugar cane as cattle feed. *World Animal Review*, **27**, 7–12.

Priestley, D. A. (1986). *Seed Aging: Implications for Seed Storage and Persistence in the Soil*. Ithaca, NY: Cornell University Press.

Puentes, C. (1974). Trials on the use of kenaf as a forage plant. *Revista Agrotecnica de Cuba*, **6**, 3–8.

Purcell, D. L. & Stubbs, W. (1965). Cattle help sheep on improved pastures. *Queensland Agricultural Journal*, **91**, 330–3.

Quesenberry, K. H. & Ocumpaugh, W. R. (1980). Crude protein, IVOMD, and yield of stockpiled limpograsses. *Agronomy Journal*, **72**, 1021–4.

Quinn, L. R., Mott, G. O., Bisschoff, W. V. A. & Freitas, L. M. M. de. (1970). In *Proceedings of the XIth International Grassland Congress*, pp. 832–5. St Lucia: University of Queensland Press.

Quirk, M. F., Bushell, J. J., Jones, R. J., Megarrity, R. G. & Butler, K. L. (1988). Live-weight gains on leucaena and native grass pastures after dosing cattle with rumen bacteria capable of degrading DHP, a ruminal metabolite from leucaena. *Journal of Agricultural Science*, **111**, 165–70.

Rahardjo, A., Patuan, L. P. S., Sulistioningsih, Prijarto, H., Gurawan, C. & Suharto, Ig. (1981). Preliminary study of the potency of agricultural waste and agroindustrial waste as animal feedstuff. In *Proceedings of the First ASEAN Workshop on Technology of Animal Feed Production Utilizing Food Waste Materials*, Bandung, Indonesia.

Ranjhan, S. K. & Chadhokar, P. A. (1984). Effective utilization of agro-industrial by-products for animal feeding in Sri Lanka. *World Animal Review*, **50**, 45–51.

Rapp, A., Murray-Rust, D. H., Christiansson, C. & Berry, L. (1972). Soil erosion and sedimentation in four catchments near Dodoma, Tanzania. *Geografiska Annaler*, **54A**, 255–318.

Reece, P. H. & Campbell, B. L. (1986). The use of $_{137}$Cs for determining soil erosion differences in a disturbed and non-disturbed semi-arid ecosystem. In *Rangelands: A Resource under Siege*, ed. P. J. Joss, P. W. Lynch & O. B. Williams, pp. 294–5. Canberra: *Australian Academy of Science*.

Reed, J. D., Capper, B. S. & Neate, P. J. H. (ed.) (1988). *Plant Breeding and the Nutritive Value of Crop Residues*. Addis Ababa: International Livestock Centre for Africa.

Rees, M. C. & Minson, D. J. (1976). Fertilizer calcium as a factor affecting the voluntary intake, digestibility and retention time

of pangola grass (*Digitaria decumbens*) by sheep. *British Journal of Nutrition*, **36**, 179–87.

Rees, M. C., Minson, D. J. & Smith, F. W. (1974). The effect of supplementary and fertilizer sulphur on voluntary intake, digestibility, retention time in the rumen, and site of digestion of pangola grass in sheep. *Journal of Agricultural Science*, **82**, 419–22.

Reis, A. R. & Garcia, R. (1989). Effect of ammonia level, time of exposure to ammonia and period of storage on chemical composition and digestibility of two tropical grasses hay. In *Proceedings XVI International Grassland Congress*, **2**, 1001–2. Versailles: Association Française pour la Production Fourragère.

Reis, A. R., Garcia, R., Gomide, J. A., Obeid, J. A. (1985). [Effects of cutting regimens on total non-structural carbohydrate content of *Brachiaria decumbens* Stapf.] *Revista da Sociedade Brasileira de Zootecnia*, **14**, 522–8.

Reul, R. H. (1979). Productive potential of wild animals in the tropics. *World Animal Review*, **32**, 18–24.

Reynolds, S. G. (1981). Grazing trials under coconuts in Western Samoa. *Tropical Grasslands*, **15**, 3–10.

Reynolds, S. G. (1988)., *Pastures and Cattle Under Coconuts*. FAO Plant Production and Protection Paper 91. Rome: Food and Agriculture Organisation of the United Nations.

Ribeiro, H. M., Pizarro, E. A., Rodriguez, N. M. & Viana, J. de A. C. (1980). Perennial soyabean hay. I. Production and storage. *Arquivos da Escola de Veterinária da Universidade Federal de Minas Gerais*, **32**, 435–43.

Richards, J. H. (1984). Root growth response to defoliation in two *Agropyron* bunchgrasses: field observations with an improved root periscope. *Oecologia*, **64**, 21–5.

Richards, J. H. & Caldwell, M. M. (1985). Soluble carbohydrates, concurrent photosynthesis and efficiency in regrowth following defoliation: a field study with *Agropyron* species. *Journal of Applied Ecology*, **22**, 907–20.

Richards, J. H., Mott, J. J. & Ludlow, M. M. (1986). Developmental morphology and tiller dynamics. In *CSIRO Division of Tropical Crops and Pastures Annual Report 1985–86*, pp. 61–2. Brisbane, Australia: CSIRO.

Rickert, K. G. (1970). Some influences of straw mulch, nitrogen fertiliser and oat companion crops on establishment of Sabi panic. *Tropical Grasslands*, **4**, 71–5.

Rickert, K. G. (1974). Green panic establishment as influenced by straw mulch and moisture. *Queensland Journal of Agricultural and Animal Sciences*, **31**, 105–17.

Rickert, K. G. & Humphreys, L. R. (1970). Effects of variation in density and phosphate application on growth and chemical composition of Townsville stylo (*Stylosanthes humilis*). *Australian Journal of Experimental Agriculture and Animal Husbandry*, **10**, 442–9.

Ridpath, M. G., Taylor, J. A. & Tulloch, D. G. (1985). Nature as a model. In *Agro-research for the Semi-Arid Tropics*, ed. R. C. Muchow, pp. 419–34. Brisbane: University of Queensland Press.

Riewe, M. E. (1961). Use of the relationship of stocking rate to gain of cattle in an experimental design for grazing trials. *Agronomy Journal*, **53**, 309–13.

Riewe, M. E., Smith, J. C., Jones, J. H. & Holt, E. C. (1963). Grazing production curves. 1. Comparison of steer gains on gulf rye grass and tall fescue. *Agronomy Journal*, **55**, 367–9.

Rika, I. K., Nitis, I. M. & Humphreys, L. R. (1981). Effects of stocking rate on cattle growth, pasture production and coconut yield in Bali. *Tropical Grasslands*, **15**, 149–57.

Ritchie, J. T. (1972). Model for predicting evaporation from a row crop with incomplete cover. *Water Resources Research*, **8**, 1204–13.

Ritchie, J. T. & Burnett, E. (1971). Dryland evaporative flux in a subhumid climate: II. Plant influences. *Agronomy Journal*, **63**, 56–62.

Rivero, R., Pérez, G., Sosa, N. & Combellas, J. (1984). Supplementation of sorghum silage for growing heifers and milking cows. *Tropical Animal Production*, **9**, 114–21.

Riveros, F. (1970). Productivity of *Desmodium intortum* (Mill.) Urb. and *Setaria sphacelata* (Schum.) Stapf. in relation to defoliation. Ph.D. thesis, University of Queensland.

Robbins, G. B. (1984). Relationships between productivity and age since establishment of pastures of *Panicum maximum* var. *trichoglume*. Ph.D. thesis, University of Queensland.

Robbins, G. B. & Bushell, J. J. (1983). A physical model of beef cattle production using inputs of sown and native pasture, leucaena and crops. *Animal Production in Australia*, **15**, 13–15.

Robbins, G. B., Bushell, J. J. & Butler, K. L. (1987). Decline in plant and animal production from ageing pastures of green panic (*Panicum maximum* var. *trichoglume*). *Journal of Agricultural Science*, **108**, 407–17.

Robbins, G. B. & Faulkner, G. B. (1983). Productivity of six ryegrass (*Lolium* spp.) cultivars grown as irrigated annuals in the Burnett district of south-east Queensland. *Tropical Grasslands*, **17**, 49–54.

Robertson, A. D. & Humphreys, L. R. (1976). Effects of frequency of heavy grazing and of phosphorous supply on an *Arundinaria ciliata* association oversown with *Stylosanthes humilis*. *Thai Journal of Agricultural Science*, **9**, 181–8.

Robertson, A. D., Humphreys, L. R. & Edwards, D. G. (1976). Influence of cutting frequency and phosphorus supply on the production of *Stylosanthes humilis* and *Arundinaria pusilla* at Khon Kaen, north-east Thailand. *Tropical Grasslands*, **10**, 33–9.

Robinson, D. W. (1967). Analysis of liveweight loss in beef cattle on native pastures in the Kimberleys, Western Australia. *Journal of the Australian Institute of Agricultural Science*, **33**, 218–19.

Rodel, M. G. M. (1971). Pasture management at Henderson. *Rhodesian Agriculture Journal*, **68**, 114–15.

Rodel, M. G. M. & Boultwood, J. N. (1981a). The effect of applied nitrogen on herbage yields of intensively grazed star grass. *Zimbabwe Agricultural Journal* **78**, 103–7.

Rodel, M. G. W. & Boultwood, J. N. (1981b). The rate of water infiltration into the soil under grass pastures grazed intensively by steers. *Zimbabwe Agricultural Journal*, **78**, 223–4.

Rodrigues, L. R. de A., Mott, G. O., Veiga, J. B., & Ocumpaugh, W. R. (1986). Tillering and morphological characteristics of dwarf elephantgrass under grazing. *Pesquissa Agropecuária Brasileira*, **21**, 1209–18.

Roe, R. (1987). Recruitment of *Astrebla* spp. in the Warrego region of south-western Queensland. *Tropical Grasslands*, **21**, 91–2.

Roe, R. & Allen, G. H. (1945). Studies on the Mitchell grass association in south-western Queensland. 2. The effect of grazing on the Mitchell grass pasture. *C.S.I.R. Bulletin*, **185**.

Rojas, O. D., McDowell, L. R., Moore, J. E., Martin, F. G. &

Ocumpaugh, W. R. (1987). Mineral concentration of tropical grasses as affected by age of regrowth. *Tropical Grasslands*, **21**, 8–14.

Román, M. A. (1979). Effect of nitrogen on forage and meat production with Bermuda grass cv. Coastcross 1, *Cynodon dactylon*, under seasonal conditions. Thesis, Universidad Autónoma Chapingo, Mexico.

Romero, A. & Siebert, B. D. (1980). Seasonal variations of nitrogen and digestible energy intake of cattle on tropical pasture. *Australian Journal of Agricultural Research*, **31**, 393–400.

Rose, C. W. (1988). Research progress on soil erosion processes and a basis for soil conservation practices. In *Soil Erosion Research Methods*, ed. R. Lal, pp. 119–39. Iowa: Soil and Water Conservation Society.

Rosenthal, G. A. & Janzen, D. H. (1979). *Herbivores. Their Interaction with Secondary Plant Metabolites*. New York: Academic Press.

Round, P. J., Mellor, W. & Hibberd, M. J. (1982). The effect of age and season of introduction on the liveweight performance of steers in the wet tropics. *Tropical Animal Production*, **7**, 43–9.

Rouquette, F. M., Matocha, J. E. & Duble, R. L. (1973). Recycling and recovery of nitrogen, phosphorus and potassium. II. Under grazing conditions with two stocking rates. *Journal of Environmental Quality*, **2**, 129–32.

Roverso, E. A., Cunha, P. G. da, Silva, D. J. da, & Montagnini, M I. (1975). Sorghum and Napier grass silage for feeding pregnant cows in winter. *Boletim de Indústria Animal*, **32**, 249–56.

Roxas, D. B., Obsioma, A. R., Castillo, L. S., Lapitan, R. M., Mamongan, V. G. & Juliano, B. O. (1983). Chemical composition and *in vitro* digestibility of straw from different varieties of rice. *Philippines Society of Animal Science Abstracts*, 20th Annual Convention, 10.

Ruby, E. S. & Young, V. A. (1953). The influence of intensity and frequency of clipping on the root system of brownseed paspalum. *Journal of Range Management*, **6**, 94–9.

Ruiz, R., Cairo, J., Martinez, R. O. & Herrera, R. S. (1981). Milk production of cows grazing Coast Cross No. 1 Bermuda grass (*Cynodon dactylon*.). 2. Sward structure and productive potential. *Cuban Journal of Agricultural Science*, **15**, 133–44.

Ruiz, R. & Rowe, J. B. (1980). Intake and digestion of different parts of the banana plant. *Tropical Animal Production*, **5**, 253–6.

Runcie, K. V. (1960). The utilization of grass by strip grazing and zero grazing with dairy cows. In *Proceedings of the VIII International Grassland Congress*, pp. 644–7. Hurley, UK: British Grassland Society.

Ruthenberg, H. (1980). *Farming Systems in the Tropics*. Oxford: Clarendon Press.

Saadullah, M.,, Haque, M. & Dolberg, F. (1980). Treating rice straw with animal urine. *Tropical Animal Production*, **5**, 273–7.

Saeki, T. (1960). Interrelationships between leaf amount, light distribution and total photosynthesis in a plant community. *Botanical Magazine (Tokyo)*, **73**, 155–63.

Salih, Y. M., McDowell, L. R., Hentges, J. F., Mason, R. M., Jr, & Conrad, J. H. (1983). Mineral status of grazing beef cattle in the warm climate region of Florida. *Tropical Animal Health and Production*, **15**, 245–51.

Samarakoon, S. P. (1987). The effects of shade on quality, dry matter yield and nitrogen economy of *Stenotaphrum secundatum* compared with *Axonopus compressus* and *Pennisetum clandestinum*. M.Agr.Sc. thesis, University of Queensland.

Sampaio, E. V. S. B., Beaty, E. R. & Ashley, D. A. (1976). Bahiagrass regrowth and physiological aging., *Journal of Range Management*, **29**, 316–9.

Sanchez, P. A. & Tergas, L. E. (1979). *Pasture Production in Acid Soils of the Tropics*. Cali, Colombia: CIAT.

Sandford, S. (1983). The development experience. In *Pastoral Systems Research in Sub-Saharan Africa*, pp. 11–21. Addis Ababa: ILCA.

Sandland, R. L. & Jones, R. J. (1975). The relation between animal gain and stocking rate in grazing trials: an examination of published theoretical models. *Journal of Agricultural Science*, **85**, 123–8.

Santana, H., Cáceres, O. & Rivero, L. (1986). Nutritive value of proteinaceous forage plants. I. Sunflowers (*Helianthus annuus*). *Pastos y Forrajes*, **9**, 155–60.

Santhirasegaram, K. (1967). Report of the Agrostologist–1966. *Ceylon Coconut Quarterly*, **18**, 57–73.

Santillan, R. A. (1983). Response of a tropical legume-grass association to systems of grazing management and levels of phosphorus fertilization. Ph.D. thesis, University of Florida.

Sarmiento, G. (1984). *The Ecology of Neotropical Savannas*. Cambridge, Mass.: Harvard University Press.

Sastradipradja, D. (1981). Feedingstuffs from the residues of the agricultural industry. In *Proceedings of the First ASEAN Workshop on the Technology of Animal Feed Production*, ed. O. B. Liang & A. T. A. Korossi, pp. 1–20. Bandung: ASEAN Committee on Science and Technology.

Saunders, G. W. (1968). Funnel ants (*Aphaenogaster* spp., Formicidae) as pasture pests in north Queensland. II. Control. *Bulletin of Entomological Research*, **59**, 281–90.

Savory, A. (1983). The Savory grazing method or holistic resource management. *Rangelands*, **5**, 155–9.

Scateni, W. J. (1966). Effect of variation in stocking rate and conservation in the productivity of sub-tropical pastures. In *Proceedings of the Xth International Grassland Congress*, pp. 947–51. Helsinki, Finland.

Scattini, W. J. (1984). Hay conservation for cattle on winter-grazed green panic (*Panicum maximum* var. *trichoglume* pasture in south-eastern Queensland. *Australian Journal of Experimental Agriculture and Animal Husbandry*, **24**, 20–5.

Schroder, V. C. (1970). Soil temperature effects on shoot and root growth of pangolagrass, slenderstem digitgrass, Coastal bermudagrass, and Pensacola bahia grass. *Proceedings of the Soil Crop Science Society of Florida*, **30**, 241–5.

Scott, J. D. (1951). *A Contribution to the Study of the Problems of the Drakensberg Conservation Area*. Science Bulletin of Department of Agriculture 324. Pretoria, South Africa.

Senra, A., Hardy, C., & Munoz, E. (1981). Management of grasslands for milk production and preservation of grass surplus. *Cuban Journal of Agricultural Science*, **15**, 237–54.

Serrao, E. A. S., Falesi, I. C., de Veiga, J. B. & Neto, J. F. T. (1979). Productivity of cultivated pastures on low fertility soils in the Amazon of Brazil. In *Pasture Production in Acid Soils of the Tropics*, ed. P. A. Sánchez & L. E. Tergas, pp. 195–225. Cali, Colombia: CIAT.

Shaw, N. H. (1978). Superphosphate and stocking rate effects on a native pasture oversown with *Stylosanthes humilis* in central

coastal Queensland. 2. Animal production. *Australian Journal of Experimental Agriculture and Animal Husbandry*, **18**, 800–7.

Shaw, N. H., Elich, T. W., Haydock, K. P. & Waite, R. B. (1965). A comparison of 17 introductions of *Paspalum* species and naturalised *P. dilatatum* under cutting at Samford, south-eastern Queensland. *Australian Journal of Experimental Agriculture and Animal Husbandry*, **5**, 423–32.

Sheldrick, R. D. & Goldson, J. R. (1978). The results of a forage-system exercise at Kitale. *Technical Report, Pasture Research Project, Ministry of Agriculture, Kenya*, **23**.

Shelton, H. M. & Humphreys, L. R. (1975). Undersowing of *Oryza sativa* with *Stylosanthes guianensis*. III. Nitrogen supply. *Experimental Agriculture*, **11**, 103–11.

Shelton, H. M., Humphreys, L. R. & Batello, C. (1987). Pastures in the plantations of Asia and the Pacific: performance and prospect. *Tropical Grasslands*, **21**, 159–68.

Shelton, H. M. & Wilaipon, B. (1984). Establishment of two *Stylosanthes* species in communal grazing areas of northeast Thailand. *Tropical Grasslands*, **18**, 180–5.

Shimojo, M. & Goto, I. (1989). Effect of sodium silicate on forage digestion with rumen fluid of goats or cellulose using culture solutions adjusted for pH. *Animal Feed Science and Technology*, **24**,173–7.

Siebert, B. D. & Field, J. B. F. (1975). Reproductive activity in beef heifers following post-weaning feeding on spear grass hay alone or with supplements. *Australian Journal of Experimental Agriculture and Animal Husbandry*, **15**, 12–16.

Siebert, B. D. & Hunter, R. A. (1977). Prediction of herbage intake and liveweight gain of cattle grazing tropical pastures from the composition of the diet. *Agricultural Systems*, **2**, 199–208.

Siebert, B. D. & Hunter, R. A. (1982). Supplementary feeding of grazing animals. In *Nutritional Limits to Animal Production from Pastures*, ed. J. B. Hacker, pp. 409–26. Farnham Royal, UK: Commonwealth Agricultural Bureaux.

Siebert, B. D. & Kennedy, P. M. (1972). The utilization of spear grass (*Heteropogon contortus*). 1. Factors limiting intake and utilization by cattle and sheep. *Australian Journal of Agricultural Research*, **23**, 35–44.

Siewerdt, L. (1980). Mecanizacao eficiente na producao de feno. *Informe Agropecuário Belo Horizonte*, **6(64)**, 23–8.

Silanikove, N., Cohen, O., Levanon, D., Kipnis, T. & Kugenheim, Y. (1988). Preservation and storage of green panic (*Panicum maximum*) as moist hay with urea. *Animal Feed Science & Technology*, **20**, 87–96.

Sillar, D. I. (1967). Effect of shade on growth of Townsville lucerne (*Stylosanthes humilis* H.B.K.). *Queensland Journal of Agriculture and Animal Science*, **24**, 237–40.

Silva, A. B. & Serrao, E. A. S. (1985). Research results on pasture spittlebugs in the Brazilian Amazon region. *Proceedings XV International Grassland Congress*, preprint 7–0–12.

Simao Neto, M. (1985). Recovery, viability and potential dissemination of pasture seed passed through the digestive tract of ruminants. Ph.D. thesis, University of Queensland.

Simao Neto, M. & Jones, R. M. (1986). The effect of storage in cattle dung on viability of tropical pasture seeds. *Tropical Grasslands*, **20**, 180–3.

Skerman, R. H. & Humphreys, L. R. (1973). The effect of temperature during flowering on seed formation of *Stylosanthes humilis*. *Australian Journal of Agricultural Research*, **24**, 317–24.

Skerman, P. J. (1977). *Tropical Forage Legumes*. FAO Plant

Production Protection Series 2. Rome: Food and Agriculture Organization of the United Nations.

Skovlin, J. M. (1971). Ranching in East Africa: A case study. *Journal of Range Management*, **24**, 263–70.

Smith, A. D. (1965). Determining common use grazing capacities by application of the key species concept. *Journal of Range Management*, **18**, 196–201.

Smith, A. & Allcock, P. J. (1985). The influence of species diversity on sward yield and quality. *Journal of Applied Ecology*, **22**, 185–98.

Smith, C. A. (1961). The utilisation of *Hyparrhenia* veld for the nutrition of cattle in the dry season. II. Veld hay compared with *in situ* grazing of the mature forage, and the effects of feeding supplementary nitrogen. *Journal of Agricultural Science*, **57**, 311–17.

Smith, C. A. (1967). Pasture management. *CSIRO Division of Tropical Pastures Annual Report 1966–67*, pp. 25–6.

Smith, C. A. & Hodnett, G. E. (1962). Compensatory growth of cattle on the grasslands of Northern Rhodesia. *Nature*, **195**, 919–20.

Smith, M. A. & Whiteman, P. C. (1985a). Animal production from rotationally-grazed natural and sown pastures under coconuts at three stocking rates in the Solomon Islands. *Journal of Agricultural Science*, **104**, 173–80.

Smith, M. A. & Whiteman, P. C. (1985b). Grazing studies on the Guadalcanal Plains, Solomon Islands. 3. Comparison of existing mixtures with koronivia (*Brachiaria humidicola*) and with natural pastures. *Journal of Agricultural Science*, **104**, 181–9.

Soldevila, M. (1980). Effect of rotation length in Pangola pastures upon the liveweight gain of growing Holstein heifers. *Journal of Agriculture of the University of Puerto Rico*, **64**, 243–6.

Soneji, S. V., Musangi, R. S. & Olsen, F. J. (1971). Digestibility and feed intake investigations at different stages of growth of *Brachiaria ruziziensis*, *Chloris gayana* and *Setaria sphacelata* using Corriedale wether sheep. 1. Digestibility and voluntary intake. *East African Agricultural and Forestry Journal*, **37**, 125–8.

Sousa, J. C. de, Conrad, J. H., Blue, W. G., Ammerman, C. B. & McDowell, L. R. (1981). Interrelationship between mineral in the soil, forage plants and animal tissue. 3. Manganese, iron and cobalt. *Pesquisa Agropecuária Brasileira*, **16**, 739–46.

Sousa, J. C. de, Conrad, J. H., Blue, W. G. & McDowell, L. R. (1979). Interrelationships among mineral levels in soil, forage plants and animal tissues. 1. Calcium and phosphorus. *Pesquisa Agropecuária Brasileira*, **14**, 387–95.

Sousa, J. C. de, Conrad, J. H., Mott, G. O., McDowell, L. R., Ammerman, C. B. & Blue, W. G. (1982). Interrelationship between minerals in the soil, forage plants and animal tissue in the north of Mato Grosso. 4. Zinc, magnesium, sodium and potassium. *Pesquisa Agropecuária Brasileira*, **17**, 11–20.

Spain, J., Pereira, J. M. & Gualdron, R. (1985). A flexible grazing management system proposed for the advanced evaluation of associations of tropical grasses and legumes. *Proceedings of the XV International Grassland Congress*, pp. 1153–5. Nishinasuno, Japan: Japanese Society of Grassland Science.

Spedding, C. R. W. (1971). *Grassland Ecology*. London: Oxford University Press.

Spencer, R. R., Akin, D. E. & Rigsby, L. L. (1984). Degradation of potassium hydroxide-treated 'Coastal' Bermudagrass stems at two stages of maturity. *Agronomy Journal*, **76**, 819–24.

Sproule, R. J., Shelton, H. M. & Jones, R. M. (1983). Effect of

summer and winter grazing on growth habit of Kenya white clover (*Trifolium semipilosum*) cv. Safari in a mixed sward. *Tropical Grasslands*, **17**, 25–30.

Squires, V. R. (1982). Behaviour of free-ranging livestock on native grasslands and shrublands. *Tropical Grasslands*, **16**, 161–70.

Steele, K. W. & Vallis, I. (1988). The nitrogen cycle in pastures. In *Advances in Nitrogen Cycling in Agricultural Ecosystems*, ed. J. R. Wilson, pp. 274–91. Wallingford, UK: CAB International.

Steinke, T. D. (1975). Effect of height of cut on translocation of ^{14}C-labelled assimilates in *Eragrostis curvula* (Schrod.) Nees. *Proceedings of the Grassland Society of Southern Africa*, **10**, 41–7.

Steinke, T. D. & Booysen, P. de V. (1968). The regrowth and utilization of carbohydrate reserves of *Eragrostis curvula* after different frequencies of defoliation. *Proceedings of the Grassland Society of Southern Africa*, **3**, 105–10.

Stephens, D. W. & Krebs, J. R. (1986). *Foraging Theory*. Princeton, New Jersey: Princeton University Press.

Stobbs, T. H. (1969b). The effect of grazing management upon pasture productivity in Uganda. II. Grazing frequency. *Tropical Agriculture*, **46**, 195–200.

Stobbs, T. H. (1969a). The effect of grazing management upon pasture productivity in Uganda. III. Rotational and continuous grazing. *Tropical Agriculture*, **46**, 293–301.

Stobbs, T. H. (1973a). The effect of plant structure on the intake of tropical pastures. 1. Variation in the bite size of grazing cattle. *Australian Journal of Agricultural Research*, **24**, 809–19.

Stobbs, T. H. (1973b). The effect of plant structure on the intake of tropical pastures. 2. Differences in sward structure, nutritive value, and bite size of animals grazing *Setaria anceps* and *Chloris gayana* at various stages of growth. *Australian Journal of Agricultural Research*, **24**, 821–9.

Stobbs, T. H. (1974a). Components of grazing behaviour of dairy cows on some tropical and temperate pastures. *Proceedings of the Australian Society of Animal Production*, **10**, 299–302.

Stobbs, T. H. (1974b). Rate of biting by Jersey cows as influenced by the yield and maturity of pasture swards. *Tropical Grasslands*, **8**, 81–6.

Stobbs, T. H. (1975). A comparison of Zulu sorghum, bulrush millet and white panicum in terms of yield, forage quality and milk production. *Australian Journal of Experimental Agriculture and Animal Husbandry*, **15**, 211–18.

Stobbs, T. H. (1977a). Seasonal changes in the preference by cattle for *Macroptilium atropurpureum* cv. Siratro. *Tropical Grasslands*, **11**, 87–92.

Stobbs, T. H. (1977b). Short-term effects of herbage allowance on milk production, milk composition and grazing time of cows grazing nitrogen-fertilized tropical grass pasture. *Australian Journal of Experimental Agriculture and Animal Husbandry*, **17**, 892–8.

Stobbs, T. H. (1978). Milk production, milk composition, rate of milking and grazing behaviour of dairy cows grazing two tropical grass pastures under a leader and follower system. *Australian Journal of Experimental Agriculture and Animal Husbandry*, **18**, 5–11.

Stobbs, T. H. & Imrie, B. C. (1976). Variation in yield, canopy structure, chemical composition, and *in vitro* digestibility within and between two *Desmodium* species and interspecific hybrids. *Tropical Grasslands*, **10**, 99–106.

Stocker, G. C. & Sturtz, J. D. (1966). The use of fire to establish

Townsville lucerne in the Northern Territory. *Australian Journal of Experimental Agriculture & Animal Husbandry*, **6**, 277–9.

Stür, W. W. & Humphreys, L. R. (1987). Seed production in *Brachiaria decumbens* and *Paspalum plicatulum* as influenced by system of residue disposal. *Australian Journal of Agricultural Research*, **38**, 869–80.

Stür, W. W. & Humphreys, L. R. (1988a). Defoliation and burning effects on the tillering of *Brachiaria decumbens*. *Journal of Applied Ecology*, **25**, 273–7.

Stür, W. W. & Humphreys, L. R. (1988b). Burning and cutting management and the formation of seed yield in *Brachiaria decumbens*. *Journal of Agricultural Science*, **110**, 669–72.

Sturtz, J. D. & Parker, G. V. (1974). Cattle liveweight changes on fodder rolls and standing hay on Townsville stylo/native grass. In *Proceedings of the Australian Society of Animal Production*, **10**, 344–8.

Stuth, J. W., Brown, J. R., Olson, P. D. & Araujo, M. R. (1985). Influence of stocking rate on the plant–animal interface in rotationally grazed mixed swards. In *Proceedings of the XVth International Grassland Congress*, pp. 1148–50. Nishi-nasuno, Japan: The Japanese Society of Grassland Science.

Sugimoto, Y., Hirata, M. & Ueno, M. (1987a). Energy and matter flows in bahiagrass pasture. IV. Nitrogen flow under rotational grazing system by Holstein heifers. *Journal of Japanese Society of Grassland Science*, **32**, 313–20.

Sugimoto, Y., Hirata, M. & Ueno, M. (1987b). Energy and matter flows in bahiagrass pasture. V. Excreting behaviour of Holstein heifers. *Journal of Japanese Society of Grassland Science*, **32**, 8–14.

Sugimoto, Y., Hirata, M. & Ueno, M. (1987c). Energy and matter flows in bahiagrass pasture. VI. Nitrogen excretion in dung and urine by Holstein heifers. *Journal of Japanese Society of Grassland Science*, **33**, 121–7.

Sullivan, G. M., Stokes, K. W., Farris, D. E., Nelsen, T. C. & Cartwright, T. C. (1980). Transforming a traditional forage/-livestock system to improve human nutrition in tropical Africa. *Journal of Range Management*, **33**, 174–8.

Sumanto, Santosa, Petheram, R. J., Perkins, J., Nana & Rusastra, W. (1987). An agro-economic profile of Padamulya village, Subang – with emphasis on drought animal rearing. *DAP Project Bulletin, James Cook University of North Queensland, Townsville*, **4**, 2–31.

Sumberg, J. E. (1983). 'Leuca-fence': A living fence for sheep using *Leucaena leucocephala*. *ILCA Newsletter*, **2**, 5.

Sumberg, J. E., McIntyre, J., Okali, C. & Alta-Krah, A. (1987). Economic analysis of alley farming with small ruminants. *ILCA Newsletter*, **28**, 2–5.

Taerum, R. (1970). A study of root and shoot growth on three grass species in Kenya. *East African Agricultural and Forestry Journal*, **36**, 155–70.

Tainton, N. M. & Booysen, P. de V. (1965). Growth and development in perennial veld grasses. 1. *Themeda triandra* tillers under various systems of defoliation. *South African Journal of Agricultural Science*, **8**, 93–110.

Taylor, C. A. (1985). Multispecies grazing research overview (Texas). In *Proceedings of a Conference on Multispecies Grazing*, ed. F. H. Baker & R. K. Jones, pp. 65–83. Morrilton, Arkansas: Winrock International Institute for Agricultural Development.

Taylor, R. D. & Walker, B. H. (1978). Comparisons of vegetation use and herbivore biomass on a Rhodesian game and cattle ranch. *Journal of Applied Ecology*, **15**, 565–81.

Teitzel, J. K., Monypenny, J. R. & Rogers, S. J. (1986). Tropical pastures in the farming system: case studies of modelling integration through linear programming. In *Proceedings of the Third Australian Conference on Tropical Pastures*, ed. G. J. Murtagh & R. M. Jones, occasional publication 3, pp. 101–9. St. Lucia. Australia: Tropical Grassland Society of Australia.

Tergas, L. E., Paladines, O., Kleinheisterkamp, I. & Velásquez, J. (1984). Animal productivity from *Brachiaria decumbens* alone and with complementary grazing of *Pueraria phaseoloides*, *Tropical Animal Production*, **9**, 1–11.

Tessema, S. & Emojong, E. E. (1984). Utilization of maize and sorghum stover by sheep and goats after feed-quality improvement by various treatments and supplements. *East African Agricultural and Forestry Journal*, **44**, 408–15.

Thiago, L. R. L. S., Leibholz, J. M. L. & Kellaway, R. C. (1981). Effect caustic soda treatment and protein supplementation of *Paspalum dilat* :*um* hay on intake and liveweight gain of cattle. *Pesquisa Agropecuária Brasileira*, **16**, 751–6.

Thomas, C. (1977). The effect of level of concentrate feeding on voluntary intake and performance by steers given silage of *Chloris gayana* during the dry season and their subsequent growth to slaughter. *East African Agricultural and Forestry Journal*, **42**, 328–36.

Thornley, J. H. M. (1972). A balanced quantitative model for shoot:root ratios in vegetative plants. *Annals of Botany*, **36**, 431–41.

Thornton, D. D. & Harrington, G. N. (1971). The effect of different stocking rates on the liveweight gain of Ankole steers on natural grassland in Western Uganda. *Journal of Agricultural Science*, **76**, 97–106.

Thornton, R. F. & Minson, D. J. (1973). The relationship between apparent retention time in the rumen, voluntary intake, and apparent digestibility of legume and grass diets in sheep. *Australian Journal of Agricultural Research*, **24**, 889–98.

Tierney, T. J., Evans, J. & Taylor, W. J. (1983). Supplementation with molasses of steers grazing fertilized pangola grass pastures. *Tropical Grasslands*, **17**, 156–63.

Tierney, T. J. & Goward, E! (1983a). Utilization of wet heath on the coastal lowlands of south-east Queensland with beef steers grazing at three stocking rates on Pangola grass (*Digitaria decumbens*) with two rates of applied nitrogen fertilizer. 1. Animal and pasture production. *Tropical Grasslands*, **17**, 132–8.

Tierney, T. J. & Goward, E. (1983b). Utilization of wet heath on the coastal lowlands of south-east Queensland with beef steers grazing at three stocking rates on pangola grass (*Digitaria decumbens*) with two rates of applied nitrogen fertilizer. 2. Pasture chemical composition and soil changes. *Tropical Grasslands*, **17**, 145–51.

Tierney, T. J. & Taylor, W. J. (1983). Productivity of a beef breeder herd grazing fertilized pangola grass in the wet coastal lowlands of south-east Queensland. *Tropical Grasslands*, **17**, 97–105.

Till, A. R. (1981). Cycling of plant nutrients in pasture. In *Grazing Animals*, ed. F. H. W. Morley, 33–53. Amsterdam: Elsevier.

Tjandraatmadja, M. (1989). The microbiology and nutritive value of tropical silages. Ph.D. thesis, University of Queensland.

Toledo, J. M., Giraldo, H. & Spain, J. M. (1987). Effecto del pastoreo continuo y el metodo de siembra en la persistencia de la asociacion *Andropogon gayanus/Stylosanthes capitata. Pasturas Tropicales Boletin*, **9**, 16–23.

Torres, R. de A., Simao Neto, M., Novaes, L. P. & Souza, R. M. de (1982). Effect of stocking rate and silage supplementation on the growth of dairy cattle on molasses grass pasture. *Pesquisa Agropecuária Brasileira*, **17**, 479–88.

Tothill, J. C. (1969). Soil temperatures and seed burial in relation to the performance of *Heteropogon contortus* and *Themeda australis* in burnt native woodland pastures in eastern Queensland. *Australian Journal of Botany*, **17**, 269–75.

Tothill, J. C. (1971). Grazing, burning and fertilizing effects on the regrowth of some woody species in cleared open forest in south-east Queensland. *Tropical Grasslands*, **5**, 31–4.

Tothill, J. C. (1974). Experiences in sod-seeding Siratro into native speargrass pastures on granite soils near Mundubbera. *Tropical Grasslands*, **8**, 128–31.

Tothill, J. C. & Mott, J. J. (eds.) (1985). *Ecology and Management of the World's Savannas*. Canberra: Australian Academy of Science.

Troughton, A. B. (1957). *The Underground Organs of Herbage Grasses*. Bulletin 44. Farnham Royal: Commonwealth Agricultural Bureaux.

Tucker, V. C., O'Grady, P., Smith, R. A. D. & Byford, I. (1972). Effect on milk yield and composition of using hay and silage conservation for dairy cows grazing Glycine–green panic pastures. *Australian Journal of Dairy Technology*, **27**, 144–8.

Tudsri, S., Watkin, B. R., Chantkam, S., Chu, A. C. P. & Forde, B. J. (1989). Effect of first year grazing management on *Stylosanthes hamata* cv. Verano production at Muak Lek, Saraburi, Thailand. *Tropical Grasslands*, **23**, 35–42.

Uchida, S. & Kitamura, Y. (1987a). Silage making from tropical pasture plants grown in south western islands of Japan. I. Effects of various treatments at ensiling on quality of Rhodes grass and Napier grass silages. *Journal of Japanese Society of Grassland Science*, **32**, 369–74.

Uchida, S. & Kitamura, Y. (1987b). Silage making from tropical pasture plants grown in south western islands of Japan. I. Effects of mixing Stylo and Siratro with Rhodesgrass on quality of silages. *Journal of Japanese Society of Grassland Science*, **32**, 375–80.

Ulloa, J. A., Watkins, K. L. & Craig, W. M. (1985). Use of urea as a source of ammonia in treating Coastal Bermudagrass hay. *Nutrition Reports International*, **32**, 901–8.

UNESCO/UNEP/FAO (1979). *Tropical Grazing Land Ecosystems*. Natural Resources Research **16**. Paris: UNESCO.

Valdes, G., Molina, A. & Garcia, R. (1988). Effect of the stocking rate, segregation and supplementation in bovine cattle fed non-irrigated Coast cross 1 bermuda grass. *Cuban Journal of Agricultural Science*, **22**, 149–55.

Vallis, I. (1983). Uptake by grass and transfer to soil of nitrogen from ^{15}N-labelled legume materials applied to a Rhodes grass pasture. *Australian Journal of Agricultural Research*, **34**, 367–76.

Vallis, I. & Jones, R. J. (1973). Net mineralisation of nitrogen in leaves and leaf litter of *Desmodium intortum* and *Phaseolus atropurpureus* mixed with soil. *Soil Biology and Biochemistry*, **5**, 391–8.

Vallis, I., Peake, (the late) D. C. I., Jones, R. K. & McCown, R. L. (1985). Fate of urea-nitrogen from cattle urine in a pasture-crop sequence in a seasonally dry tropical environment. *Australian Journal of Agricultural Research*, **36**, 809–17.

Van der Plank, J. E. (1968). *Disease Resistance in Plants*. New York: Academic Press.

Varma, A., Yadav, B. P. S. & Sampath, K. T. (1987). An ensilage technology for the tribal farmers of North-Eastern Hills Region. *Indian Journal of Animal Sciences*, **57**, 1306–9.

Vasquez, C. M. & Lao, O. (1978). Influence of the number of plots on milk production in an intensive system of exploitation. In *Primer Seminario Cientifico Tecnico, Provincia de Las Tunas*, pp. 68–72. Havana, Cuba: Estacion Central de Pastos.

Veitia, J. L. & Marquez, J. R. (1973). Digestibility of fresh Pangola grass (*Digitaria decumbens*) and Rhodes grass hay (*Chloris gayana*) at three cutting intervals. *Cuban Journal of Agricultural Science*, **7**, 23–37.

Velásquez, G. J. A. & González, J. E. (1972). The nutritive value of groundnut (*Arachis hypogaea*) straw. *Agronomía Tropical*, **22**, 287–90.

Vélez, C. A. & Escobar, L. G. (1970). Effects of seasonal N application to Para grass on beef production. *Acta Agronomia*, **20**, 65–90.

Vera, R. R., Pizarro, E. A., Martins, M. & Viana, J. A. C. (1983). Yield and quality of tropical legumes during the dry season: *Galactia striata* (Jacq.) Urb. *Proceedings of the XIV International Grassland Congress*, pp. 786–8. Boulder, Colorado, USA: Westview Press.

Verma, N. C. & Mojumdar, A. B. (1985). Studies on quality and feeding value of maize + cowpea silage. *Indian Journal of Nutrition and Dietetics*, **22**, 194–6.

Vicente-Chandler, J., Rivera-Brenes, L., Caro-Costas, R. R., Rodriquez, J. P., Bonita, E. & racia, W. (1953). *The management and utilization of the forage rops of Puerto Rico*. Bulletin 116, University of Puerto Rico Agriculture Experiment Station.

Vieira, J. M. (1985). The effect of type of pasture and stocking rate on pasture characteristics, animal behaviour and production. Ph.D. Thesis, University of Queensland.

Vijchulata, P., Chipadpanich, S. & McDowell, L. R. (1983). Mineral status of cattle raised in the villages of central Thailand. *Tropical Animal Pr luction*, **8**, 131–7.

Vilela, D., Cardosa, R. M., Silva, J. F. C. da, & Gomide, J. A. (1980). Effect of concentrate supplementation on nutrient intake and milk production of cows on molasses grass (*Melinis minutiflora* Beauv.) pasture. *Revista da Sociedade Brasileira de Zootecnia*, **9**, 214–32.

Voisin, A. (1959). *Grass Productivity*. London: Crosby Lockwood.

Waidyanatha, U. P. de S., Wijesinghe, D. S. & Stauss, R. (1984). Zero-grazed pasture under immature Hevea rubber: productivity of some grasses and grass–legume mixtures and their competition with Hevea. *Tropical Grasslands*, **18**, 21–6.

Walker, B. (1968). Grazing experiments at Ukiriguru, Tanzania. II. Comparisons of rotational and continuous grazing systems on natural pastures of hardpan soils using an 'extra-period latin square change over' design. *East African Agriculture and Forestry Journal*, **34**, 235–44.

Walker, B. (1969). Effects of feeding hay on the early wet season weight loss of cattle in western Tanzania. *Experimental Agriculture*, **5**, 53–7.

Walker, B. (1977). Productivity of *Macroptilium atropurpureum* cv. Siratro pastures. *Tropical Grasslands*, **11**, 79–86.

Walker, B. (1980). Effects of stocking rate on perennial tropical legume grass pastures. Ph.D. thesis, University of Queensland.

Walker, B. & Scott, G. D. (1968). Grazing experiments at

Ukirigura, Tanzania. 1. Comparisons of rotational and continuous grazing systems on natural pastures of hardpan soils. *East African Agriculture and Forestry Journal*, **34**, 224–34.

Wanapat, M. & Devendra, C. (eds.) (1985). *Relevance of Crop Residues as Animal Feeds in Developing Countries*. Bangkok: Funny Press.

Wanapat, M. & Topark-Ngarm, A. (1985). *Proceedings of the XV International Grassland Congress*, pp. 959–60. Nishi-nasuno, Japan: Japanese Society of Grassland Science.

Ward, H. K. & Cleghorn, W. B. (1970). The effects of grazing practices on tree regrowth after clearing indigenous woodland. *Rhodesian Journal of Agricultural Research*, **8**, 57–65.

Warren, S. D., Thurow, T. L., Blackburn, W. H. & Garza, N. E. (1986). The influence of livestock trampling under intensive rotation grazing on soil hydrologic characteristics. *Journal of Range Management*, **39**, 491–5.

Watkin, B. R. & Clements, R. J. (1978). The effects of grazing animals on pastures. In *Plant Relations In Pastures*, ed. J. R. Wilson, pp. 273–89. Melbourne: CSIRO.

Watson, S. E. & Whiteman, P. C. (1981a). Grazing studies on the Guadacanal Plains, Solomon Islands. 2. Effects of pasture mixtures and stocking rate on animal production and pasture components. *Journal of Agricultural Science*, **97**, 353–64.

Watson, S. E. & Whiteman, P. C. (1981b). Animal production from naturalized and sown pastures at three stocking rates under coconuts in the Solomon Islands. *Journal of Agricultural Science*, **97**, 669–76.

Weinmann, H. (1961). Total available carbohydrates in grasses and legumes. *Herbage Abstracts*, **31**, 255–61.

Westoby, M. (1980). Elements of a theory of vegetation dynamics in arid rangelands. *Israel Journal of Botany*, **28**, 169–94.

Weston, E. J., Harbison, J., Leslie, J. K., Rosenthal, K. M. & Mayer, R. J. (1981). *Assessment of the Agricultural and Pastoral Potential of Queensland*. Agriculture Branch Technical Report 27. Brisbane: Queensland Department of Primary Industries.

Whiteman, P. C. (1970). Seasonal changes in growth and nodulation of perennial tropical pasture legumes in the field. II. Effects of controlled defoliation levels on nodulation of *Desmodium intortum* and *Phaseolus atropurpureus*. *Australia Journal of Agricultural Research*, **21**, 207–14.

Whiteman, P. C. (1980). *Tropical Pasture Science*. Oxford: Oxford University Press.

Whiteman, P. C., Halim, N. R., Norton, B. W. & Hales, J. W. (1985). Beef production from three tropical grasses in southeastern Queensland. *Australian Journal of Experimental Agriculture*, **25**, 481–8.

Whyte, R. O. (1968). *Grasslands of the Monsoon*. London: Faber.

Wilaipon, N., Aitken, R. L. & Hughes, J. D. (1981). The use of apical tissue analysis to determine the phosphorus status of *Stylosanthes hamata* cv. Verano. *Plant & Soil*, **59**, 141–6.

Wilaipon, B., Gigir, S. A. & Humphreys, L. R. (1979). Apex, lamina and shoot removal effects on seed production and growth of *Stylosanthes hamata* cv. Verano. *Australian Journal of Agricultural Research*, **30**, 293–306.

Wilaipon, B., Gutteridge, R. C. & Chutikul, K. (1981). Undersowing upland crops with pasture legumes. 1. Cassava with *Stylosanthes hamata* cv. Verano. *Thai Journal of Agricultural Science*, **14**, 333–7.

Wilaipon, B. & Humphreys, L. R. (1981). Influence of grazing on the seed production of *Stylosanthes hamata* cv. Verano. *Thai Journal of Agricultural Science*, **14**, 69–81.

Wilaipon, P. & Humphreys, L. R. (1976). Grazing and mowing effects on the seed production of *Stylosanthes hamata* cv. Verano. *Tropical Grasslands*, **10**, 107–11.

Wildin, J. H. & Chapman, D. G. (1987). *Ponded Pasture Systems-Capitalising on Available Water*. Rockhampton, Australia: Queensland Department of Primary Industries.

Wilkins, R. J. (1982). Improving forage quality by processing. In *Nutritional Limits to Animal Production from Pastures*, ed. J. B. Hacker, pp. 389–408. Farnham Royal, UK: Commonwealth Agricultural Bureaux.

Wilkinson, J. M. (1983a). Silages made from tropical and temperate crops. Part 1. The ensiling process and its influence on feed value. *World Animal Review*, **45**, 36–45.

Wilkinson, J. M. (1983b). Silages made from tropical and temperate crops. Part 2. Techniques for improving the nutritive value of silage. *World Animal Review*, **46**, 35–46.

Wilkinson, P. R. (1964). Pasture spelling as a control measure for cattle ticks in southern Queensland. *Australian Journal of Agricultural Research*, **15**, 822.

Wilkinson, S. R. & Lowrey, R. W. (1973). Cycling of mineral nutrients in pasture ecosystems. In *Chemistry and Biochemistry of Herbage*, ed. G. W. Butler & R. W. Bailey, pp. 247–315. London: Academic Press.

Williams, C. H. (1977). Trace metals and superphosphate: Toxicity problems. *Journal of Australian Institute of Agricultural Science*, **43**, 99–109.

Wilson, G. P. M. & Hennessy, D. W. (1977). The germination of excreted kikuyu grass seed in cattle dung pats. *Journal of Agricultural Science*, **88**, 247–9.

Wilson, J. B. (1988). A review on the control of shoot:root ratio, in relation to models. *Annals of Botany*, **61**, 433–49.

Wilson, J. R. (1975). Influence of temperature and nitrogen on growth, photosynthesis and accumulation of non-structural carbohydrate in a tropical grass, *Panicum maximum* var. *trichoglume*. *Netherlands Journal of Agricultural Science*, **23**, 48–61.

Wilson, J. R. (1976a). Variation of leaf characteristics with level of insertion on a grass tiller. Development rate, chemical composition and dry matter digestibility. *Australian Journal of Agricultural Research*, **27**, 343–54.

Wilson, J. R. (1976b). Variation of leaf characteristics with level of insertion on a grass tiller. II. Anatomy. *Australian Journal of Agricultural Research*, **27**, 355–64.

Wilson, J. R. (1983). Effects of water stress on *in vitro* dry matter digestibility and chemical composition of herbage of tropical pastures species. *Australian Journal of Agricultural Research*, **34**, 377–90.

Wilson, J. R., Anderson, K. L. & Hacker, J. B. (1989). Dry matter digestibility *in vitro* of leaf and stem of buffel grass (*Cenchrus ciliaris*) and related species and its relation to plant morphology and anatomy. *Australian Journal of Agricultural Research*, **40**, 281–91.

Wilson, J. R., Brown, R. H. & Windham, W. R. (1983). Influence of leaf anatomy on the dry matter digestibility of C_3, C_4 and C_3/C_4 intermediate types of *Panicum* species. *Crop Science*, **23**, 141–6.

Wilson, J. R., Catchpoole, V. R. & Weier, K. L. (1986). Stimulation of growth and nitrogen uptake by shading a rundown green panic pasture on brigalow clay soil. *Tropical Grasslands*, **20**, 134–43.

Wilson, J. R. & Hattersley, P. W. (1983). *In vitro* digestion of bundle sheath cells in rumen fluid and its relation to the suberized lamella and C_4 photosynthetic type in *Panicum* species. *Grass Forage Science*, **38**, 219–23.

Wilson, J. R. & Mannetje, L.'t. (1978). Senescence, digestibility, and carbohydrate content and buffel grass and green panic leaves in swards. *Australian Journal of Agricultural Research*, **29**, 503–16.

Wilson, J. R. & Minson, D. J. (1980). Prospects for improving the digestibility and intake of tropical grasses. *Tropical Grasslands*, **14**, 253–9.

Wilson, J. R. & Ng, T. T. (1975). Influence of water stress on parameters associated with herbage quality of *Panicum maximum* var. *trichoglume*. *Australian Journal of Agricultural Research*, **26**, 127–36.

Wilson, J. R. & Sandland, R. L. (1976). The relationship between growth and nitrogen and phosphorus content of green panic and kikuyu grass. *Plant and Soil*, **44**, 341–58.

Wilson, J. R., and Wong, C. C. (1982). Effects of shade on some factors influencing nutritive quality of green panic and Siratro pastures. *Australian Journal of Agricultural Research*, **33**, 937–49.

Wilson, L. L., Fisher, D. D., Katsigianis, T. S. & Baylor, J. E. (1981). Mineral composition of tropical forages and metabolic blood profiles of grazing cattle and sheep on calcium-dominant Caribbean soils. *Tropical Agriculture* (Trin.), **58**, 53–62.

Winks, L., Lamberth, F. C. & O'Rourke, P. K. (1977). The effect of a phosphorus supplement on the performance of steers grazing Townsville stylo-based pasture in north Queensland. *Australian Journal of Experimental Agriculture and Animal Husbandry*, **17**, 357–66.

Winks, L. & O'Rourke, P. K. (1977). Some observations on feeding salt to steers grazing native pasture in North Queensland. *Journal of Australian Institute of Agricultural Science*, **43**, 76–7.

Winks, L. & O'Rourke, P. K. & McLennan, S. R. (1982). Liveweight of steers supplemented with molasses, urea and sulfur in northern Queensland. *Australian Journal of Experimental Agriculture and Animal Husbandry*, **22**, 252–7.

Winter, W. H. (1987). Using fire and supplements to improve cattle production from monsoon tallgrass pastures. *Tropical Grasslands*, **21**, 71–81.

Winter, W. H., Edye, L. A. & Williams, W. T. (1977). Effects of fertilizer and stocking rate on pasture and beef production from sown pastures in northern Cape York Peninsula. 2. Beef production and its relation to blood, faecal and pasture measurements. *Australian Journal of Experimental Agriculture and Animal Husbandry*, **17**, 187–96.

Winter, W. H., Siebert, B. D. & Kuchel, R. E. (1977). Cobalt deficiency of cattle grazing improved pastures in northern Cape York Peninsula. *Australian Journal of Experimental Agriculture and Animal Husbandry*, **17**, 10–15.

Wong, C. C., Rahim, H. & Sharudin, M. A. Mohd. (1985). Shade tolerance potential of some tropical forages for integration with plantations. 1. Grasses. *MARDI Research Bulletin*, **13**, 225–47.

Wong, C. C. & Wilson, J. R. (1980). Effects of shading on the growth and nitrogen content of green panic and Siratro in pure and mixed swards defoliated at two frequencies. *Australian Journal of Agricultural Research*, **31**, 269–86.

Woolford, M. K. (1985). The silage fermentation. In *Microbiology of Fermented Foods*, vol. 2, ed. B. J. B. Wood, pp. 85–112. London: Elsevier Applied Science Publishers.

World Commission on Environment and Development. (1987). *Our Common Future*. Oxford: Oxford University Press.

Xande, A. (1978). Ensilage of grass, a conservation technique for obviating the food shortage of ruminants during the dry season. 1. Theoretical and practical aspects – particularities of tropical forages. *Nouvelles Agronomiques des Antilles et de la Guyane*, **4**, 63–80.

Yadava, P. S. & Kakati, L. N. (1985). Seasonal variation in herbage accumulation, net primary productivity and system transfer functions in an Indian grassland. In *Ecology and Management of the World's Savannas*, ed. J. C. Tothill & J. J. Mott, pp. 273–6. Canberra: Australian Academy of Science.

Yazman, J. A., McDowell, R. E., Cestero, H., Arroyo Aguilú, J. A., Rivera Anaya, J. D., Soldevila, M. & Román Garcia, F. (1982). Efficiency of utilization of tropical grass pastures by lactating cows with and without supplement. *Journal of Agriculture of the University of Puerto Rico*, **66**, 200–22.

Zandstra, H. C. (1983). Research on crop animal systems. In *Proceedings Crop–Livestock Research Workshop*, pp. 14–37. Los Banos: International Rice Research Institute.

Index

volatile fatty acids 166

water 44, 91, 125, 126
  runoff 29
  stock 21, 23

weeds 66, 82, 84, 85, 86, 119, 163, 165, 167, 169
wilting 144
wool production 119

zinc 35, 96

Printed in the United States
By Bookmasters